ELEVATOR TECHNOLOGY

ELLIS HORWOOD SERIES IN TRANSPORTATION
Series Editor: F. T. BARWELL, Emeritus Professor of Mechanical Engineering, University College, Swansea

The development of transport technology depends on the interaction of scientific advance with social and economic problems. For example, the Maglev transit device, as recently introduced at Birmingham Airport, was rendered possible by the use of semi-conductors in advanced electronic controls and the application of modern control theory. The justification for its development was partly environmental, notably the elimination of noise and vibration, and partly economic, i.e., un-manned operation and reduced maintenance cost.

The series treats various transport modes and devices from the functional viewpoint with particular emphasis on the solution of design problems from the economic viewpoint.

Jessop, A.	**Bayesian Decision Methods in Transport Planning and Operations**
Pursall, B. R.	**Underground Transport Systems Technology**
Wright, D. S.	**Freight Transport Calculations**

VERTICAL TRANSPORTATION Section
Editor: Dr. G. C. Barney, Director of Network Unit, University of Manchester Computer Centre and Research Director, Lift Design Partnership.

This important series of books covering the field of lift/elevator and passenger conveyors is published in collaboration with The International Association of Elevator Engineers.

The series will present authoritative texts by internationally-renowned experts on all aspects of vertical passenger transportation, which fields include elevators or lifts, escalators and other passenger conveyors. By this means, engineers and others in the industry will have the benefit of the dissemination of knowledge, aiding the training and education of future generations and encouraging and enabling research and development into new advanced systems.

Barney, G. C.	**Elevator Technology 1**
Janovsky, L.	**Elevator Mechanical Design**

ELEVATOR TECHNOLOGY

Editor:
G. C. BARNEY, B.Sc., M.Sc., Ph.D., C.Eng., F.I.E.E.
Director of Networking
University of Manchester Regional Computer Centre
and Research Director,
Lift Design Partnership, London

Published for
THE INTERNATIONAL ASSOCIATION OF ELEVATOR ENGINEERS
by

ELLIS HORWOOD LIMITED
Publishers · Chichester

First published in 1986 by
ELLIS HORWOOD LIMITED
Market Cross House, Cooper Street,
Chichester, West Sussex, PO19 1EB, England
The publisher's colophon is reproduced from James Gillison's drawing of the ancient Market Cross, Chichester.

Distributors:
Australia and New Zealand:
JACARANDA WILEY LIMITED
GPO Box 859, Brisbane, Queensland 4001, Australia

Canada:
JOHN WILEY & SONS CANADA LIMITED
22 Worcester Road, Rexdale, Ontario, Canada

Europe and Africa:
JOHN WILEY & SONS LIMITED
Baffins Lane, Chichester, West Sussex, England

North and South America and the rest of the world:
Halsted Press: a division of
JOHN WILEY & SONS
605 Third Avenue, New York, NY 10158, USA

© 1986 International Association of Elevator Engineers/Ellis Horwood Limited

British Library Cataloguing in Publication Data
Elevator technology.
1. Elevators
I. Barney, George C.
II. International Association of Elevator Engineers
621.8′77   TJ1370

Library of Congress Card No. 86–15285

ISBN 0–7458–0072–6 (Ellis Horwood Limited)
ISBN 0–470-20702–7 (Halsted Press)

Phototypeset in Times by Ellis Horwood Limited
Printed in Great Britain

COPYRIGHT NOTICE
All Rights Reserved. No part of this publication may be reproduced, stored in a retrieval system, or transmitted, in any form or by any means, electronic, mechanical, photo-copying, recording or otherwise, without the permission of Ellis Horwood Limited, Market Cross House, Cooper Street, Chichester, West Sussex, England.

Table of contents

Introduction xv

Part 1 — Theoretical design aspects of lift traffic

1 Traffic Design
   Dr G. C. Barney, University of Manchester, Manchester, UK
   1. Introduction 3
   2. Derivation of the round trip time of a single car 4
      2.1 Common method 4
      2.2 Equation-based method 4
      2.3 Example 1 6
   3. Limitations and assumptions 8
      3.1 Average highest reversal floor 8
      3.2 Unequal interfloor heights 8
      3.3 Arrival process 8
      3.4 Unequal floor populations 9
      3.5 Commentary 9
   4. Actual passenger arrival rates 10
      4.1 Calculation of interval and car load for a desired
          passenger arrival rate 10
      4.2 Example 2 11
      4.3 Commentary 11
   5. Computer-aided traffic analysis of lift systems 12
      5.1 Computer-aided design 12
      5.2 Up-peak CAD analysis 12
      5.3 Down-peak CAD analysis 14
      5.4 Interfloor CAD analysis 16
      5.5 Example 3 18
      5.6 Commentary 20
   6. Conclusion 20
   7. Acknowledgement 20
   8. References 20

2 Statistical evaluation of traffic design formulae
   Dr N. A. Alexandris, Piraeus Graduate School of Industrial Studies, Athens, Greece
   1. Introduction 22
   2. Passengers' arrival process 23

Table of contents

    3. Qualitative measures of multi-car lift systems in up-peak performance....24
        3.1 Mean terminal queue size....25
        3.2 Mean passenger waiting time....26
        3.3 Average number of busy lifts....27
    4. Generalized formulae for the average highest reversal floor and expected number of stops....28
        4.1 Evaluation of H....30
        4.2 Evaluaton of S....31
        4.3 Simplification of generalized formulae....32
    5. Conclusions....35
    6. References....35

3 A search for an index of lift traffic performance
Dr M. Wareing, UMIST, Manchester, UK
    1. Specification for an index of lift traffic performance....36
    2. Lift traffic signalling (background)....37
    3. Performance indices using conventional traffic signalling....37
        3.1 Definition....37
        3.2 Simulation results....37
        3.3 Definition of performance indices....39
    4. An index based in extended traffic signalling....40
        4.1 Definition....40
        4.2 Simulation analysis....42
    5. Conclusions and discussion....43
    6. References....44

Part 2 — Practical aspects of lift traffic....

4 UK practices of handling lift traffic in hospitals
Mr P. H. Beard, Wessex Regional Health Authority, Winchester, UK
    1. Lift traffic in hospitals....47
    2. Nucleus system....48
    3. TRAM/NETSYS traffic modelling and network analysis....48
        3.1 TRAM....48
        3.2 NETSYS traffic communication network analysis....50
        3.3 Assessment of service....50
        3.4 Feedback....50
    4. Control systems....51
    5. Quality of service....52
        5.1 Improving existing lifts....52
        5.2 Schemes under consideration....52
        5.3 User opposition....52
    6. Criticism....53
    7. Acknowledgement....53

5 The traffic performance of lifts: Principles and practice
Dr E. M. McKay, Rubicon Technical Services, Berkhamsted, UK
    1. Introduction....54
    2. Traffic performance concepts....55
        2.1 Conventional design....55
        2.2 A more generalized approach....55
    3. The data logger....56
    4. Case study data analysis and rules....56
    5. Potential traffic design procedures....59
    6. Acknowledgements....61
    7. References....61

6 Simulation and data logging
Eng. A. Lustig, S. Lustig Consulting Engineers, Tel Aviv, Israel
1. Introduction 62
2. Simulation 62
   2.1 Lift simulation program (LSP) 63
3. The data logger 63
4. Simulation and data logging 67
   4.1 Comparison of data logger and simulation results 69
   4.2 Modernization 71
5. Summary 71

7 Practising lift simulations
Ir E. H. Allaart, Tebodin, The Hague, Netherlands
1. Introduction 72
2. AMRO Bank project 73
3. Simulation procedure 74
4. Other traffic requirements 76
   4.1 Trains and lifts 76
   4.2 Basement service bunching 78
5. Conclusion 80

Part 3 — Mechanical design

8 A new design of compensating cable
Mr Richard Laney and **Mr William McCallum,** Siecor Corporation, Rodney Mount, USA
1. General principles 83
2. Dense aggregate compensating cable 85
3. Installation 87
4. Economics and future outlook 88
5. Experience 89

9 A means of minimizing the effect of steel member torsion in elevator travelling cable
Mr Alfred Garshick, BIW Cable Systems, Boston, USA
1. Introduction 90
2. Current practice 91
3. Test programme 91
   3.1 Standard strength members 91
   3.2 New design strength members 94
   3.3 Performance in control cable with new steel core 94
4. Conclusion 99

10 The problem of application of plastic lining in friction drives of lifts
Dr Wojciech Cholewa and **Dr Józef Hansel,** University of Mining and Metallurgy, Cracow, Poland
1. Introduction 101
2. Effect of lining on the life of ropes 101
3. Operational suitability of lining material 102
   3.1 Tests of friction factor 103
   3.2 Tests of wear resistance during rope slip 106

3.3 Tests of resistance to altering pressures 106
4. Conclusions 106
5. References 108

Part 4 — Control systems

11 Use of proportional valves for hydraulic elevators
Mr Eduard Hadorn, Beringer Hydraulik, Neuheim, Switzerland
1. Introduction 111
2. The requirements for an ideal elevator driven by a hydraulic system . . 111
3. Description 112
   3.1 Valve type LRV 112
   3.2 The electronic control card 112
   3.3 Adjustment and test 115
4. Final summary 115

12 A new elevator controlled by inverter
Mr E. Watanabe, Mr H. Kamaike, Mr S. Suzuki, Mr T. Ishii and **Mr S. Yokota,** Mitsubishi Electric Corporation, Inazawa, Japan
1. Introduction 116
2. Structure of the VVF controlled elevators 117
   2.1 General comparison 117
   2.2 VVVF control device for high speed and very high speed elevators 117
   2.3 VVVF control device for low and medium speed elevators 118
3. Advantages 120
   3.1 Power consumption 120
   3.2 Power supply capacity 122
   3.3 Reduction on bearing load of the elevator machine-room 124
4. Conclusion 125

13 Variable voltage and frequency elevator drive
Dr Jalal T. Salihi, United States Elevator Corporation, Spring Valley, USA
1. Basic components 126
2. Operation of induction motor from a variable frequency
   and voltage source 127
3. Controlled slip mode of operation in induction motor 130
4. Current or torque control 132
5. Speed control 132
Appendix: Analysis of torque current control scheme 134

14 Microprocessors in elevator controls
Dr-Ing. Joris Schroeder, Schindler Management AG, Lucerne, Switzerland
1. Introduction 136
2. Positioning controls 138
3. Drive controls 139
4. Supervisory or group controls 140
5. Indicator and communication systems 141
6. Diagnostic techniques 141
7. General man/machine interface 143
8. Conclusion 143

Part 5 — Elevatoring buildings

15 Top/down sky lobby lift design
Mr James W. Fortune, Lerch, Bates and Associates Inc., Littleton, USA
1. Introduction 149
2. Review of concepts and terminologies 149
3. A model for a top/down sky lobby 150
4. Shuttle service 151
5. Conclusions 152

16 Bush House: Lift for the world
Mr Michael Godwin, Lift Design Partnership, London, UK
1. Introduction 154
2. The experience 156
3. Problems encountered and problems solved 157
4. Listening to lifts.. 159
5. Some unexpected problems 160
6. Reliability and performance standards 161
7. Conclusion 162
Appendix: Specification detail of centre block lifts at Bush House complex 162
    A.1 Original installation 162
    A.2 Modernized lifts 162

17 Theoretical performance versus actual measurement
Mr V. Quentin Bates, Lerch, Bates and Associates, Littleton, USA
1. Introduction 165
2. A 'hindsight' review 166
3. Conclusions 172

18 Elevators for existing residential buildings
Mr Hans Westling, Swedish Council for Building Research, Stockholm, Sweden
1. Introduction 174
2. Background 175
3. Inexpensive 176
4. Functional 178
5. Economical use of space 178
6. Safety 179
    6.1 Safety requirements 179
    6.2 Main data for the lifts 180
    6.3 Drive system ....
7. Speedy installation 181
8. The market 181
9. Pilot projects 181
10. Experience gained and remaining problems 182
11. Acknowledgements 186

Part 6 — Lift management

19 The energy consumption of elevators
Dr.-Ing. Joris Schroeder, Schindler Management AG, Lucerne, Switzerland
1. Introduction 189
2. The energy consumption plot 189
3. The typical trip 190

4. The energy equation..........................193
5. Elevator energy consumption in a building..........194
6. Conclusion..................................195

20 Towards improvements in lift maintenance
Mr A. M. Godwin, Lift Design Partnership, London, UK
1. Background..................................196
2. Quality of products and imperfect reliability............197
3. Objects of maintenance..........................198
   3.1 The theory..............................198
   3.2 Preventive maintenance: What does it mean?......199
4. Performance and reliability measurement...............200
5. Service contracts presently offered...................201
6. A new concept: Performance guaranteed maintenance (PGM) contracts 201
   6.1 Factor 1 — Call-backs........................202
   6.2 Factor 2 — Down-time.......................202
   6.3 Factor 3 — Planned maintenance activities..........202
   6.4 Factor 4 — Intensity of use....................203
   6.5 Factor 5 — Satisfactory group control..............203
7. An example of a PGM contract.....................204
8. Conclusion..................................206

21 Remote monitoring of lifts
Dr J. R. Beebe, Lift Innovantions Ltd., Bolton, UK
1. Introduction.................................207
2. Lift-in-service indication.........................208
   2.1 Scale of the problem........................208
   2.2 Nature of the problem.......................208
   2.3 A practical solution.........................209
   2.4 Lift-in-service indicator......................209
3. Remote signalling.............................210
   3.1 Considerations............................210
   3.2 Solutions................................210
   3.3 Experience...............................213
4. Trends for the future...........................213
5. Conclusion..................................215
6. References..................................215

Part 7 — Standards and safety

22 International standardization in the lift industry
Eng. F. de Crouy–Chanel, Otis Elevator International, Paris, France
1. Introduction.................................219
2. Standardization work within ISO/TC 178..............220
   2.1 The ISO participants........................220
   2.2 The organisation..........................220
   2.3 Progress of the work........................220
3. Highlights of the ISO activity.....................221
4. Standardization work within CEN/TC 10...............225
   4.1 The CEN members.........................225
   4.2 The organization..........................225
   4.3 Progress of the work........................226
   4.4 Coordination with ISO......................226
5. Conclusion..................................227

23 Safety gear and European standards
Eng. Carlo Distaso, IGV/ELEVATORI, Milan, Italy
1. Introduction 228
2. Progessive safety gears 230
   2.1 Physical-chemical effects 230
   2.2 Theoretical considerations 231
3. Hydraulic buffer 232
   3.1 Description and basic theory 232
   3.2 Further analysis 233
4. Conclusions 235
5. References 235

24 Elevator safety
Mr D. A. Swerrie, Principal Safety Engineer, Elevagor Unit, San Francisco, California, USA
1. Introduction 236
2. History 237
3. The pendulum swings 239
4. The answer to safety? 241
5. Closing remarks 242
6. Reference 243

25 Explosion protection of electric hoist controls
Dipl.-Ing. Paul Schick, R. Stahl Switchgear, D.7118, Kunzelsau, West Germany
1. Explosion hazard and protection 244
2. Constructional requirements and classifications for explosion protected apparatus in hazardous areas 245
3. Types of protection and installation 246
4. Explosion protection of electrical apparatus for lifts 247
5. Summary 251
6. References 251

26 Availability, reliability and safety of lifts and their components
Dipl.-Ing. H. Streng, Consultant, Stuttgart, West Germany
1. Introduction 252
   1.1 Safety 252
   1.2 Reliability and availability 253
2. Reliability and probability of failures 253
3. Probability of failures 253
4. Safety and availability 254
5. Practical benefits from the analysis of accident reports 255
6. Measures to avoid injury to persons, material and property . . 255
7. Requirements in safety codes 257
8. Installation design and component selection 258
9. Lift components and safety devices 258
10. Other factors 259
11. Conclusion 260
12. References 260

Part 8 — Innovation

27 Latest development in lift technology
Mr M. Kaakinen, KONE Lift Group, Helsinki, Finland
1. Introduction 265

Table of contents

    2. The lift round trip time 266
       2.1 The elements of the lift round trip time 266
       2.2 Lift cycle time 267
    3. Control systems 268
       3.1 System principles 268
       3.2 Operating principles 269
    4. Examples of microcomputer lift controllers 270
       4.1 Low rise system 270
       4.2 Medium rise system 270
       4.3 High rise system 272
       4.4 Lift monitoring and command system 273
    5. Drive systems 273
       5.1 Controlled a.c.-drives/stator voltage control or
           eddy current braking 274
       5.2 d.c.-drive with static converter 274
       5.3 Controlled a.c.-drives with voltage and frequency control 274
    6. Examples of drives 275
    7. Automatic doors 276
       7.1 Door gear 276
       7.2 Passenger detection 278
    8. Summary 278

28 'Skytrak': A new era of passenger transport?
Mr Michael Godwin, Lift Design Partnership, London, UK
    1. Introduction 279
    2. The technology 280
    3. Safety 280
    4. Power system 282
    5. Conclusions 284

Part 9 — Escalators

29 The kinematics and dynamics of escalator steps and safety
Eng. Dr Lubomir Janovsky, Technical University, Prague, Czechoslovakia
    1. Introduction 289
    2. Kinematics and dynamics of the steps 290
       2.1 Methods of calculation 290
       2.2 Technical data and fundamentals of calculation 292
    3. Case studies 293
       3.1 Case 1: Emergency brake failure 293
       3.2 Case 2: Emergency brake operation 299
    4. Conclusions 301

30 Heavy duty escalators for public service
Dipl.-Ing. Emil Braun, Orenstein and Koppel AG, Dortmund, West Germany
    1. Capital investment and corrosion protection 303
    2. Operational readiness and safety 306
    3. Safety devices 309
    4. Step band speed and driving properties 310

31 Escator safety in the UK
Mr B. G. James, Health and Safety Executive, Bootle, UK
    1. Introduction 312
    2. Current position 312

3. Falling of persons. 313
4. Trapping of persons 313
5. Passenger safety 314
6. Action taken within the UK. 315
7. Conclusion. 316

32 Wheelchair escalator
Mr E. Watanabe, Mr S. Yokota, Mr M. Yonemoto and Mr M. Asano, Mitsubishi Electric Corporation, Inazawa, Japan
1. Introduction 318
2. Outline of wheelchair escalator. 319
3. Basic structure 319
4. Operating procedure 322
5. Conclusion. 324

33 Spiral escalator
Mr S. Goto, Mr H. Nakatani, Mr T. Kaida, Mr M. Tomidokoro and Mr R. Saito, Mitsubishi Electric Corporation, Inazawa, Japan
1. Introduction 326
2. Structure and drive principle 326
   2.1 Overall structure 326
   2.2 Drive mechanism. 328
   2.3 Steps. 328
   2.4 Step chains 331
   2.5 Truss. 331
3. Conclusion. 331

Part 10 — Epilogue

34 Practising engineering, we each make a difference
Mr V. Quentin Bates, Lerch, Bates and Associates, Littleton, USA
1. Ingenuity. 335
2. Two heads are better than one 337
3. One more floor 341
4. In conclusion 342

Index of contributors 343

Index 352

THE INTERNATIONAL ASSOCIATION OF ELEVATOR ENGINEERS
The rapid growth of the building industry and associated technologies demands a parallel growth in the field of vertical transportation.

Both the freight and passenger elevator industry need qualified engineers in the field whose knowledge keeps pace with technological and scientific developments.

The International Association of Elevator Engineers (IAEE) will provide a forum for the exchange of information and experience vital for training and education.

The aims of the IAEE are summarized as:

1. The definition and promotion of vertical transportation engineering as a specialized profession;
2. The promotion of international standards;
3. The encouragement of higher professional standards throughout the industry;
4. The promotion of scientific and technological expertise;
5. The co-operation between members.

Further information regarding membership, consitutions etc. can be obtained from the IAEE Secretariat, 10 rue Massena, 06000, Nice, France.

ELEVCON 86 and the associated exhibition ELEVATORS 86 are organised by Sher Group, 10 rue Massena, 06000, Nice France.

Introduction

This book contains the written contributions of presentations made at the 1st International Convention for Elevator Engineering, ELEVCON 86, held 11–13 February 1986 at the Acropolis, Nice, France.

The speakers at ELEVCON 86 came from twelve different countries and represent a wide range of topics and areas of expertise. These proceedings are a unique record of their presentations and in itself provides an educational experience for elevator engineers and others. This is the beginning of a series in Elevator Technology.

A conference organizer cannot determine the content of the presentations and hence what is offered by the author defines the quality and topic range for the conference. When it comes to arranging the written record some sequence can be developed in order to present to readers a more cohesive entity. This procedure is not without effort and sometimes means quite extensive revision of the manuscripts submitted.

The book is divided into ten parts. Parts 1 and 2 deal with the starting points to lift provision namely the theoretical and practical aspects of handling lift traffic. The following Parts 3 and 4 discuss practical problems in the mechanical and control areas. Part 5 presents the ways in which consultants bring design into reality.

Management topics are described in Part 6 with an emphasis on modern methods and technology with Part 7 defining some of the problems of maintaining safety. This part illustrates how safe a transportation system a lift system has become. Part 8 allows the imagination to run into the future with descriptions of new techniques and inventions for the future. The associated transportation area of escalators is covered extensively in Part 9. Finally Part 10 suggests elevator engineers should come together to share ideas — which is the purpose of this book.

G. C. Barney
Februrary 1986

Part 1
Theoretical design aspects of lift traffic

1
Traffic design

Dr G. C. Barney, University of Manchester, Manchester, UK

ABSTRACT

The sizing of lift systems to serve the demands of a building's population has interested the lift community since the 1920s. The methods used were somewhat rough and ready. However, by the 1970s a recognized method of calculation had evolved for up-peak traffic sizing based on the mathematical determination of the H, S and P parameters (average highest reversal floor, average number of stops and average number of passengers).

In the 1970s digital computer simulation techniques evolved which allowed specific traffic and lift installations to be examined. These computer packages utilize proven mathematical methods.

This chapter reviews traffic design and presents modern design methods.

1. INTRODUCTION

The traffic sizing of lift systems is conventionally carried out by determining a value for the round trip time (RTT) of a single lift car serving an up-peak traffic condition. Then knowing or assuming a number of lift cars (L) in a group, a value for the up-peak interval (UPPINT) of successive car arrivals at the main terminal (MT) is found. The traditional up-peak handling capacity (UPPHC) over a five minute period can then be calculated assuming cars fill with passengers (P) to 80% of the contract capacity (CC). If this handling capacity slightly exceeds or is equal to the anticipated passenger arrival rate (λ) then the design is complete and a value for the percentage population (% POP) served can be stated. Mathematically:

$$\text{UPPINT} = \text{RTT}/L \qquad (1)$$

$$\text{UPPHC} = \frac{300}{\text{UPPINT}} \times 0.8 \times \text{CC} \qquad (2)$$

$$\text{\% POP} = \frac{\text{UPPHC}}{\text{POPULATION}} \times 100\% \qquad (3)$$

The starting point for this procedure is the derivation of the round trip time for a single car.

2. DERIVATION OF THE ROUND TRIP TIME FOR A SINGLE CAR

2.1 Common Method

A common method used by consultants and lift manufacturers and found in Codes of Practice and textbooks is to calculate the RTT as two components; a standing time and a running time, namely:

Standing time

    Time to load P passengers at MT
    Time to open and close doors at MT
    Time to open and close doors at S stops
    Time to unload P passengers at S stops
        Subtotal standing time $= (A)$

Running time

    Time to start, accelerate, run at contract speed, decelerate and level car at $(S+1)$ stops.
    Time to pass remaining floors at contract speed to top floor.
    Time for express run down from top floor to MT
        Subtotal running time $= (B)$

therefore

$$\text{RTT} = (A) + (B) \qquad (4)$$

This method is effective, but not elegant and does depend on certain assumptions and suffers from a number of possible errors.

2.2 Equation-based Method

Consideration of Fig. 1 shows that a round trip is comprised of a number of components:

(a) Time to transfer P passengers into (t_p) and out (t_u) of the lift cars.
(b) Time to open (t_o) and close (t_c) the car doors $(S+1)$ times.
(c) Time to accelerate, run at contract speed, decelerate and level car $(S+1)$ times.

Ch. 1] Traffic design

(d) Time to travel past remaining floors at contract speed to highest floor (*H*).
(e) Time for express run from highest floor (*H*) to MT.

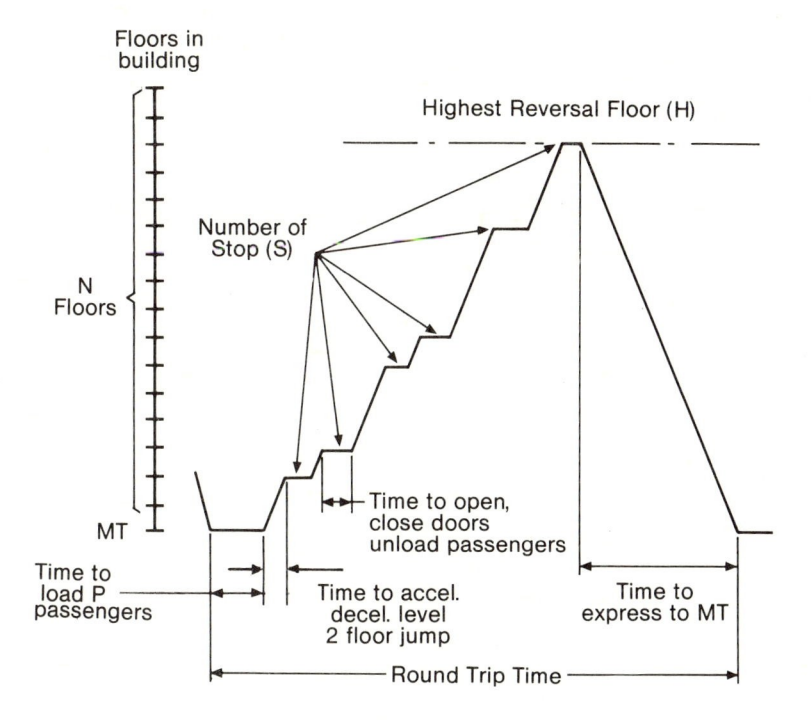

Fig. 1 — Components of a round trip for a single car.

therefore

$$\text{RTT} = (a) + (b) + (c) + (d) + (e) \tag{5}$$

$$= Pt_p + Pt_u + (S+1)(t_o + t_c) + (S+1)t_f(1) + (H-S)t_v + (H-1)t_v \tag{6}$$

which simplifies to:

$$\text{RTT} = 2HT_v + (S+1)t_s + 2P\,t_p \tag{7}$$

where:

$$t_v = \frac{df}{v}$$

df is a standard interfloor height and v is the contract speed.

$$t_s = t_f(1) + t_c + t_o - t_v$$

where $t_f(1)$ is a single-floor jump time

$$t_p = \frac{t_p + t_u}{2}$$

P is taken as 80% of contract capacity

S and H can be calculated statistically (Bassett Jones, 1923; Gaver and Powell 1971) from:

$$S = N\left[1 - \left(1 - \frac{1}{N}\right)^P\right]; \quad H = N - \sum_{i=1}^{N-1}\left(\frac{i}{N}\right)^P \qquad (8)$$

Values of H and S can be tabulated (Table 1) for a range of car sizes and all reasonable numbers of floors per building zone.

2.3 Example 1
Given:

$N = 16;\ CC = 16;\ L = 5;\ POP = 1000$

$t_v = 1;\ t_f(1) = 5;\ t_o = 1.5;\ t_c = 2.5;\ t_p = 1.2$ (times in seconds)

$P = 12.8\ (16 \times 0.8)$

$\left.\begin{array}{l} H = 15.3 \\ S = 9.0 \end{array}\right\}$ Table 1

$t_s = 5 + 1.5 + 2.5 - 1 = 8.0$

therefore

$$\begin{aligned}
\text{RTT} &= 2 \times 15.3 \times 1 + (9+1)\,8 + 2 \times 12.8 \times 1.2 \\
&= 30.6 + 80.0 + 30.7 \\
&= 141.3 \text{ seconds}
\end{aligned}$$

$\text{UPPINT} = 141.3/5 = 28.3$ seconds

Traffic design

Table 1 — Table calculated by conventional formulae of average highest reversal floor (*H*) and expected number of stops (*S*) for buildings with 5–24 floors above the main terminal using British Standard Code car contract capacities. Cars assumed to carry 80% of contract capacity (shown in brackets)

H
S

Number of floors served above main terminal	\multicolumn{6}{c}{Car contract capacity (80% capacity value is shown in brackets)}					
	6 (4.8)	8 (6.4)	10 (8.0)	12 (9.6)	16 (12.8)	20 (16.0)	24 (19.2)
5	4.6 / 3.3	4.7 / 3.8	4.8 / 4.2	4.9 / 4.4	4.9 / 4.7	5.0 / 4.9	5.0 / 4.9
6	5.4 / 3.5	5.6 / 4.1	5.7 / 4.6	5.8 / 5.0	5.9 / 5.4	5.9 / 5.7	6.0 / 5.8
7	6.2 / 3.7	6.5 / 4.4	6.6 / 5.0	6.7 / 5.4	6.8 / 6.0	6.9 / 6.4	6.9 / 6.6
8	7.1 / 3.8	7.4 / 4.6	7.5 / 5.3	7.6 / 5.8	7.8 / 6.6	7.9 / 7.1	7.9 / 7.4
9	7.9 / 3.9	8.2 / 4.8	8.4 / 5.5	8.6 / 6.1	8.7 / 7.0	8.8 / 7.6	8.9 / 8.1
10	8.7 / 4.0	9.1 / 4.9	9.3 / 5.7	9.5 / 6.4	9.7 / 7.4	9.8 / 8.1	9.9 / 8.7
11	9.6 / 4.0	10.0 / 5.0	10.2 / 5.9	10.4 / 6.6	10.6 / 7.8	10.7 / 8.6	10.8 / 9.2
12	10.4 / 4.1	10.8 / 5.1	11.1 / 6.0	11.3 / 6.8	11.5 / 8.1	11.7 / 9.0	11.8 / 9.7
13	11.2 / 4.1	11.7 / 5.2	12.0 / 6.1	12.2 / 7.0	12.5 / 8.3	12.6 / 9.4	12.7 / 10.2
14	12.1 / 4.2	12.6 / 5.3	12.9 / 6.3	13.1 / 7.1	13.4 / 8.6	13.6 / 9.7	13.7 / 10.6
15	12.9 / 4.2	13.4 / 5.4	13.8 / 6.4	14.0 / 7.3	14.3 / 8.8	14.5 / 10.0	14.7 / 11.0
16	13.7 / 4.3	14.3 / 5.4	14.7 / 6.5	14.9 / 7.4	15.3 / 9.0	15.5 / 10.3	15.6 / 11.4
17	14.5 / 4.3	15.2 / 5.5	15.86 / 6.5	15.8 / 7.5	16.2 / 9.2	16.4 / 10.6	16.6 / 11.7
18	15.4 / 4.3	16.0 / 5.5	16.5 / 6.6	16.8 / 7.6	17.1 / 9.3	17.4 / 10.8	17.5 / 12.0
19	16.2 / 4.3	16.9 / 5.6	17.4 / 6.7	17.7 / 7.7	18.1 / 9.5	18.3 / 11.0	18.5 / 12.3
20	17.0 / 4.4	17.8 / 5.6	18.2 / 6.7	18.6 / 7.8	19.0 / 9.6	19.3 / 11.2	19.4 / 12.5
21	17.9 / 4.4	18.6 / 5.6	19.1 / 6.8	19.5 / 7.9	19.9 / 9.8	20.2 / 11.4	20.4 / 12.8
22	18.7 / 4.4	19.5 / 5.7	20.0 / 6.8	20.4 / 7.9	20.9 / 9.9	21.1 / 11.5	21.3 / 13.0
23	19.5 / 4.4	20.4 / 5.7	20.9 / 6.9	21.3 / 8.0	21.8 / 10.0	22.1 / 11.7	22.43 / 13.2
24	20.3 / 4.4	21.62 / 5.7	21.8 / 6.9	22.2 / 8.0	22.7 / 10.1	23.0 / 11.9	23.2 / 13.4

$$\text{UPPHC} = \frac{300}{28.3} \times 12.8 = 135.7 \text{ persons/5 minute}$$

$$\text{\% POP} = \frac{135.7}{1000} = 13.6\%$$

Note the significance of middle term of the RTT equation: any time saved in stopping/starting a lift is more effective than (say) doubling the contract speed.

3. LIMITATIONS AND ASSUMPTIONS

Although the equation based method is effective and elegant it does depend on certain assumptions and limitations some of which apply to the Common Method.

3.1 Average Highest Reversal Floor (H)

The common method assumes the top floor as the average highest reversal floor. This is incorrect as Table 1 indicates. Fig. 2 which plots the error shows that a small car serving a large number of floors attracts a very high error in H. As H can be calculated easily this error is unnecessary.

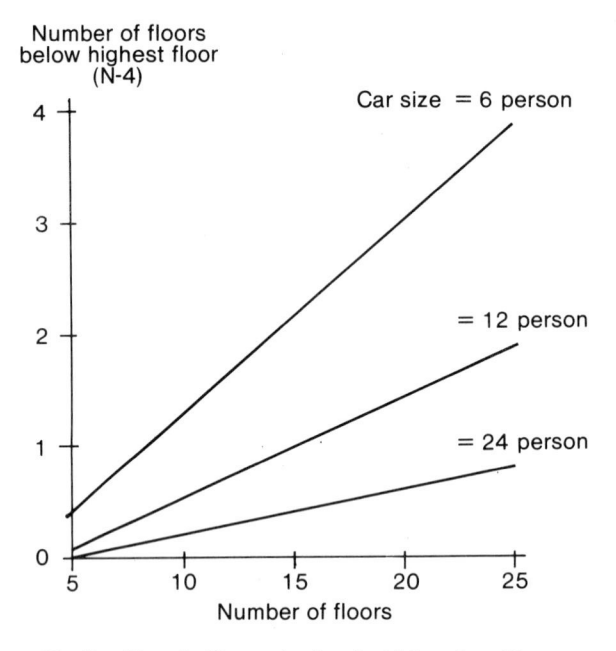

Fig. 2 — Error in H assuming it to be highest floor N.

3.2 Unequal Interfloor Heights, Interfloor Jumps Greater than One

It is assumed that contract speed is reached in a single floor jump over a standard interfloor height. This is untrue for cars running at contract speeds over 1.75 m/s with an interfloor height of 3.3 m, as contract speed is not reached in a single floor jump. Also most buildings have some interfloor heights larger than others, e.g. lobby, service, conference, restaurant, etc. Fig. 3 shows such a situation and indicates a correction method.

3.3 Arrival Process

Normally passengers are assumed to arrive according to a rectangular probability distribution function but a Poisson pdf has been suggested. If values for H and S are calculated for a Poisson pdf (Barney and Dos Santos,

Traffic design

```
Given: N=16; H=15.3;
S=9.0; df=3.3 m; t_f(1)=5.0;
t_f(2)=6.1
```

Height of building to H
(15.3)3.3+(2−1)3.3
+2(1.5−1)3.3=57 m

Average distance between stops
$$= \frac{H}{S} = \frac{57}{9} = 6.3 \text{ m}$$

Average number of floors between stops
$$= \frac{6.3}{df} = \frac{6.3}{3.3} = 1.9$$

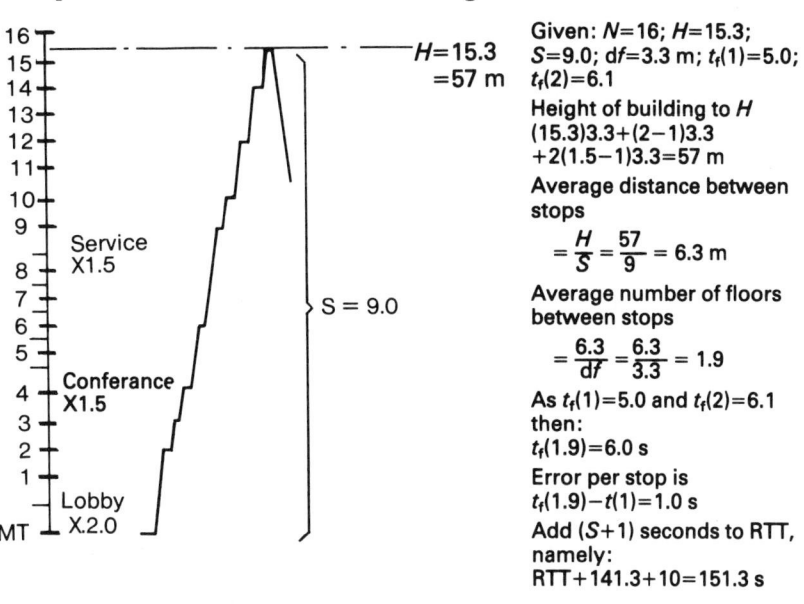

As $t_f(1)=5.0$ and $t_f(2)=6.1$ then:
$t_f(1.9)=6.0$ s

Error per stop is
$t_f(1.9)−t(1)=1.0$ s

Add $(S+1)$ seconds to RTT, namely:
RTT+141.3+10=151.3 s

Fig. 3 — Correction to RTT for unequal interfloor heights and interfloor jumps greater than one.

1985, Table 2.14) it will be found that for specific systems values for H and S are lower in value than those calculated assuming a rectangular pdf. In fact, tests of other suitable probability functions, such as Erlangian and exponential, produce values between the rectangular and Poisson figures. It can then be concluded therefore that the rectangular values for H and S *provide a pessimistic value of RTT with some built-in spare capacity.*

3.4 Unequal Floor Populations
Building floors generally are not equally populated. Using statistical techniques new formulae for H and S can be obtained:

$$S = N - \sum_{i=1}^{N}\left(1-\frac{U_i}{U}\right)^P ; \quad H = N - \sum_{j=1}^{N-1}\left(\sum_{i=1}^{j}\frac{U_i}{U}\right)^P \tag{9}$$

Where U_i is the population of floor i and U is the total building population.

It is not possible to tabulate such a vast range of possibilities. However, the effect of unequal populations is not unfavourable as Fig. 4 shows the value of S decreases and the values of H generally decreases except for an unusual population distribution.

3.5 Commentary
An elegant equation-based method has been presented, which includes the derivation of the average highest reversal floor and which is not sensitive to the passenger arrival process. In addition corrections are possible to cater

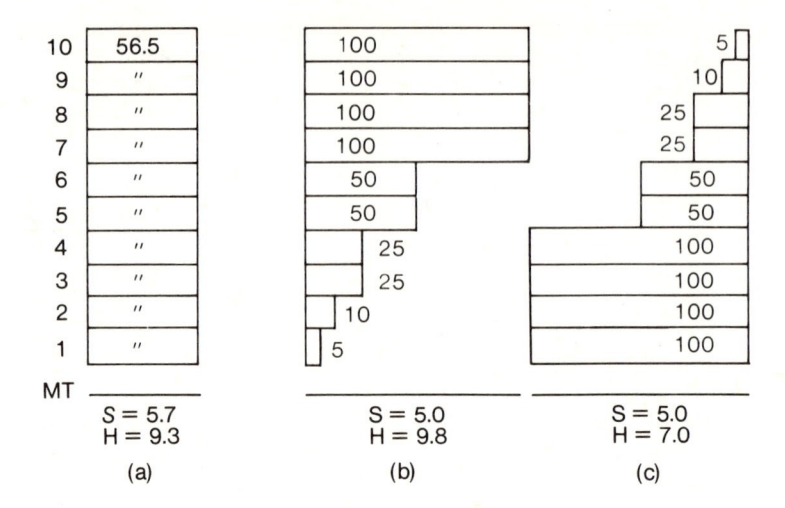

Fig. 4 — Effect on *H* and *S* of unequal floor populations. (a) Equal floor populations; (b) unequal floor populations (unlikely); (c) unequal floor populations (likely).

for unequal floor populations, unequal interfloor heights and the inability of drive systems to reach contract speed in a single floor jump. The effects of the control system have not been discussed or why 80% car loadings seem appropriate.

4. ACTUAL PASSENGER ARRIVAL RATES

The RTT equation assumes in the calculation that cars will load to 80% of contract capacity. The value obtained for UPPHC also makes the same assumption. But this is only correct if the arrival rate (λ) exactly equals UPPHC! What happens if either too few passengers or too many passengers arrive to load the cars 80% contract capacity on average? The answer is the car load will alter and with it the average number of stops, the average highest reversal floor, the RTT and the interval.

4.1 Calculation of Interval and Car Load for a Defined Passenger Arrival Rate

Suppose that the lift designer selects a desired value for car interval (INT). This is not an uncommon practice as interval is a quality of service parameter. If the passenger arrival rate is λ persons per second then the number of passengers (*P*) per car is:

$$P = \lambda \text{INT} \qquad (10)$$

Now given values for t_v, t_s and t_p; *H* and *S* can be obtained from Table 1 (by interpolation if necessary) and a value for RTT obtained. By selecting a suitable number of cars (*L*), the interval can be calculated. Should this value

Traffic design

for interval differ from that in eq (10) then a new value must be selected until the initial and final values of interval converge. (This is an iterative procedure.)

4.2 Example 2

Using the given data and results from Example 1 suppose 120 persons arrive for service during up-peak.

A suitable initial value of interval could be:

$$INT_{initial} = UPPINT \times \frac{\lambda}{UPPHC} \qquad (11)$$

$$= 28.3 \times \frac{120}{136} = 25.0 \text{ seconds}$$

Table 2 illustrates the procedure. Compare the 80% load values of UPPINT(28.3s) and UPPHC (136) obtained in Example 1 with the actual traffic demand values of INT (24.1s) and λ (120).

Table 2 — Iterative calculation of interval (INT)

Parameter	Trial 1	Trial 2	
Initial interval (INT) (s)	25.0	24.2	INTi
P	10.0	9.7	
H	15.0	15.0	
S	7.6	7.4	
RTT	122.8	120.5	
L	5	5	
Final interval (INT) (s)	24.6	24.1	INTf
OK?	NO	YES	
*New initial interval (s)	24.2	—	
Load	—	61%	

*New initial interval = INTi − 2(INTi − INTf)

4.3 Commentary

Using the procedure outlined it is possible to exactly estimate car intervals and loads for any passenger arrival rate. In Example 2 the effect of passenger arrivals altering from 120 to 136 persons per 5 minutes can be clearly seen. What is not known of course, is passenger average waiting time (AWT).

5. COMPUTER-AIDED TRAFFIC ANALYSIS OF LIFT SYSTEMS

5.1 Computer-aided Design

Computer-aided Design (CAD) enables engineers and designers to use computers to assist in the decision of complex engineering problems. CAD involves three phases; input, computation and output. In the case of CAD for lift systems these phases comprise:

Input phase: Enter: building data; lift system data; passenger data.
Computation phase: Perform a discrete digital simulation of the lift system (time consuming).
Output phase: Examine: graphical output; spatial plots; car load; car interval; passenger waiting time; percentiles of waiting time.
Examine: numerical tables

At UMIST, University of Manchester, a CAD suite called LSD (Lift Simulation and Design) has been used to simulate 10 000 lift systems. Some twelve control systems have been emulated. Results will now be presented obtained by using the LSD suite simulating up-peak, down-peak and interfloor traffic conditions using the five different control policies listed in Table 3.

5.2 Up-peak CAD Analysis

Fig. 5 shows the results of a large number of simulations of lift systems operating under up-peak traffic conditions. Each individual simulation is depicted by a cross. For each simulation the car load and car interval is calculated by the method outlined in section 4. Thus the horizontal axis plots percentage car load. The vertical axis plots a performance figure as a normalized value of passenger average waiting time (AWT) divided by the evaluated interval (E:INT). The value of AWT has been obtained from the simulation of each lift system and is an exact figure for all individual passenger waiting times.

Examination of Fig. 5 indicates that up to 60% car loads the average performance figure is 50% (i.e. AWT is about 0.5 INT, as tradition has it). At 60% car loads some lift systems indicate a performance figure of 100% (i.e AWT = INT). Up to 80% car loads few runs exceed the 100% performance figure, but for car loads over 90% few lift systems can provide a performance figure below 100%.

Fig. 6 is an idealized version of Fig. 5. The curve that emerges is of a classic exponential form found in the design of telephone systems, water supply and computer systems. It shows that at the standard design value of 80% car loading, the performance figure is 85% with a range of from 40 to 130%. This range arises owing to the randomness of passenger arrivals, passenger destinations and the running of the cars. Investigation of the

Table 3 — Definition of lift control system policies (Barney and Dos Santos, *Elevator Traffic, Analysis, Design and Control* (Peter Peregrinus, London, 1985)

Control system name (mnemonic)	Description	Nearest proprietary system
Fixed sectoring (Birdirectional) (FSO)	Divides a building zone into fixed sectors equal to number of lifts each sector deals with both up and down landing calls.	Otis VIP260
Fixed sectoring (Priority timed) (FS4)	Divides a building zone into separate fixed up-sectors equal to the number of lifts minus one and separate fixed down-sectors equal to the number of lifts. each sector is timed through six stages of priority.	Westinghouse Mark 4
Dynamic sectoring (DS)	Divides a building zone into a variable number of sectors defined by car positions. Each sector contains a variable number of floors.	Schindler Aconic
Call allocation (ACA) (Closs, 1970)	Sectorless building zones. Landing calls allocated to cars on the basis of a cost function uses special call registration panel on landing.	UMIST Research
Computer Group Control (CGC) (Lim, 1983)	Divides a building into dynamic sectors of non-contiguous floors. Allocates landing calls on the basis of a cost function. Self-tunes estimation parameters.	ILMC System 990

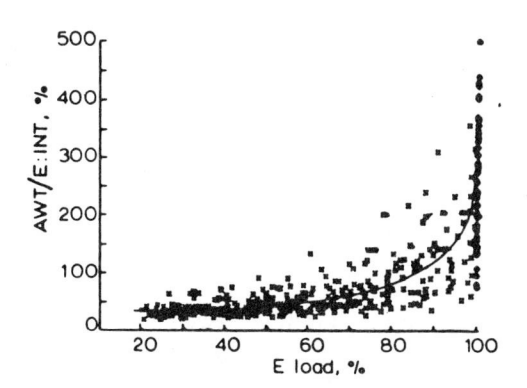

Fig. 5 — Performance figures using expected interval versus load for the collective algorithm.

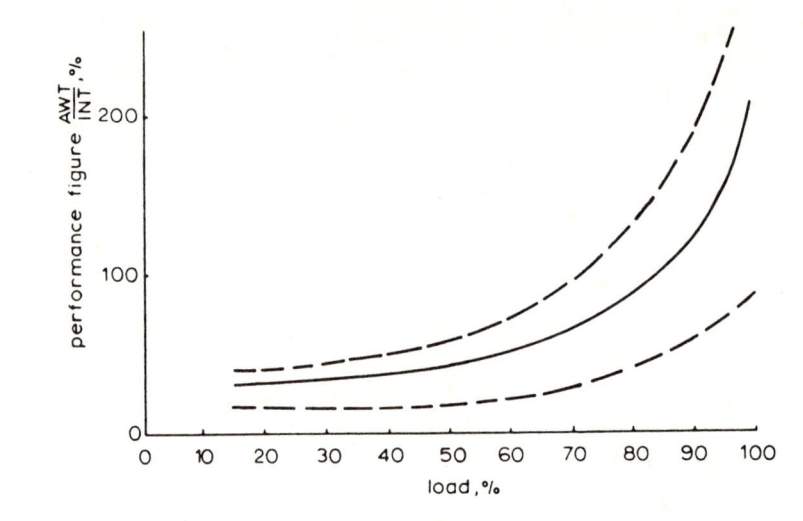

Fig. 6 — Up-peak performance. ——— average values; — — — possible range of values.

effect of the control policy on up-peak performance showed no visible differences. This result can be implied as most control systems can only return empty cars to the main terminal once the up-peak traffic condition has been detected.

Fig. 6 can be tabulated as Table 4.

Table 4 — Tabulation of Fig. 6.

Percent car load	30	40	50	60	70	75	80	85	90	95
AWT/INT	0.32	0.35	0.40	0.50	0.65	0.74	0.85	1.01	1.30	1.65

5.3 Down-peak CAD Analysis

The best way of characterizing down-peak passenger demand (α) is to relate it to UPPHC, namely:

$$\alpha = \frac{\lambda}{\text{UPPHC}} \times 100\% \tag{12}$$

where λ is the down-peak passenger arrival rate over a 5-minute period. In the same way as for the up-peak case many simulations of lift systems serving a down-peak traffic pattern were made.

Fig. 7 provides an idealized graph of the performance figure against a range of down-peak demand. A straight line average has been drawn following the relationship:

Traffic design

$$DNAWT = 0.85 \, \alpha UPPINT \quad (13)$$

Thus once a value for UPPINT is known, it is possible to estimate passenger average waiting times for various down-peak traffic loads. Fig. 7 indicates

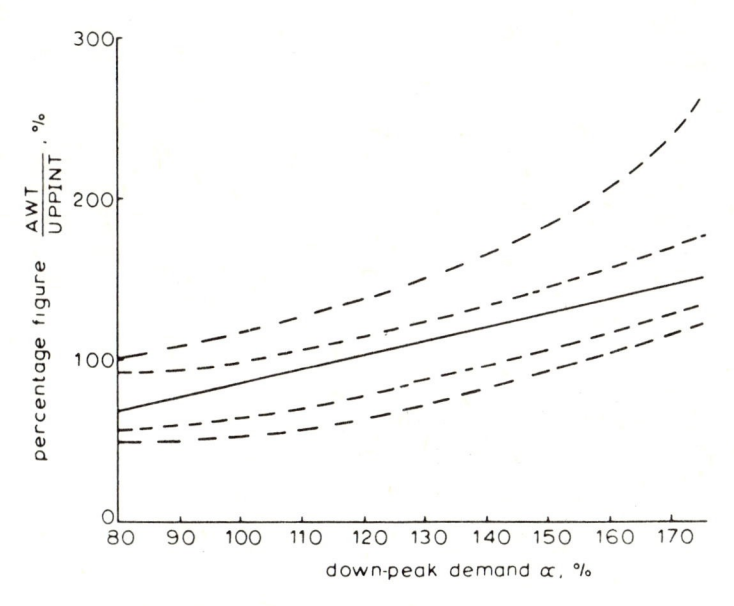

Fig. 7 — Down-peak performance figure. —— overall probable limit of values; ——— possible limit of values owing to influence of control algorithms.

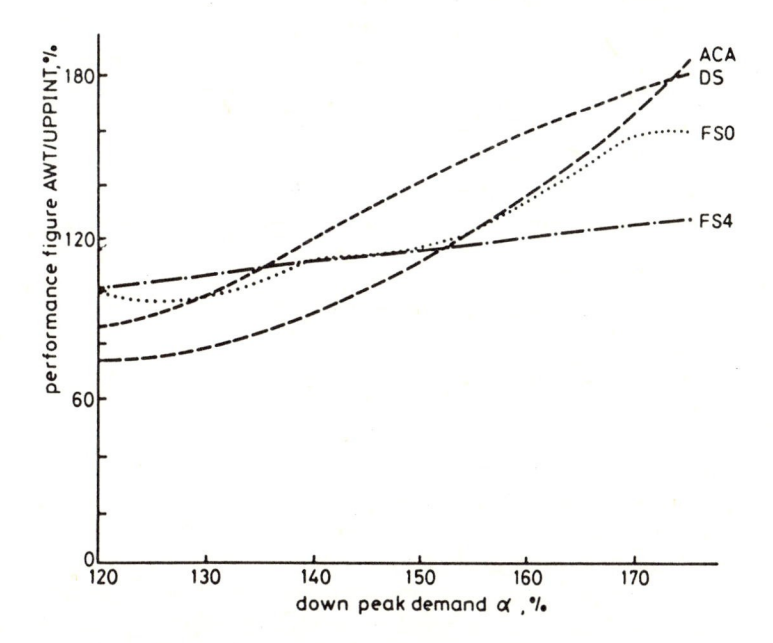

Fig. 8 — Down-peak performance of CGC algorithm.

possible ranges for the effects of randomness and control policies; Fig. 8 gives more detail of the latter effect.

It is interesting to note that algorithm FS4 performs very consistently for all levels of demand, only being outclassed by ACA up to demands of 150%. It can be assumed that the priority timing feature in FS4 provides this 'even handed' characteristic. Algorithm ACA 'knows' passenger destinations (which in down-peak is obviously the MT) and its ability to minimize AWT operates well up to medium load levels.

Only analysis of the down-peak situation are shown in Fig. 9. For a wide range of down-peak demand the ratio of down peak stops to up peak stops is 0.5 and the ratio of down-peak interval to up-peak interval is 0.67. The former result is useful when calculating down peak RTT and intervals. The latter result, however, confirms the generally held belief that every lift system has a handling capacity for down-peak traffic 50% greater than for up-peak traffic.

5.4 Interfloor CAD Analysis

Let the interfloor demand (β) be:

$$\beta = \frac{\lambda}{\text{UPPHC}} \times 100\% \tag{13}$$

where λ is the interfloor passenger arrival rate over a 5-minute period.

It is more common to think of the proportion of a building's population who will use a lift each hour. For example, a value of $\beta = 0.2$ for a 15% up-peak arrival rate indicates that 36% of the building population will use the lifts every hour. Interfloor traffic demands over $\beta = 0.25$ can be considered very heavy. Fig. 10 shows the results of extensive simulations of the interfloor traffic pattern. A straight line average has been drawn following the relationship:

$$\text{IFAWT} = (1.784\beta + 0.22)\text{UPPINT} \tag{14}$$

The use of eqn (14) can be made to estimate passenger average waiting times for various levels of interfloor traffic loads.

The effect of the control policy under interfloor traffic is depicted in Fig. 11. The ACA algorithm is the best as passenger destinations are always known and the self-tuning algorithm CGC performs only a little worse. The DS algorithm which is a non-computer based algorithm offers a very similar performance, which is not surprising as it was designed to deal with interfloor traffic. The fixed sectoring algorithms are significantly worse in performance.

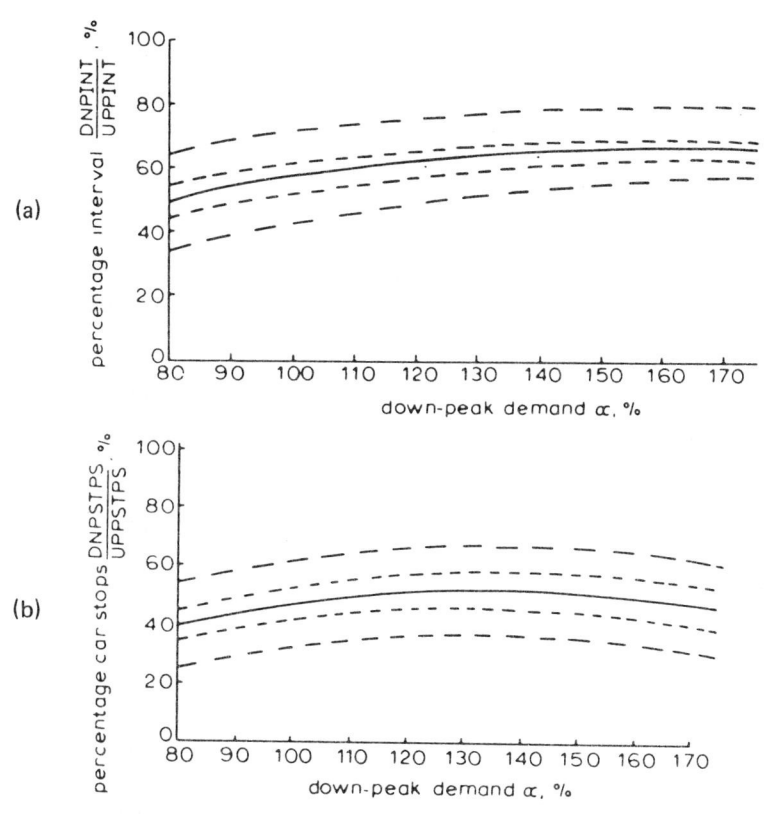

Fig. 9 — (a) Down-peak percentage interval. (b) Down-peak percentage car stops. — — overall probable limit of values; — — — — possible limit of values owing to influence of control algorithms.

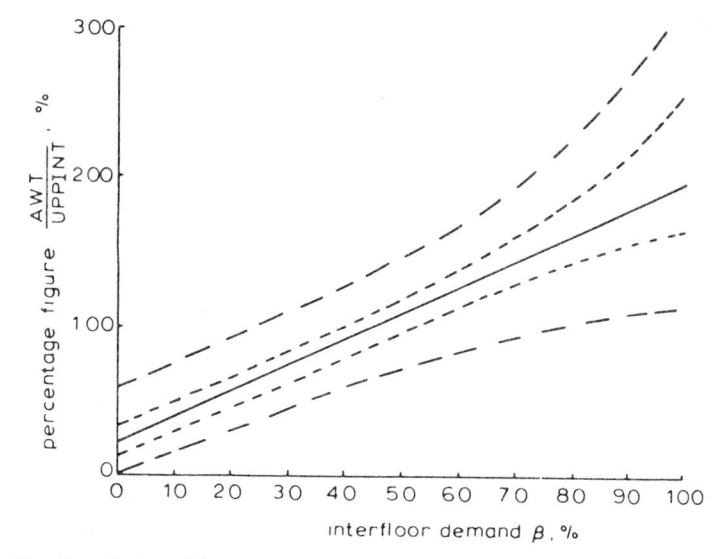

Fig. 10 — Balanced interfloor performance figure. — — overall probable limit of values; — — — possible limit of values owing to influence of control algorithms.

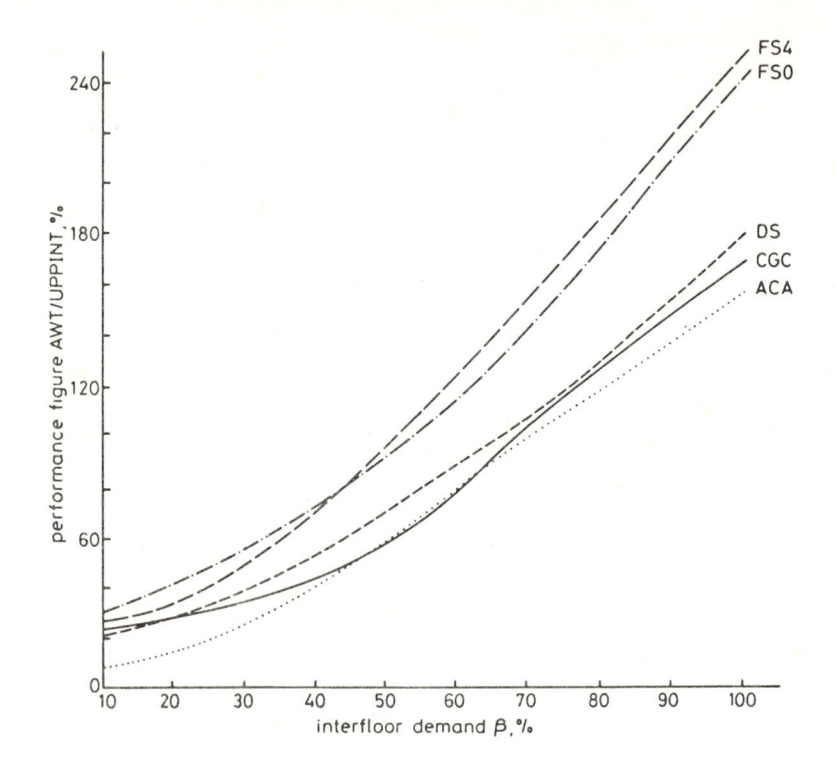

Fig. 11 — Interfloor performance of CGC algorithm.

Fig. 12 shows the number of stops made by each car per minute under the range of traffic demands. The curve shows a limitation at about 4.5 stops/car/minute. This limit is entirely due to the floor to floor cycle time (door operations and flight time) and passenger transfers. It is interesting to note that Bedford's (1966) criteria of a busy system as 2.25 stops/car/minute corresponds to a value of β equal to 16% about 25–30% of the building population using the lifts every hour.

5.5 Example 3
Using the given data and results from Examples 1 and 2 suppose down-peak demand is 180 persons per 5 minutes and interfloor demand is 25 persons per 5 minutes. Determine average waiting time for each traffic condition.

(a) Up-peak:

UPPINT = 28.3 s; UPPHC = 136 (Example 1)

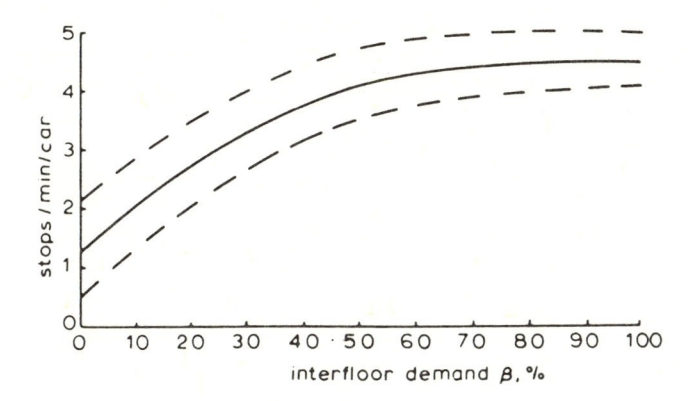

Fig. 12 — Balanced interfloor numbers of stops per car per minute. ––– probable limit of values.

$$\text{INT} = 24.1 \text{ s}; \text{ Load} = 61\% \text{ (Example 2)}$$

$$\frac{\text{AWT}}{\text{INT}} = 0.52 \text{ for a load of } 61\% \text{ (Table 4)}$$

therefore

$$\text{UPPAWT} = 0.52 \times 24.1 = 12.5 \text{ s}$$

(b) Down-peak:

$$= \frac{180}{136} = 1.32 \text{ (eq (12))}$$

therefore

$$\text{DNAWT} = 0.85 \times 1.32 \times 28.3 = 31.8 \text{ s (eq (13))}$$

(c) Interfloor:

$$\beta = \frac{25}{136} = 0.18 \text{ (eq (13))}$$

therefore

$$\text{IFAWT} = (1.74 \times 0.18 + 0.22) \, 28.3 = 15.1 \text{ s (eq (14))}$$

Thus:

$$\text{UPPAWT} = 12.5 \text{ s}; \text{ DNAWT} = 31.8 \text{ s}; \text{ IFAWT} = 15.1 \text{ s}$$

Given the method outlined, it is now possible to determine the effect of one lift going out of service. For example, suppose the system calculated above had four lifts in service instead of five. The results would be:

$$\lambda = 120;\ \text{UPPINT} = 35.5\ \text{s};\ \text{UPPHC} = 109;$$
$$\text{UPPAWT} = 70\ \text{s};\ \text{DNAWT} = 49.5\ \text{s}\ \text{IFAWT} = 21.9\ \text{s}$$

Significantly the reduced system cannot cope with up-peak.

5.6 Commentary

Computer analysis has shown a better understanding of how lifts work, namely:

(i) Car loadings of 80% under up-peak conditions should not be exceeded.
(ii) Passenger AWTs can be estimated for up-peak
(iii) Control policies have little effect during up-peak
(iv) A lift system has 50% more capacity during down-peak over up-peak
(v) A lift system serving down-peak traffic has 50% of the stops likely during up-peak
(vi) Passenger AWTs can be estimated for down-peak
(vii) Control policies are significant during down-peak; those with priority timing perform well
(viii) Control policies are significant during interfloor traffic with dynamic sectoring and self-turning systems performing well.

6. CONCLUSION

The traffic sizing of lift systems will always be concerned with averages and probabilities. Specific systems are unlikely to be average and more likely each building population will behave in odd and disturbing ways. However, provided the lifts installed are not grossly undersized the building population will always adapt to what they are given, albeit with some complaints. The information presented here obtained by extensive computer analysis gives a better understanding to lift traffic design and should enable certain pitfalls to be avoided.

7. ACKNOWLEDGEMENT

The author acknowledges with thanks permission granted by the Institution of Electrical Engineers, London to use materials from the book *Elevator Traffic*.

8. REFERENCES

Barney, G. C. and Dos Santos, S. M. (1985). *Elevator Traffic, Analysis, Design and Control* (Peter Peregrinus, London).

Bedford, R. J. (1966). Lift traffic recordings and analysis, *GEC J.*, **33** (2), 69–77.

Closs, G. G. (1970). The computer control of passenger traffic, PhD Thesis, University of Manchester Institute of Science and Technology.

Gaver, D. P. and Powell, B. A. (1971). Variability in round trip times for an elevator car during up-peak, *Transpn. Res.*, **5** (4), 301–307.

Jones, Bassett (1923). The probable number of stops made by an elevator, *GE Rev.*, **26** (8), 583–587.

Lim, S. H. (1983). A computer based lift control algorithm, PhD thesis, University of Manchester Institute of Science and Technology.

2

Statistical evaluation of traffic design formulae

Dr N. A. Alexandris, Piraeus Graduate School of Industrial Studies, Athens, Greece.

ABSTRACT

Statistical analysis has shown that the most general traffic condition of interfloor traffic with unequal floor population can provide analytical formula to obtain values for mean highest reversal floor (H) and expected number of stops (S) in lift systems. These values are used consequently to evaluate the round trip time (RTT), which is the fundamental design criterion to determine lift performance. The formulae for H and S are derived assuming that the arrivals at the main terminal and each of the upper floors follow a Poisson distribution. Some statistical evidence that suggests that this is not an unreasonable assumption will be presented. Also, using the same assumption the average queue size in the main terminal, the mean passenger waiting times, the average number of busy lifts aand the passengers' delay times are evaluated considering a multi-car lift system as a multi-channel bulk service queuing problem.

1. INTRODUCTION

In the first lift installations in multi-storey buildings the main considerations taken into account were the position and the space occupied by the lift wells. The number of lifts to be installed and their control were based on empirical methods, which have proved adequate only for designing lift installations for buildings with a small number of floors and low traffic. Thus research started to achieve improvements by using more sophisticated methods, taking into account the way the incoming passengers seek service, the randomness of people approaching the lifts and the random character of passengers' destinations. The modelling of lift systems has moved along two courses: mathematical models and computer simulation models.

Here, some lift traffic design formulae derived using statistics, probability and queuing theory, are presented and the possibility of how a mathematical investigation can help designers of lift installations is examined.

2. PASSENGERS' ARRIVAL PROCESS

It is a difficult task to obtain data on the arrival pattern of passengers approaching a lift system at the main terminal of a multi-level building. Observation is the only procedure that can be employed. Because of the lack of any effective automatic method, which could be used to record the number of arriving passengers, the use of observers must avoid disrupting the normal passenger activity arising mainly from a suspicion of being watched. However, a continuous presence of observers causes a familiarity and a relaxation of this suspicion.

Observations can be made, either over a consecutive number of days, or on the same day of a consecutive number of weeks. This latter procedure is preferred as traffic patterns are most likely to be similar on the same day in a week, provided that there are no seasonal changes and the dates do not fall on special days (i.e. holidays, transport strikes, severe weather).

Thus few surveys have been made of the frequency distribution of passenger arrivals at the main terminal of multi-level buildings. To correct the situation three buildings (Table 1) have been surveyed on selected

Table 1 — Buildings used in Poisson survey tests

Building	N (floors)	L (cars)	CC (persons)	Population (persons)
1	15	3	10	397
2	20	4	20	1657
3	24	8	24	1845

days and selected periods. As a Poisson distribution has been found in other areas of pedestrian movement; the main concern of the survey was to determine whether the assumption that the passenger arrivals at the main terminal of a multi-level building follow a Poisson distribution, holds in the real world.

The procedure which has been followed was:

(i) find the observed (actual) frequencies of passenger arrivals;
(ii) estimate the mean arrival rate;
(iii) estimate the variance;

(iv) find the theoretical frequencies of passenger arrivals;
(v) estimate 'chi-squared';
(vi) compare the values obtained for chi-squared with standard statistical tables of chi-squared for Poisson distributions;
(vii) postulate a 'goodness of fit' as poor, fair or good.

Having tabulated and plotted (Fig. 1) the acquired data certain conclusions can be made.

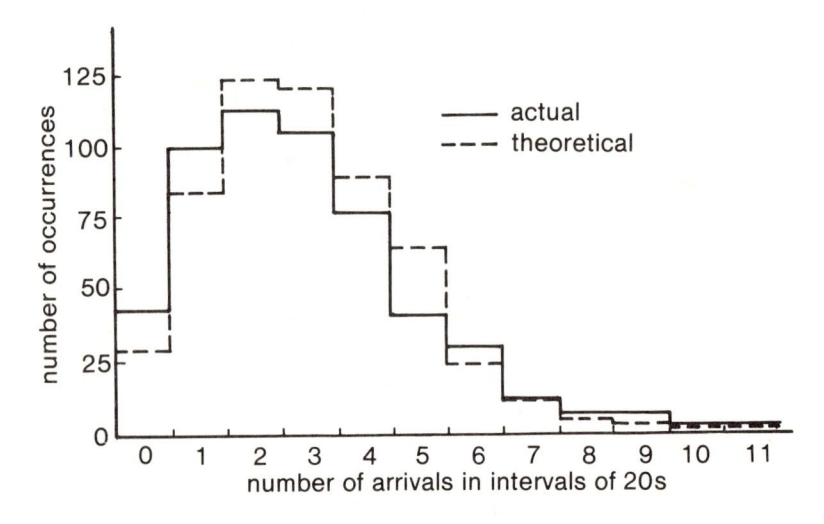

Fig. 1 — Comparison between actual arrival distribution and Poisson arrival distribution for Building 2.

(a) Comparison of the observed and theoretical values calculated for the mean and variance showed a *Poisson fit to be reasonable*.
(b) The chi-squared goodness-of-fit tests gave evidence that the Poisson arrival rate assumption *at least cannot be rejected*.

Thus, although there may be other theoretical distributions which may better accommodate the data (but taking into account the resulting complexity for more complicated distributions, and that the traffic pattern will not be exactly the same each day, owing to factors outside and inside the building) the Poisson distribution must be considered as a *good approximation to the actual empirical distribution*.

3. QUALITATIVE MEASURES OF MULTI-CAR LIFT SYSTEMS IN UP-PEAK PERFORMANCE

The qualitative study of lift systems performance requires a complicated mathematical analysis. This analysis considers a multi-car lift system as a multi-channel bulk service queuing problem with the passengers' service in batches by each lift car and has as a primary aim to derive those steady state probabilities, which are necessary to evaluate:

(i) The size of queue created at the main terminal of a multi-level building.
(ii) The time passengers have to wait at the main terminal until they enter the lift car.
(iii) The mean number of active lifts.

All of which characterize system utilization.

The analysis given here treats lift systems comprising more than one lift car, but it is restricted to the up-peak traffic condition only, and is based upon the following assumptions.

(i) Only the up-peak traffic pattern is to be considered. (This exists when the dominant traffic flow is in the upward direction with all or the majority of passengers entering the lift system at the main terminal of the building.)
(ii) Lifts do not leave without passengers.
(iii) Lifts return to the main terminal, even when there are no calls.
(iv) There is no limit to the length of the queue of passengers waiting for service.
(v) Passenger arrivals obey the Poisson process.
(vi) There are no priorities, passengers use whichever lift becomes free.
(vii) The queue service discipline is first in, first out.
(viii) Service is by batches of size no greater than the contract capacity of the lifts.
(ix) The service time for each batch is exponentially distributed.

None of these assumptions is unreasonable as most proprietary controllers operate as assumption (ii), (iii), (vi) and (viii) and passengers' behaviour often obeys assumptions (v) and (vii).

The formulae derived for each qualitative measure after a complex mathematical process are not given here. Only the curves resulting from these formulae are presented in each case.

3.1 Mean Terminal Queue Size

Figure 2 shows two sets of graphs for the mean queu size ($\bar{q}$) (i.e. the average number of passengers awaiting service) in the main terminal for the number of lifts $L = 1, \ldots, 8$ and $C = 6$ and $C = 24$ car capacities. Operational curves for car capacities 8, 10, 12, 16 and 20 are not shown since they are of similar shape.

Note that both axes of Fig. 2 have been normalized for the purposes of comparison. The vertical axis $\bar{q} (LC)^{-1}$ represents the normalized average number of queuing passengers woth LC the total lift system handling cpacity. The axis R represents the facility utilization or percentage load.

Examining Fig. 2 for $L = 1$ shows that the queue length at the main terminal grows at an alarming rate when the facility utilization factor

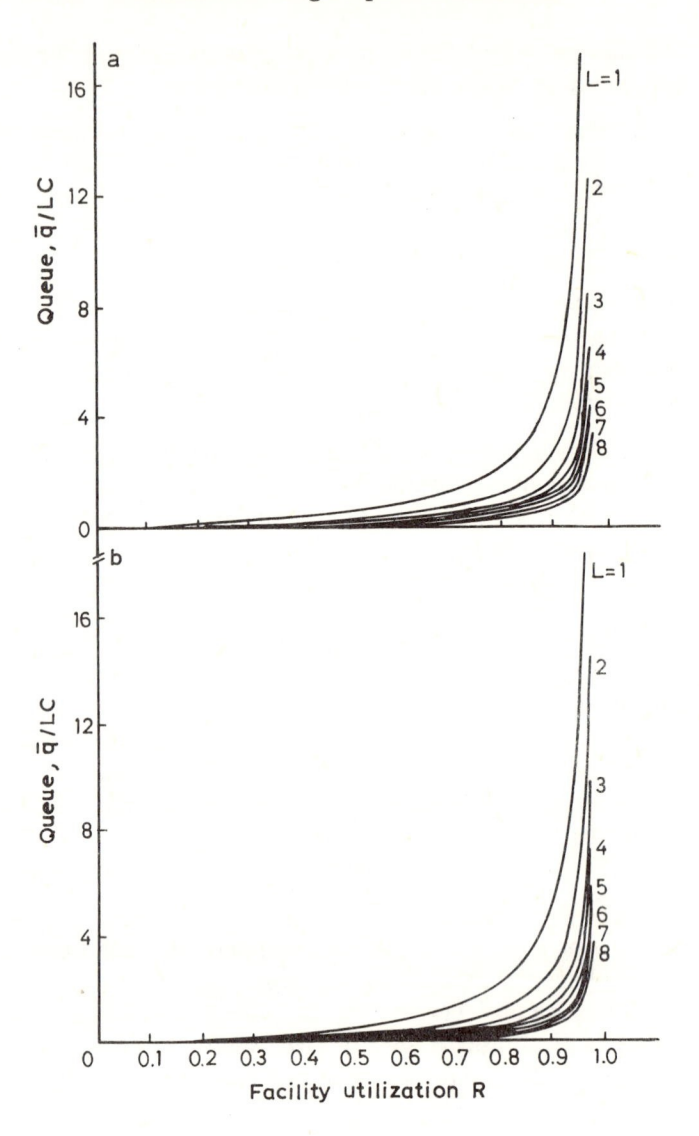

Fig. 2 — Mean queue length in terminal for multi-lifts, (a) $C = 24$; (b) $C = 6$.

becomes greater than 80%. Increasing the number of lifts (L) allows higher utilization factors before queue lengths begin to grow rapidly. But in all cases there is an upper limit at about 90% loading.

3.2 Mean Passenger Waiting Time
Figure 3 includes the two special cases of car capacities of $C = 6$, $C = 24$ (other car capacities lead to similar graphs), for the mean passenger waiting time ($\bar{w}$) in the main terminal, versus the facility utilization factor (R).

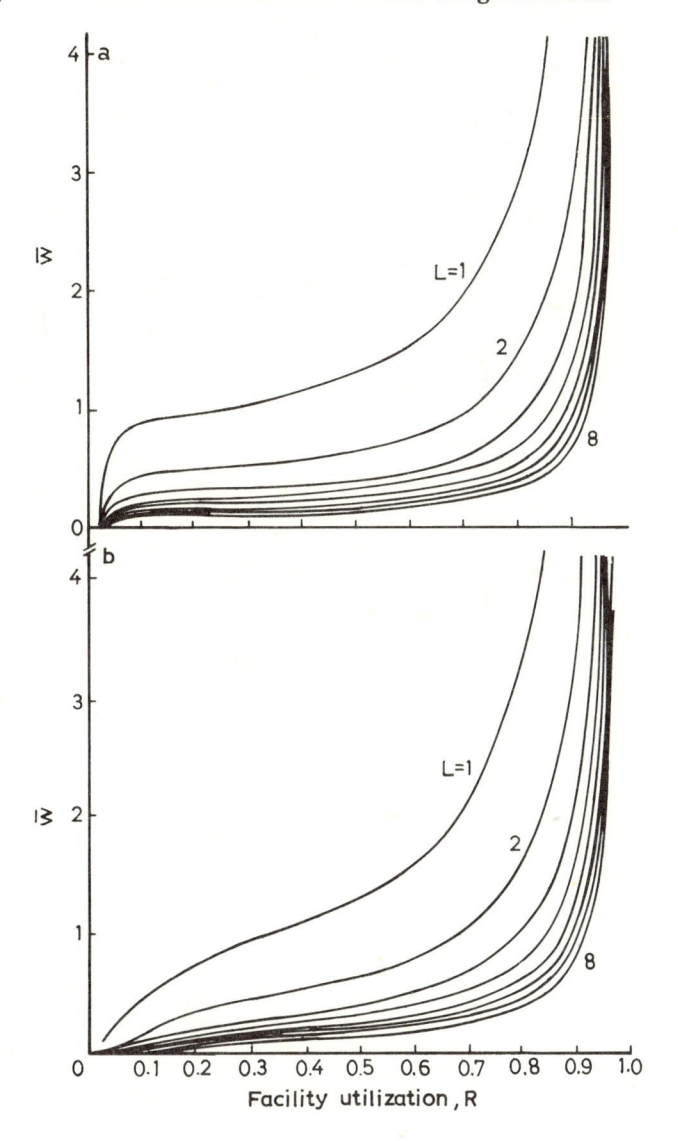

Fig. 3 — Mean passenger waiting times, $\bar{w}$ (a) $C = 24$; (b) $C = 6$.

Similar to the average queue length graphs, the mean waiting time grows rapidly at about 80% facility utilization, and slightly better results are achieved through increasing the number of lift cars L.

3.3 Average Number of Busy Lifts

Figure 4 shows to sets of graphs for the average number of busy lifts ($\bar{L}$).

Each graph shows 12 curves corresponding to the car contract capacity ($C = 2, 4, 6, \ldots, 24$) and for lift systems of one and eight cars (multi-lift systems with numbers of cars between one and eight exhibit similar curves to Fig. 4).

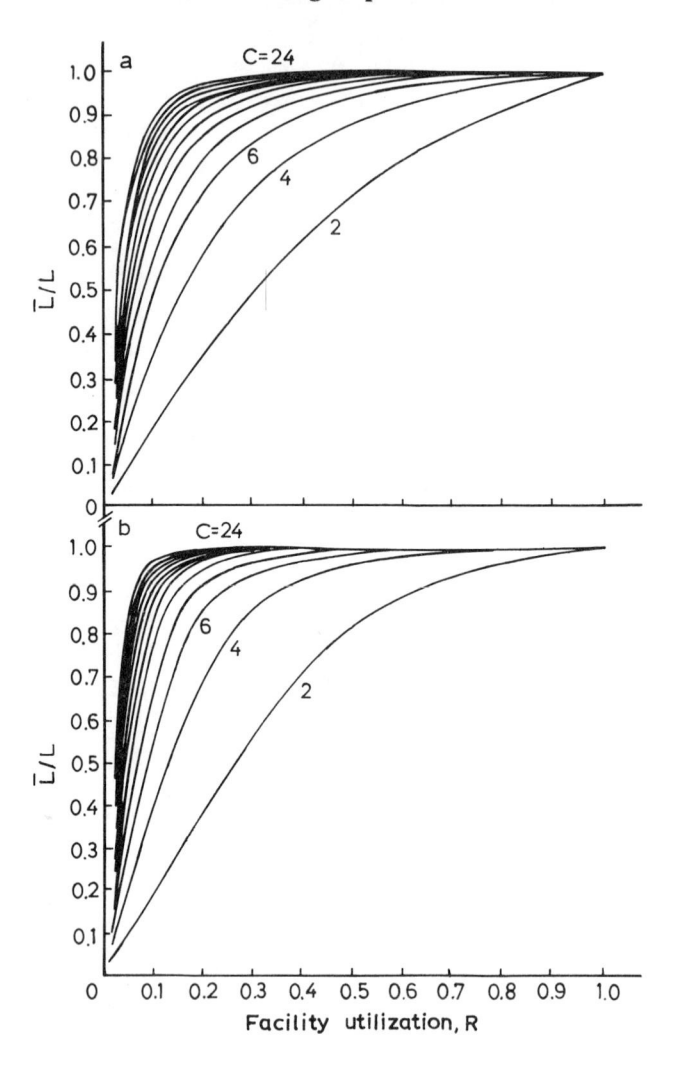

Fig. 4 — Average number of active lifts (a) $L = 1$; (b) $L = 8$.

From these curves it can be seen that almost all of the lifts become busy at quite low values of facility utilization. As the car capacity increases lifts become busy for the same facility utilization factor R, the change becoming limited both for large car capacities and values of R greater than 40%.

4. GENERALIZED FORMULAE FOR THE AVERAGE HIGHEST REVERSAL FLOOR AND EXPECTED NUMBER OF STOPS

The fundamental design criterion to determine lift performance is the evaluation of the round trip time (RTT), which is defined as the time in seconds for a single car trip around the building from the time the first passenger enters the car until the doors reopen after the car has returned to the main terminal of the building. The RTT can be expressed as a function of the highest reversal floor (H), which is the highest floor serviced by the lift car during each trip, the number of passengers carried on each trip (P) and the lift dynamics. The lift dynamics are essentially the car door opening times, inter-floor travel time and lift contract speed. Expressions for these variable as well as the RTT have been developed by variety of workers. In particular Barney and Dos Santos (1977) have developed a simple expression for the RTT as:

$$\text{RTT} = 2Ht_\upsilon + (S+1)t_s + 2Pt_p \qquad (1)$$

where S is the mean number of stops made above the ground floor, P is the average number of passengers carried each trip, t_p is the average time for a passenger to enter or leave the lift car, t_s is the time associated with each stop, and t_υ is the time taken to travel between two floors at contract speed (i.e. t_υ = interfloor distance x (contract speed)$^{-1}$).

Equation (1) can only be evaluated when values can be assigned to h, s and P. Usually the design is carried out for up-peak traffic only with equal floor population. Formulae for the case of unequal floor populations can be derived for up-peak and the two common probability distribution functions, uniform and Poisson. But analysis of down-peak and inter-floor traffic is difficult analytically.

Statistical analysis has shown (Alexandris, 1977) that the most *general traffic condition* of interfloor traffic with unequal floor demand and population can provide formulae.

Consider a building with:

N floors above the main terminal,
U the total building population above the main terminal,
U_i, U_j the population of floors i, j,
λ_i the rate of passenger arrivals of the ith floor,
T the lift cycle time.

Assume that:

(i) the probability of no calls from the ith floor to the jth floor is given by

$$\text{pr}_{ij} = e^{-\lambda T} \frac{U_j}{U} \qquad (2)$$

for $i, j = 0, 1, 2, \ldots, N$.
(ii) $\text{pr}_{ii} = 1$, that is there are no calls registered from one floor to itself.
(iii) No passenger goes to the main terminal from an upper floor *except* to leave the building.

All these assumptions are realistic for a practical lift system.

4.1 Evaluation of H

Considering the general traffic condition:
The ith floor ($i=1,2,\ldots,N$) will be the highest reversal floor, either because passengers enter the lift at the main terminal (event A_0) or because passengers enter the lift at any other floor (events $A_1, A_2, \ldots, A_N$).

Assuming that passengers enter the lift car at the ith floor, the probability π_{ij}, that the jth floor will be the highest reversal floor is given by:

$$\pi_{ij} = (1 - p_{ij}) \sum_{k=j+i}^{N} \prod p_{ik} = \mathrm{pr}(A_i) \qquad (3)$$

for $i = 1, 2, \ldots, N$, and $j = i, i+1, \ldots, N$, and pr_{ij} is defined by eqn (2).

As the events $A_0, A_1, \ldots, A_N$ are not mutually exclusive, Poincaré's formula can be applied:

$$\mathrm{pr}\left(\sum_{k=0}^{N} A_k\right) = \sum_{k=0}^{N} \mathrm{pr}(A_k) - \sum_{\substack{k_1, k_2 = 0 \\ k_1 < k_2}}^{N} \mathrm{pr}(A_{k_1}, A_{k_2})$$

$$+ \sum_{\substack{k_1, k_2, k_3 = 0 \\ k_1 < k_2 < k_3}}^{N} \mathrm{pr}(A_{k_1} A_{k_2} A_{k_3})$$

$$+ \ldots + (-1)^N \mathrm{pr}(A_1, \ldots, A_N) \qquad (4)$$

Assuming that the events $A_0, A_1, \ldots, A_N$ are independent, i.e. the arrivals at each floor are independent of each other then eqn (4) simplifies:

$$\mathrm{pr}\left(\sum_{k=0}^{N} A_k\right) = \sum_{k=0}^{N} \mathrm{pr}(A_k) - \sum_{\substack{k_1, k_2 = 0 \\ k_1 < k_2}}^{N} \mathrm{pr}(A_{k_1})\mathrm{pr}(A_{k_2})$$

$$+ \sum_{\substack{k_1, k_2, k_3 = 0 \\ k_1 < k_2 < k_3}}^{N} \mathrm{pr}(A_{k_1})\mathrm{pr}(A_{k_2})\mathrm{pr}(A_{k_3})$$

$$+ \ldots + (-1)^N \sum_{k_1=1}^{N} \mathrm{pr}(A_{k_i}) \qquad (5)$$

Setting:

$$\mathrm{pr}\left(\sum_{k=0}^{N} A_k\right) = \pi_j$$

Ch. 2]   **Statistical evaluation of traffic design formulae**   31

as the probability that the *j*th floor is the highest reversal floor then:

$$\pi_j = \sum_{k=0}^{N} \pi_{kj} - \sum_{\substack{k_1,k_2=0 \\ k_1<k_2}}^{N} \pi_{k_1j}\pi_{k_2j} + \sum_{\substack{k_1k_2k_3=0 \\ k_1<k_2<k_3}}^{N} \pi_{k_1j}\pi_{k_2j}\pi_{k_3j}$$

$$+ \ldots + (-1)^N \prod_{k_j=1}^{N} \pi_{kj} \tag{6}$$

Hence by definition the average highest reversal floor *H* is:

$$H = \sum_{j=1}^{N} j\pi_j = \sum_{j=1}^{N} j \sum_{k=0}^{N} \pi_{kj} - \sum_{\substack{k_1,k_2=0 \\ k_1<k_2}}^{N} \pi_{k_1j}\pi_{k_2j}$$

$$+ \ldots + (-1)^N \prod_{k_1=1}^{N} \pi_{kj} \tag{7}$$

where $\pi_{ki} = (1 - \mathrm{pr}_{ki}) \sum_{m=i+1}^{N} \mathrm{pr}_{km}$.

Since $\mathrm{pr}_{kk} = 1$, then $\pi_{kk} = 0$ and eqn (7) simplifies to:

$$H = \sum_{\substack{j=1 \\ k \neq j}}^{N} j \sum_{k=0}^{N} \pi_{kj} - \sum_{\substack{k_1,k_2=0 \\ k_1<k_2 \\ k_1,k_2 \neq j}}^{N} \pi_{k_1j}\pi_{k_2j} + \ldots + (-1)^N \prod_{\substack{k_j=1 \\ k_i \neq j}}^{N} \pi_{kj} \tag{8}$$

Equation (8) thus represents a value for *H* for the case of the most general traffic condition.

4.2 Evaluations of *S*
Using the same assumptions as stated in the derivation of *H*, a similar derivation for *S* can be presented. Since pr_{ki} has been defined as the probability of no calls being registered from the *k*th floor to the *i*th floor, then the probability that at least one call has been registered from *k*th to the *i*th floor becomes:

$$W_{ki} = 1 - p_r k_i = 1 - e^{-\lambda_k T} \frac{U_j}{U} \tag{9}$$

The number of stops made by a lift car in an *N* floor building is:

$$S = S_1 + S_2 + \ldots + S_N \tag{10}$$

Where the variables S_i, $(i = 1, 2, \ldots, N)$ are assumed to be independent random variables, which take the values 1 or 0 depending on whether or not the ith floor is a stopping floor. The expected number of stops S, is given by:

$$S = E(s) = \sum_{i=1}^{N} E(S_i) \tag{11}$$

The probability that $S_i = 1$ can be evaluated using Poincaré's formula

$$p_r(S_i = 1) = \sum_{k=0}^{N} W_{ki} - \sum_{\substack{k_1, k_2 = 0 \\ k_1 < k_2}}^{N} W_{k_1 i} W_{k_2 i}$$

$$+ \ldots + (-1)^N \prod_{k_j = 0}^{N} W_{k_j i} \tag{12}$$

as the ith floor will be a stopping floor because a call has been registered from the main terminal, or the first floor, or the second floor, ..., of the Nth floor. W_{ki} is given by eqn (9).

Similarly to the derivation of H, $\text{pr}_{kk} = 1$ and $W_{kk} = 0$ (assumption (ii)) then eqn (12) becomes:

$$S = \sum_{i=0}^{N} \sum_{\substack{k=0 \\ k \neq i}}^{N} W_{ki} - \sum_{\substack{k_1, k_2 = 0 \\ k_1 < k_2 \\ k_1, k_2 \neq i}}^{N} W_{k_1 i} W_{k_2 i}$$

$$+ \ldots + (-1)^N \prod_{\substack{k_j = 0 \\ k_1 < k_2 < \ldots < k_N \\ k_1, k_2, \ldots, k_N \neq N}}^{N} W_{k_j i} \tag{13}$$

Equation (13) thus represents a value for S for the case of the most general traffic condition.

4.3 Simplification of Generalized Formulae

Equations (8) and (13) represent generalized formulae for H and S for *any* traffic condition. Some simplification of these formulae which are quite complex and require mathematical abilities and the aid of a digital computer to apply them easily and successfully are given below.

(i) Passenger arrivals at certain floors

Assume there are no arrivals at certain floors, i.e. there may be a floor population but they are inactive at the time of interest. For example, assume arrivals occur only at the main terminal and floors 4 and 5, then eqn (8) for H becomes:

$$H = \sum_{j=1}^{N} j[\pi_{0j} + \pi_{4j} + \pi_{5j} - (\pi_{0j}\pi_{4j} + \pi_{0j}\pi_{5j} + \pi_{4j}\pi_{5j}) + \pi_{0j}\pi_{4j}\pi_{5j}] \quad (14)$$

where

$$\pi_{nj} = 1 - e^{-\lambda_n T}\frac{U_j}{U} \prod_{i=j+1}^{N} e^{-\lambda_n T}\frac{U_i}{U} \qquad n = 0,4,5 \quad (15)$$

and eqn (13) for S becomes

$$S = \sum_{i=0}^{N} [W_{0i} + W_{4i} + W_{5i} + (W_{0i}W_{4i} + W_{0i}W_{5i} + W_{4i}W_{5i})$$
$$+ W_{0i}W_{4i}W_{5i}] \quad (16)$$

where

$$W_{ni} = 1 - e^{-\lambda_n T}\frac{U_i}{U} \qquad n = 0, 4, 5 \quad (17)$$

(ii) Up-peak traffic

Considerable computational simplification arises if the inter-floor traffic has a dominant flow, e.g. as in the case of up-peak, namely:

$$H = N - \sum_{j=1}^{N} \prod_{i=N-j+1}^{N} e^{-\lambda T}\frac{U_i}{U} \quad (18)$$

$$S = N - \sum_{i=1}^{N} e^{-\lambda T}\frac{U_i}{U} \quad (19)$$

Further simplifications arise if the floor populations are equal i.e.:

$$\frac{U_i}{U} = \frac{1}{N}$$

giving:

$$H = N - \sum_{j=1}^{N}(e^{\frac{-\lambda\tau}{N}})^j \qquad (20)$$

$$S = N(1 - e^{\frac{-\lambda t}{N}}) \qquad (21)$$

(iii) Down-peak traffic

As in section (ii) formulae for the down-peak traffic condition can be obtained for unequal floor populations, namely:

$$H = \frac{N(N+1)}{2}\left(\sum_{k=1}^{N} \pi_{k0} - \sum_{\substack{k_1,k_2=0 \\ k_1<k_2}}^{N} \pi_{k_1}\pi_{k_20} + \ldots + (-1)^N \prod_{k_1=0}^{N} \pi_{k_10}\right) \qquad (22)$$

where

$$\pi_{k0} = (1 - e^{-\pi_k T}) \prod_{k=k+1}^{N} e^{-\pi_k T} \qquad (23)$$

$$S = \sum_{k=1}^{N} W_{k0} - \sum_{\substack{k_1=k_2=0 \\ k_1<k_2}}^{N} W_{k_10}W_{k_20} + \ldots + (-1)^{N+1}\Pi W_{k_0} \qquad (24)$$

where

$$W_{k0} = 1 - e^{-\lambda_k T} \qquad (25)$$

Further simplifications occur for equal floor populations, i.e.:

$$\frac{U_i}{U} = \frac{1}{N}$$

giving:

$$\sum_{k=1}^{N}\left(\sum_{k_1=1}^{N} \pi_{k_10} - \sum_{\substack{k_1,k_2=0 \\ k_1<k_2}}^{N} \pi_{k_10}\pi_{k_20} + \ldots + (-1)\prod_{k_1=0}^{N} k_10\right) \qquad (26)$$

where

$$\pi_{k0} = (1 - e^{-\lambda T})(e^{-\lambda T})^{N-k} \qquad (27)$$

$$S = N - \binom{N}{2}W_0^2 + \binom{N}{3}W_0^3 + \ldots + (-1)^{N+1}W_0^N \qquad (28)$$

where

$$W_0 = 1 - e^{-\lambda T} \qquad (29)$$

5. CONCLUSIONS

This chapter provides qualitative design curves for multilift systems that had previously been determined empirically by simulation.

These curves show that the average queue lengths and average queue waiting times in the main terminal increase at an alarming rate when the lift system utilization factor R (or loading) exceeds 80%. In both cases when R is approximately 90% the system design parameters reach unacceptable values, a fact that must be taken into account when designing a specific lift system.

The curves plotted for the mean number of busy or active lifts show that almost all lift cars become active for very low values of the utilization factor R. In fact at about the 40% utilization factor all cars are busy regardless of the lift car capacity.

Also the equations derived for H and S enable any realistic traffic condition to be analysed although some of the expression are complex.

Hence, the results presented here should be of particular use for lift system designers.

5. REFERENCES

Alexandris, N. A. (1977). Statistical models in lift systems, PhD Thesis, University of Manchester.

Barney, G. C. and Dos Santos, S. M. (1977). *Lift Traffic Analysis, Design and Control*, (Peter Pereginus, London).

3
A search for an index of lift traffic performance

Dr. M. Wareing, UMIST, Manchester, UK

ABSTRACT

Most practising life engineers are aware of the need for an 'ideal' indicator of lift traffic performance, which is independent of traffic demands and system configuration. The existence of such an indicator would allow objective system comparisons and more effective performance tuning. Current response indicators require demand characteristics in order to assess performance significance in a suitable index. Unfortunataly measurable information from the lift system has been limited to conventional landing and car call signals. Both Beebe (1980) and McKay (1980) variously investigated the merits of call registration rate and car stopping rate as demand indicators but found only limited success. The conception of 'busy period' (Levey, 1977) was felt by Beebe (1980) as offering a convenient indicator. However, recent simulation studies did not confirm this idea. Further simulation investigations have been performed which utilize passenger destination information at call registration time. Although considerable advantages are available in terms of supervisory control, the extra information has not proved useful when employed in a simple demand/response index. A complex demand/response relationship is thus indicated, but an adequate analytical representation is not yet available and the work of Alexandris (1977) suggests it may be difficult to attain.

1. SPECIFICATION FOR AN INDEX OF LIFT TRAFFIC PERFORMANCE

Performance indices in conventional systems analysis relate system input to output for representing system efficiency. Several workers either explicitly or tacitly observe the requirement for an index of lift traffic performance (McKay, 1980; Lauere, 1978; Beebe, 1980; Morrison, 1981). An index of lift traffic performance should be largely independent of traffic demand since it enables comparisons between different system epochs; and with suitable normalization allows comparison between different systems. Any analysis of potential indices of lift traffic performance should thus examine:

(i) Stability in the face of varying traffic demand;
(ii) Sensitivity to disturbances of interest in system tuning;
(iii) Consistency with accepted performance indicators.

2. LIFT TRAFFIC SIGNALLING (Background)

Two methods of registering passenger service requirements have evolved (a) landing and car call buttons; and (b) landing call stations registering passenger destinations. Method (a) is well known but Method (b) would require a different passenger attitude and extra instrumention and thus has not been implemented on real systems. McKay (1976) defined system response time as a compromise indicator of system performance as viewed by users of the Method (a) signalling arrangement.

3. PERFORMANCE INDICES USING CONVENTIONAL TRAFFIC SIGNALLING

3.1 Definition

McKay (1980) and Beebe (1980) report investigations of landing call registration rates and car stopping rates as demand indicators. McKay found a weak linear demand–response relationship and Beebe confirmed this. Beebe (1980) proposed 'busy' period defined by Levey et al. (1977) as a demand indicator but problems have been found. The author was thus led to define two alternative parameters:

(i) 'Demand service distribution' (DSD) representing an estimate of the time required to flight all cars around all currently registered car and landing calls according to a basic dynamic sectoring control algorithm.
(ii) 'Lost time' (LTM) measuring time spent by lift cars in a given sampling interval whilst stationary with open doors and with registered demands.

DSD and LTM are arranged be recalculated at 1 second intervals. For a moving sampling interval of 5 minutes the dynamic measures of 'demand service distribution' and 'lost time' are then simply summations of all DSD and LTM evaluations made within the moving sampling window (ΣDSD and ΣSRT). Similarly, system response time measurements are logged from the moving sampling window in parallel with the above demand characterization and summed (ΣSRT).

3.2 Simulation results

Results presented in this chapter result from simulations of a four car 10 person car capacity system operating in a 16-floor building under six levels of interfloor traffic. In addition each simulation was carried out with partitioning of +10% on door times, −10% on drive acceleration, +50% extra population of floor 8 and disturbed supervisor parameters. These four curves and the normal curve were plotted and found to have very small variations. Hence all graphs shown are plotted for the average of all four values.

Simulation analysis of the qualities ΣLTM and ΣDSD demand and ΣSRT response indicators appeared to display close similarities in their temporal variations. Assuming a 0.75 level of significance, significant ΣDSD versus ΣSRT and ΣLTM versus ΣSRT correlations were found for interfloor traffic (β) $\leq$ 0.7 where saturation of car capacity began to prevent ΣLTM (passenger transfer time) from reflecting ΣSRT. Average values of ΣDSD, ΣLTM and ΣSRT measures were found to vary considerably with interfloor traffic demand (β). Figure 1 depicts the interfloor demand variation by calculated up-peak interval (ADSD, ALTM, ASRT). Individual values of ADSD, ALTM and ASRT were obtained from 1 hour interfloor traffic simulations.

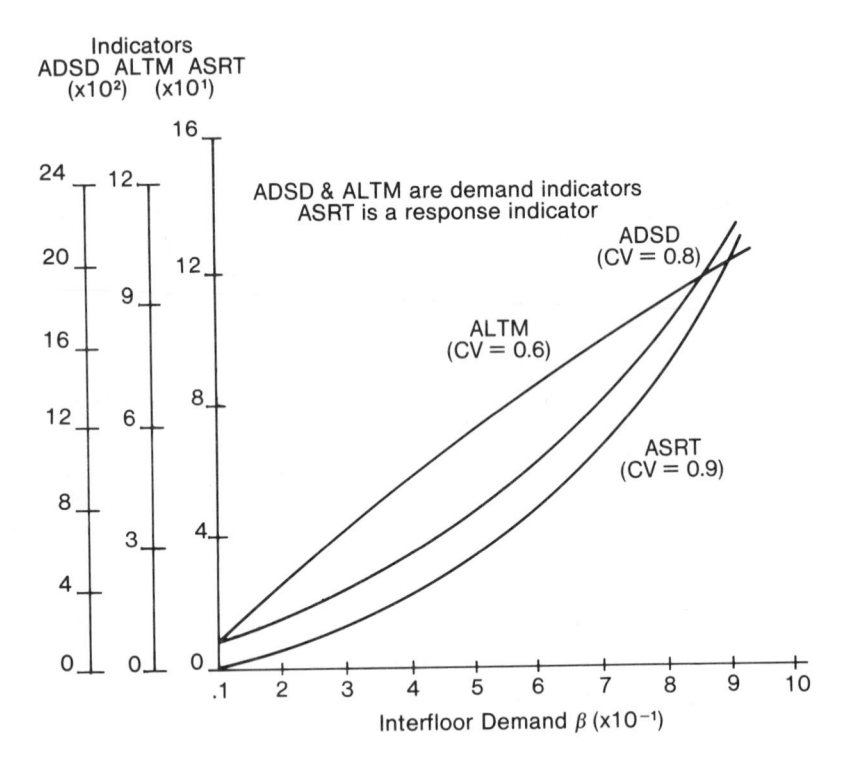

Fig. 1 — Conventional traffic information.

Figure 1 shows ADSD and ASRT with similar exponential increases although less pronounced for ADSD. Additionally ALTM appears to increase fairly linearly, which is not surprising given the dependence of ALTM on passenger transfer times. It can thus be seen that the relationship

Ch. 3]   **A search for an index of lift traffic performance**   39

of ASRT with ALTM or ADSD is definitely non-linear but dependent upon effects which are not accounted for in the correlation test assuming 0.75 levels of significance.

3.3 Definition of performance indices

With two demand indicators (ADSD, ALTM) and one response indicator (ASRT), it is necessary to consider how they may be employed in a performance index. Given the lack of supporting theory, an obvious approach is to assume a linear demand/response relationship. A simple demand/response index presents three possibilites:

$$PI1 = \frac{ADSD}{ASRT}, \quad PI2 = \frac{ALTM}{ASRT}, \quad PI3 = \frac{\frac{ADSD}{MNADSD} + \frac{ALTM}{MNALTM}}{\frac{ASRT}{MNASRT}}$$

(1)

Such indices can be expected to rise for improved performance since low values of a response would be maintained for increasing demand measure. PI3 overcomes considerable disparities in the relative magnitudes of ADSD, ALTM and ASRT by dividing by mean values obtained from the entire investigation range (MNADSD, MNALTM, MNASRT).

Figure 2 depicts interfloor demand variation of PI1, PI2 and PI3. All three performance index definitions exhibit demand variation. Coefficients of demand variation for the undistributed plots are shown on the graph.

Three important observations can be derived from these values:

(i) The size of variations suggests none of the indices will have value in terms of performance comparison applications.
(ii) The size of variations suggest none of the indices will have value in terms of performance comparison applications.
(iii) Demand variations might obscure the effects of tuning or disturbances in the system.

Investigation of (iii) was conducted by introducing four levels of disturbance. The effects of these disturbances has already been displayed in Figs. 1 and 2. A rather obvious conclusion to be drawn is the predominance of demand variations with respect to disturbance variations.

Additionally, comparing performance conclusions of PI1, PI2 and PI3 with ASRT (reflecting the traditional system response time indicators), a number of anomalies are apparent: door, drive and population disturbances are reflected by increased ASRT values for $\beta < 1.0$. PI1, PI2 and PI3 on the other hand indicate performance improvements at the same levels of demand. In view of the isolated use of ARST in the denominator terms of the performance indices, it is apparent that performance representation anomalies must be attributable to disturbance dependence of the demand

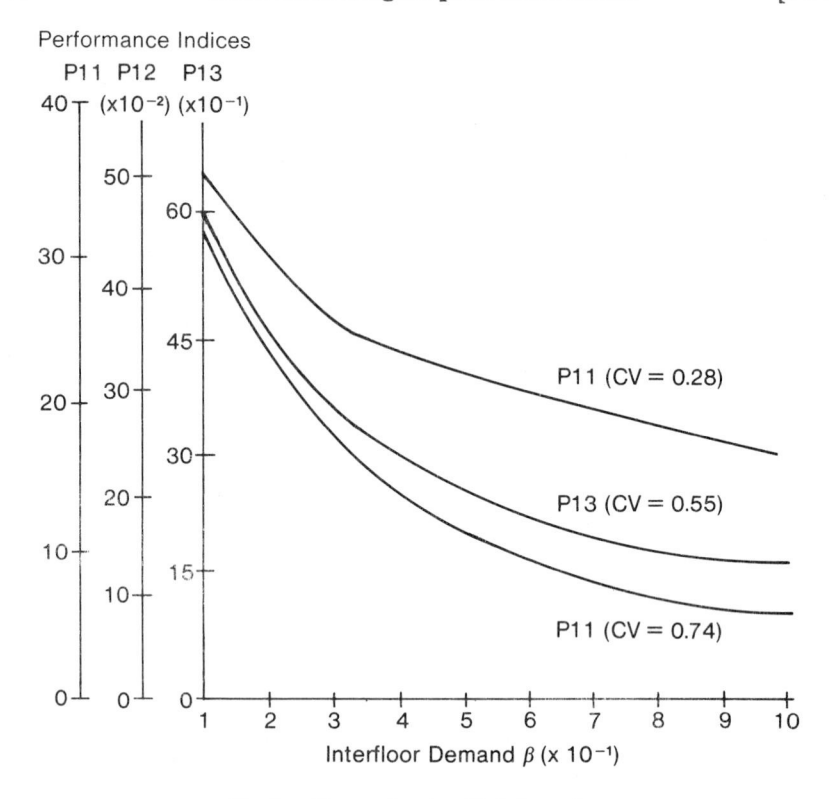

Fig. 2 — Conventional traffic information.

indicators. The failure of PI1, PI2 and PI3 to meet any of the functional requirements of a performance index (listed in section 1) suggests that further demand information must be investigated as a means of obtaining a more satisfactory performance index.

4. AN INDEX BASED IN EXTENDED TRAFFIC SIGNALLING

4.1 Definition
Extra demand information which might produce a disturbance independent demand characterization and enable a largely demand independent performance index includes:

(i) Passenger destinations at call registration time.
(ii) Number of passengers travelling to registered destinations.

Both have been observed in Section 2 to be unobtainable from conventional traffic signalling without employing an advanced signalling arrangement (e.g. 'call allocation', Close, 1970). Additionally provision of extra transducers is necessary to provide information on passenger numbers. If it is now assumed with modern technological developments that this is possible, how may it be employed in a future performance index?

Considering lift performance in terms of classical mechanics, it is possible to view traffic demand in terms of work necessary to effect passenger journeys (i.e. the product of pasenger masses with their required travelling distances). However, such a process necessitates acquiring accurate passenger masses whilst they are using the system. A convenient simplification can be obtained if all passengers are assumed to constitute 'unit' masses. Furthermore, acquisition of accurate travel distances can be obviated by defining them in terms of number of floors. Ignoring the effect of passengers in transit it is possible to define demanded work as:

$$DW = \sum_{i=1}^{NF} \sum_{j=1}^{NPQ(i)} TD(i,j) \qquad (2)$$

where $TD(i,j)$ is the requested travel distance (in floors) of passenger j, on floor i, $NPQ(i)$ is the number of passengers queueing at floor i and NF is the number of building floors.

DW represents an instantaneous measure of system demand. A dynamic indicator of demand may be defined as:

$$\text{Requested demand (RD)} = \sum_{n=1}^{LW} DW_n \qquad (3)$$

where DW_n is demanded work evaluated at time n of a moving sampling interval of length LW seconds.

Provision of extra passenger information at call registration time enables the measurement of passenger journey times. This must be considered an improvement in comparison to system response time measurements since it provides a more complete description of passenger–system interaction. A dynamic response indicator may thus be defined as:

$$TJT = \sum_{n=1}^{LW} JT_n \qquad (4)$$

where JT_n is the completed 'moving window' passenger journey time, at update n, of a moving sampling window of LW seconds.

Assuming a linear demand/response relationship the performance index may be defined as:

$$PI = \frac{ARD}{ATJT} \qquad (5)$$

where ARD is the average RD obtained from 1 hr simulated activity divided by calculated system up-peak interval, and ATJT is the average TJT obtained from 1 hr simulated activity divided by calculated system up-peak interval.

4.2 Simulation analysis

Figure 3 shows the ARD demand indicator, ATJT response indicator and PI performance index. Given the linear dependence of requested demand (RD) on passenger numbers (see eqn. (2) and (3)) it is not surprising that ARD should display a linear increase with traffic demand (β). Additionally

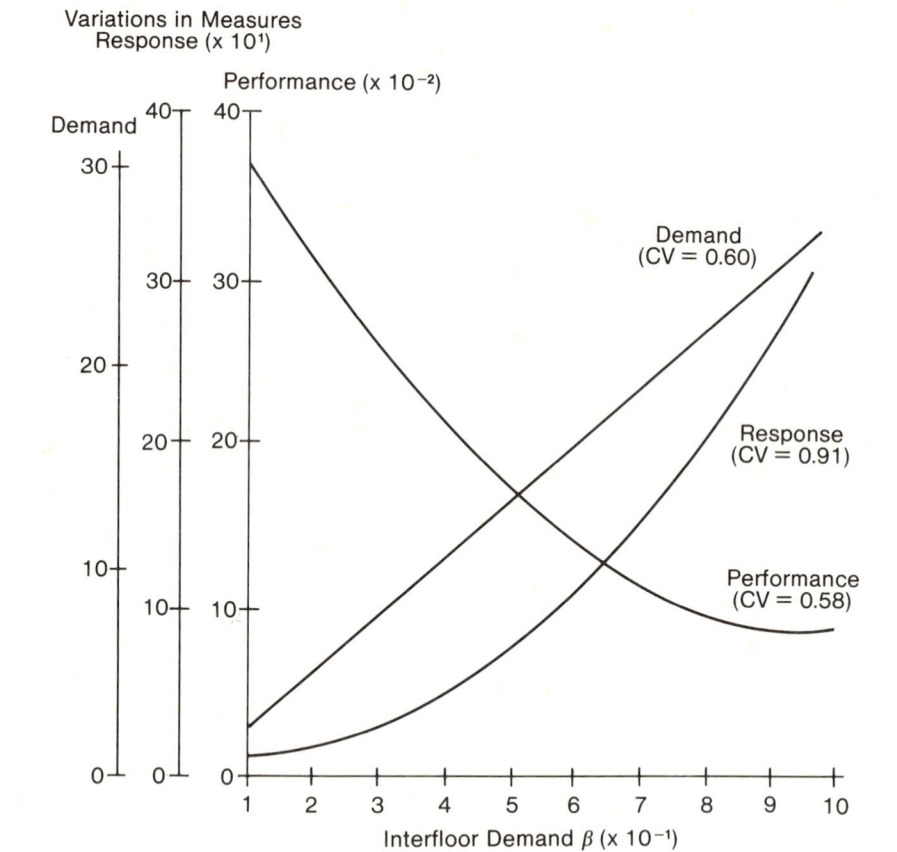

Fig. 3—Extended traffic information.

the ATJT response indicator displays an exponential increase with passenger demand (β) and reflects exponential increases in journey time reported by Moussallati (1974). Disparities of demand variation between ARD and ATJT can be seen to be responsible for the 58% variation in PI. PI does not therefore represent an improvement over performance indices based on conventional traffic signalling.

Examination of Figure 3 graphically demonstrates the overwhelming demand variation with respect to the applied disturbance variations. It is thus possible to conclude that system tuning or disturbance effects would not be discernible from demand variations existing in practical systems.

Examination of the demand and disturbance variation of ARD in Fig. 3 reveals that only 'population' disturbance has any measurable effect on ARD. All other applied disturbances exert measurable effect upon the response indicator ATJT. It is therefore possible to see how door, drive and supervisory disturbances produce consistent PI and ATJT performance representation (e.g. decreased values of ATJT reflecting concomitant increases in PI). It can now be concluded that PI offers considerable improvement over the consistency of performance indices based on conventional traffic signalling except in representing traffic disturbances.

5. CONCLUSIONS AND DISCUSSION

Simulation analysis of performance indices assuming a linear demand response relationship and utilizing conventional controller signals demonstrated shortcomings in terms of stability, sensitivity and consistency requirements (Wareing, 1985). These difficulties were attributed to problems in demand characterization. Having exhausted the number of reliable signals which may be obtained from conventional traffic signalling, further demand and response characterizations were investigated utilizing passenger numbers and destination information which is currently unavailable from concentional traffic signalling. Simulation analysis of a performance index based on extended demand and response characterizations demonstrated improved consistency with indices based on conventional traffic signalling, but similar levels of demand dependence rendered it useless in performance comparisons or tuning applications.

Having exhausted all the demand and response information that may be obtained froma a working lift system it must now be concluded that stability and sensitivity problems of the onvestigated performance index must be attributable to an incorrectly assumed linear demand–response relationship. It is therefore apparent that a more representative relationship should be arrived at from either empirical relations on a trial and error basis or to construct a more representative mathematical model of lift system behaviour. In view of the experiences of McKay (1980b), Levey *et al.* (1977) and Alexandris (1977), both courses of action are likely to be complex and difficult to achieve. A more satisfactory approach might be to utilize the currently emerging technology of expert systems. Justification for this action derives from the considerable application success afforded to expert systems in fields which lack a supporting mathematical theory (e.g. medicine, chemistry, tax law, geology). A suitably configured 'expert system' following basic lift traffic engineering heuristics, performance profiling and extended performance criteria (Lim, 1983) might then obviate the requirement for a classical form of performance index.

6. REFERENCES

Alexandris, N. (1977). 'Statistical models in lift systems', PhD Thesis, CSC, UMIST.
Beebe, J. R. (1980). 'Lift management', PhD Thesis, CSC, UMIST.
Closs, G. D. (1970). 'The computer control of passenger traffic in large lift systems', PhD Thesis, CSC, UMIST.
Lauere, R. J. (1978). The measure of elevator performance, *Elevator World*, (September), 16–19.
Levey, D., Yadin, M. and Alexandrovitz, A. (1977). Optimal control of elevators, *Int. J. Systems Sc.*, B(3), 301–320.
Lim, S. H. (1983). 'A computer based lift control algorithm', PhD Thesis, CSC, UMIST.
McKay, E. M. (1980). Lift systems in high rise flats: an exploratory study of their traffic performance, *Building and Environment*, 15, 17–25.
Morrison, E. L. (1981). Evaluating elevator performance, *Elevator World*, (April), 39–43.
Moussallati, M. Z. (1974). 'Lift performance under balanced interfloor traffic', MSc Dissertation, CSC, UMIST.
Wareing, M. (1985). 'A Search for an Index of Lift Traffic Performance', PhD Thesis, CSC, UMIST.

Part 2
Practical aspects of lift traffic

4

Practices of handling lift traffic in hospitals

Mr. P. H. Beard, Wessex Regional Health Authority, Winchester, UK

ABSTRACT

The UK National Health Service owns approximately 7000 lifts with a replacement value in excess of £200 million, and is currently installing approximately 500 lifts per annum.

In the period 1968–1980, new hospitals were usually at least six storeys high. These hospitals have sometimes proved unpopular, with high operating and maintenance costs, and the lift service has often been unsatisfactory.

Most current UK hospital building is based on a standardized building template (nucleus system) usually of two storeys, connected to a linear hospital street. This has resulted in radical changes in lift policy, with the conflicting dilemma of single lifts at intervals along the street compared with pairs of lifts at greater spacing.

The Chapter will discuss methods of measuring actual traffic in completed hospitals and comparing with predicted traffic from computer studies. Also work being progressed on control systems to improve lift service.

Finally, the author will seek to promote better understanding of the characteristics of differing types of hospital traffic, in order that hospital lift service is comparable with standards expected in other modern public buildings.

1. LIFT TRAFFIC IN HOSPITALS

Lift traffic in hospitals is unpredictable, through the variety of its types. Staff, visitors, bed cases to and from operating and therapy departments, catering supply and disposal are among its components. The provision of dedicated lifts for different traffic is expensive and, where it has been provided, is capable of being abused, particularly by staff. Generally in the UK only bed/passenger lifts are provided, and in many cases clean supply/ dirty disposal traffic is not segregated.

2. NUCLEUS SYSTEM

Most current UK hospital building is based on a standardized building template, with variations for specialized departments usually of two storeys and connected to the hospital streets as shown in Fig. 1.

In the plan, Street Section 4 contains a single bed/passenger lift intended for carrying non-ambulant patients, disabled members of staff and visitors who have difficulty using the stairs. Street Section 8 contains a goods lift, which is the same size as the bed/passenger lift but is intended for stores deliveries, linen, sterile supplies and catering traffic.

Each lift provides a standby to the other, but it is necessary to make alterations to the supply and delivery schedules to avoid interference with priority patient needs. The separation of the lifts is approximately 60 m.

The basic plan allows for a spare lift shaft, which is used as a sub-waiting area, or cleaner's room. However, it is uneconomic to retrofit lifts and the provision of a pair of lifts in Section 4 would appear to be a better investment, in spite of the loss of the adjacent staircase.

The lift provision is clearly restricted in the example shown and is not readily conducive to traffic analysis.

In many recent hospital designs, the Nucleus design has been modified to three floors, with resultant increases in vertical traffic, and a greater need for closer examination of traffic flows.

Several Nucleus hospitals have incorporated ramps as a means of achieving vertical circulation. These have been combined with a material handling system using tugs and trolleys. Such a system replaces some lifts and can be used in cases of emergency for the evacuation of patients.

3. TRAM/NETSYS TRAFFIC MODELLING AND NETWORK ANALYSIS

The TRAM/NETSYS traffic modelling an network analysis programs have been developed by the UK Department of Health and are now marketed by Cedar Design Systems.

3.1 TRAM

TRAM allows traffic modelling between hospital departments and consists of two linked modelling programs, the first of which helps the user to balance the expected patient flow through the hospital departments; and the second which permits the movements of any group of people or goods between parts of any hospital to be examined for any 15 minute period of the day or night.

The programs have been validated and are available for general use. They work from simple input of departmental sizes against an existing data-

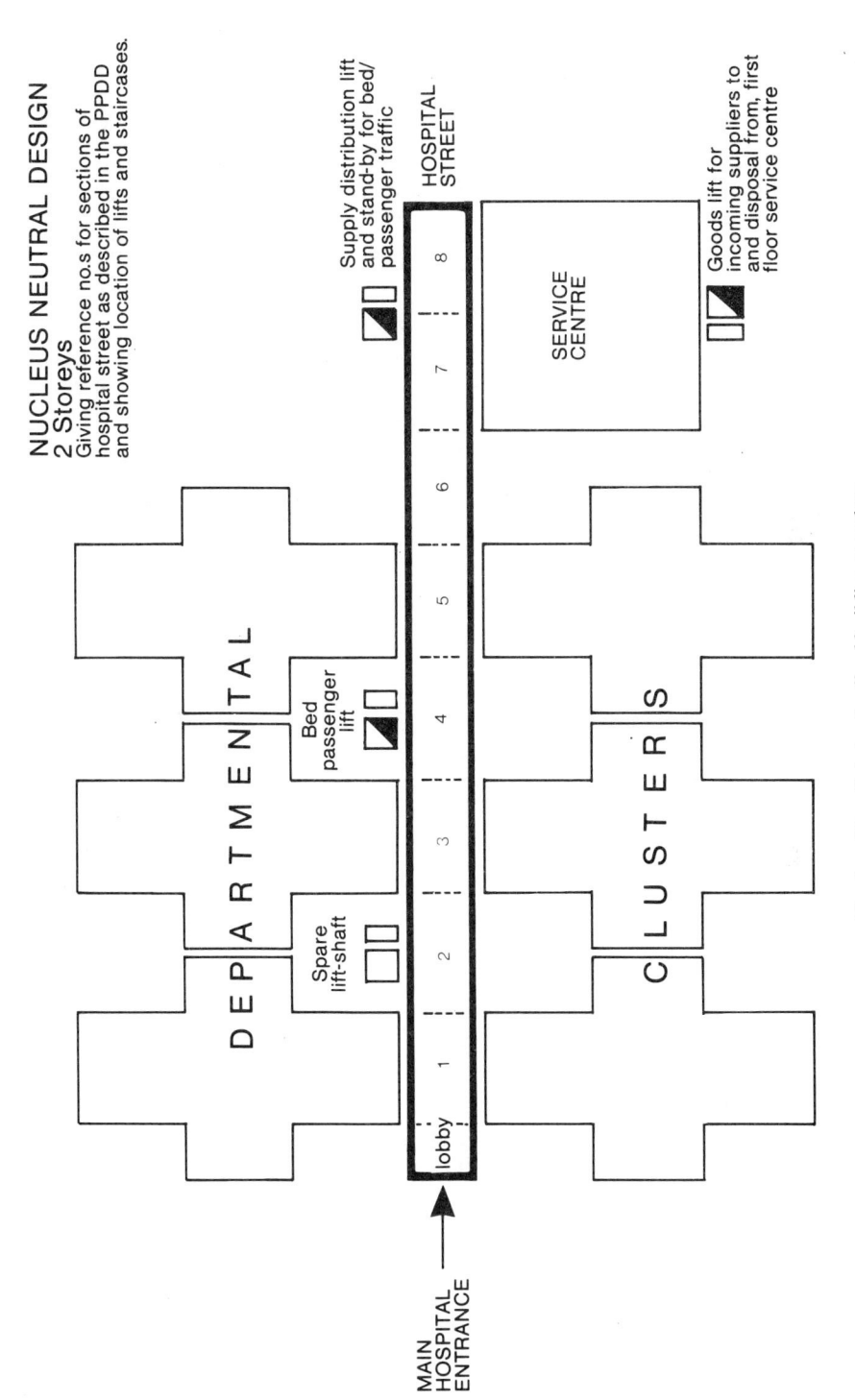

Ch. 4]   **UK practices of handling lift traffic in hospitals**   49

Fig. 1 — Hospital standardized building template.

base, any part of which can easily be altered by the user. There are 36 departments already in the data-base working on a 24-hour average day.

TRAM is a fast interactive or batch program. Simulation of various departmental mixes, which would normally take months of manual calculation, is made very easy. Any part of the hospital brief input, or of the existing data-base can easily be changed and the program re-run. Data are available for many different departments which can be applied if the user does not already have suitable data of his own.

The second TRAM modelling program is used with an input of a hospital development plan, to drive the NETSYS program.

3.2 NETSYS Traffic/Communication Network Analysis

The movement of any or all groups of people and goods on one or more links of the hospital communication network can be seen in tabular or graphical output. Simulation of traffic movements created by departmental sizes and the operational policies gives a profile of lift usage through a 24-hour day (Fig. 2). The use of corridors, lifts, stairs and ramps can be simulated for any part of the day.

3.3 Assessment of Service

The lift designer is presented with the estimation of the number of passengers (or passenger equivalents in the case of beds/trolleys with attendants) to be carried by the lift(s) in the specified period, 'say peak 5 min'. This must be equated with the handling capacity of the respective lift system, e.g:

$$\text{Handling capacity} = \frac{L \times 300 \times 0.5 \times CC}{RTT} \tag{1}$$

where L is the nuber of cars, CC is the contract capacity in persons, and RTT is the round trip time.

This assumes that the number of passengers carried each trip is 50% (say) of contract capacity.

It is expected that the average waiting time for a lift should not exceed 45 seconds as a measure of the grade of service.

3.4 Feedback

Nucleus hospitals completed at present have functioned for a relatively short period and it is premature to check whether the lift capacity will be adequate in the long term. Also, some hospitals have further phases planned which will result in heavy traffic between departments. Other bottlenecks are self-correcting as the organization adapts itself to the constraints, although causing delays elsewhere. The Nucleus policy of lifts results in a low provision compared with previous hospitals of more than two or three storeys. It is in line with the assumptions that ambulant staff and visitors will walk up one storey and down two storeys. Also, lifts will not be required for evacuation of patients during a fire except in exceptional circumstances.

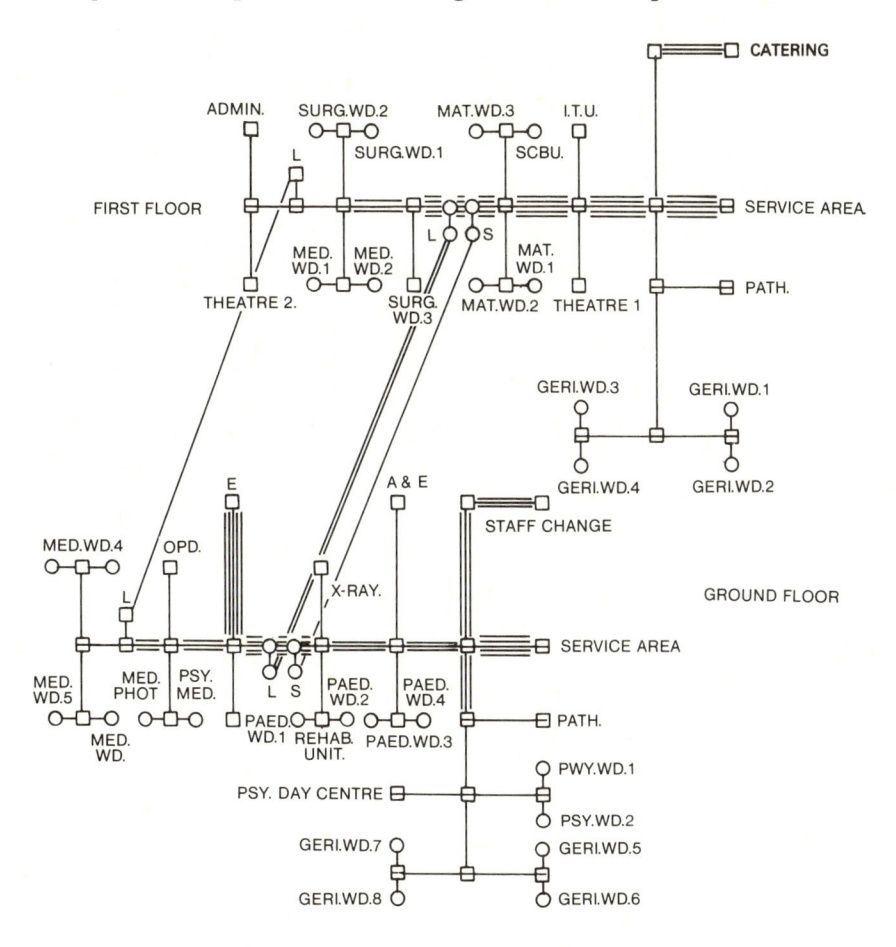

Ch. 4]   **UK practices of handling lift traffic in hospitals**   51

Fig. 2 — Example of a NETSYS output.

The result is that the lift contract may be less than 1% of the total building contract, compared with up to 10% for commercial buildings above six storeys.

4. CONTROL SYSTEMS

Single lifts serving only two or three levels are usually simple automatic push button (SAPB). Where a pair of lifts is installed, interconnected SAPB has been provided, while preset parking is sometimes used. Here, one lift is parked at each floor so that the descending lift automatically despatches the other lift to the upper floor. This system has the disadvantage that extra travel is generated. In most Nucleus hospitals, hydraulic lifts have been installed in preference to traction and it is an advantage to park these lifts at ground level. This is because automatic relevelling is provided and there will be less sinking of the lift through contraction of the oil, and leakage, when the lift is at the lower level.

5. QUALITY OF SERVICE

5.1 Improving Existing Lifts

In several cases where the lift service is unsatisfactory, surveys have been carried out by manufacturers, research organizations and others. These studies may have been performed over a period of only one or two days, which may detract from the accuracy of results since hospitals may have a variable traffic affected by operation departments and outpatient clinics etc. Many recommendations for improvements have not been carried out through lack of finance, bearing in mind that it is difficult to justify schemes with long payback periods.

Assuming that the complete replacement of the lift(s) is not possible and ignoring the appearance or décor, the only solution is to improve the existing lift drive and/or controls. For a multiple lift system, important factors are the total lift mileage and grade of service.

Reducing the total distance travelled by the lifts will directly reduce maintenance costs and energy consumption.

Now grade of service (G) is represented by:

$$G = \frac{WI}{2} + \frac{RTT}{4} \qquad (2)$$

where WI is the waiting interval and RTT the round trip time.

By reducing these times, the grade of service will be improved thereby increasing staff morale and even safety, since in an emergency a lift is likely to be available earlier than otherwise. Also, improved levelling may be achieved.

5.2 Schemes under Consideration

Schemes currently investigated by the author include:
 (i) Putting lifts into larger groups so that the grade of service is improved and 'double buttoning' is avoided. This will included microprocessor control systems with fault diagnosis coupled to the hospital building automatic system.
 (ii) a.c. variable voltage systems to replace Ward-Leonard drives to save energy.
 (iii) a.c. variable voltage drives to replace pole changers or single speed drives, for better levelling and smoother ride and to reduce brake wear.

5.3 User Opposition

Medical/nursing professionals are often opposed to larger lift groups through lack of confidence in the multi-lift system to deliver a lift quickly, without resorting to priority key arrangements. The technology is not trusted and the users do not like having 'all their eggs in one basket'. It is hoped that by demonstrating successful systems, this user-resistance will be overcome. By having all the lifts of a similar size and contract load, it is easier to group them. For example, in a hospital where it is expected that there will be only one orthopaedic patient with balkan beam attachment per week it is necessary to install a 2100 kg bed-lift. Unfortunately, this low usage is an obstacle to grouping this lift with slightly smaller lifts.

6. CRITICISM

In some hospitals, multiple lift installations have been in service for over ten years, maintained by the original manufacturer but yet often providing a poor grade of service. For various reasons, the manufacturer has not made serious proposals to the hospital authorities for, for example, an improved control system. The maker may feel that the proposals are not welcome or that there is no likelihood of money being available for any improvement.

In addition, the insurance company's surveyor has made six-monthly inspections and, while reporting defects affecting safety, has made no comments on, for example, factors which have caused excessive wear on components. The hospital has therefore had little assistance from its technical advisers in getting the best service from the lift system. As a result, the industry has undersold itself. Hopefully, this comment will go some way towards remedying his situation.

7. ACKNOWLEDGEMENT

The author would like to thank the Wessex Regional Health Authority for allowing the preparation and presentation of this chapter and the many parties who have influenced it. The opinions expressed are the author's and do not necessarily reflect the policy of the UK Department of Health and Social Security who own approximately 7000 lifts with a replacement value in excess of £200 million and is currently installing approximately 300 lifts per annum.

5

The traffic performance of lifts: principles and practice

Dr E. M. McKay, Rubicon Technical Services, Berkhamsted, UK

ABSTRACT

A passenger lift system which services its calls quickly may be said to have a good traffic performance. One which does so economically may be said to have a good traffic design. Fundamental to the whole subject of lift traffic design is the adoption of usable definitions for the terms traffic performance and traffic demand. Application of these concepts will be illustrated using results obtained from a study made by the Building Research Establishment of lifts in high-rise residential blocks of flats. Emphasis will be given to the value of empirical methods as a basis for selection of the system parameters when the only data available will be numbers of floors, bed-spaces, occupants etc. An account of the data logging system (built around a PDP-11 minicomputer) and of the methods of data analysis will also be included.

1. INTRODUCTION

A wide ranging study has been made, by the Building Research Establishment, of lifts in high-rise residential buildings. It was undertaken at the request of various departments of central government in the United Kingdom. The programme of work covered various topics which were said, or thought, to be of interest to those who live in such buildings, or to those responsible for the provision and maintenance of the lift systems. It also included studies of system reliability, vandalism and access control and tenant grievances. A major part of the work was the study (McKay, 1980b) of traffic performance. The objectives for this part of the programme were:

(i) To develop and field test a computer-based data logger to monitor lift systems in use.
(ii) To use the logger in eight high-rise residential blocks and to make a case study of the traffic and traffic performance of each system, and
(iii) To explore the potential for such detailed data as the basis for empirical traffic design procedures.

2. TRAFFIC PERFORMANCE CONCEPTS

2.1 Conventional design

Traffic performance calculations have been widely used as the basis for specifying the main traffic parameters of lift systems for office buildings. The typical procedure is a trial and error method which estimates for each trial configuration:

(i) The mean interval between car departures from the ground floor.
(ii) The number of persons carried, from the ground floor, within a specified interval — typically of five-minute duration.

The validity of this approach, and its underlying assumptions may be questioned on several counts. Perhaps more important is its very limited application, resulting from the extreme conditions of use assumed, i.e. saturated demand at the ground floor and zero demand elsewhere. These and other criticisms are discussed more fully elsewhere (McKay, 1976).

2.2 A more generalized approach

An important requirement of the investigation into lifts in high-rise residential buildings was therefore to find a new and more appropriate measure of traffic performance which can be:

(i) Applied to any passenger lift system.
(ii) Applied to any mode of traffic.
(iii) Expressed by floor or floor-group and/or by either or both directions of intended travel.
(iv) Readily obtained (directly or by derivation) from data logged automatically.
(v) Corresponds closely to a user's concept of traffic performance.

The variable proposed was System Response Time and defined as the time which elapses between the registration and cancellation of a landing cell. As is seen it satisfies completely requirements (i)–(iv). Also it is likely to constitute a major part of any user's (subjective) test as it is the longest wait-time associated with any given landing call.

If accepted as the traffic performance variable it remains only to decide on the probability of excess to be associated with the design value. For example it may be considered sufficient simply to employ the mean or median value. Alternatively the design value which might be exceeded only on 5%, or 1% of occasions, could be chosen as the index of traffic performance. 'Demand' information has value only (i), if it is required as input to a procedure for the prediction of traffic performance or (ii) is used to 'explain', to interpret the measured value of the traffic performance index. Different descriptors of demand are suggested for each application.

A demand descriptor is suggested to 'explain' experimentally obtained traffic performance data from an actual system. Also such a variable will be more useful if it corresponds only to a signal recognized by the system itself. Examples of this type might be calls and/or stops per unit time. For the prediction of traffic performance at the building design stage it makes more sense to use a procedure which requires as input only those data available at that time. Those data will be limited to building/occupancy parameters such as the numbers of floors or rooms, flats or bedspace or simply floor area etc.

3. THE DATA LOGGER

The description given in what follows is in outline only (McKay, 1980a). The data logging system comprised a PDP 11/05 minicomputer, centralized and distributed interfacing, a visual display panel, a keyboard/printer terminal and the recording medium.

A modular approach to system design permitted expansion if required of either of the initial limits (up to four cars and up to 24 floors served) although both limits were adequate for the eight systems monitored in this study.

Perhaps the most important aspect in the logger's design was that of deciding on the record format. The one adopted derived from the work of Swindells (1975). It comprised (a) time markers, at minute intervals, except for periods of lift system inactivity in excess of three minutes, (b) an event string generated at the time of closure of a car's doors, and (c) certain hourly trend data which could be optionally supressed. The event string contained the elements shown in Table 1.

In addition to signal sequence checking the program logic permitted other inferences such as door recycling at a floor, a zero response time (resulting from the car being already at the required floor) etc.

4. CASE STUDY DATA ANALYSIS AND RESULTS

Building, occupancy and lift system data pertaining to each of the eight sites are given in Table 2. With minor variations lift service provision was found to be one of two kinds; either an interconnected pair of cars with automatic return to ground floor or a pair of independently controlled cars each serving ground and alternate above-ground floors. While the number of floors per building varied little amongst the eight buildings the same cannot be said for

Table 1 — Event string definition

SEC:	seconds since last minute marker
FLR:	floor number
CAR:	car number
DIR:	direction set at time of door closure
RTU:	response time for UP landing call if any
TRD:	response time for DOWN landing call if any
DHT:	door hold-open time
DOT:	door opening time
DCT:	door closing time
ERR:	error code — generated by unexpected signal sequence
CRLF:	carriage return/line feed is event string terminator

Note. Times where indicated are output in seconds except for DOT and DCT which are in tenths of a second.

Table 2 — Building lift system details

	\multicolumn{8}{c}{Building}
	A	B	C	D	E	F	G	H
Totals above ground								
Floors	16	21	21	19	21	16	16	22
Floors served	15	21	20	19	20	15	17	21
Households	94	120	82	91	82	94	102	154
Bedspaces	244	200	222	182	222	244	272	440
Typical floor								
1 room flats		2						1
2 room flats	2	4	1	5	1	2	2	1
3 room flats	4		3		3	4	4	3
4 room flats								2
Total flats	6	6	4	5	4	6	6	7
Total bedspaces	16	10	11	10	11	16	16	20
System details								
Cars and capacity	2×8p	2×8p	2×8p	2×8p	2×8p	2×8p	2×8p	2×10p
Interconnected	No	No	No	Yes	No	No	Yes	Yes
Collective type	Down	Down	Full	Full	Full	Down	Full	Full
Home floor	—	—	—	Ground	—	—	Ground	Ground
Overal flight time	76	119	89	80	89	76	105	86
Door times:								
Hold open	5/5	4/4	5/10	12/16	6/8	3/5	10	7
Complete cycle	10/11	10/11	10/15	17/22	11/13	10/11	15	13

Notes: Rooms given exclude kitchen and bathroom. (Also taken as equal to the number of bedspaces B.)
Buildings A and F: 16th floor has 4 × 1 room flats.
Buildings C and E: 21st floor has 2 x 1 room flats.
Building D   18th and 19th floors each have 3 × 1 room flats.
Building B   1st floor unoccupied ('auxiliary' ground).
Building G   1st and 2nd floors unoccupied (storage floors).
Overall flight times = sum of flight times between terminal floors.
Two sets of door times given for systems with independent cars OR where there is a significant difference between the two cars.

the numbers of flats and bedspaces; these ranged from 82 to 154 and 182 to 440 per building, respectively. Records of lift system operation and use were obtained within the period 0630 to 2400 hours, over at least 10 weekdays and three weekends.

Initial trials were run on these data to determine the 'best' periods for detailed analysis. From these investigations an early morning and a late afternoon period, each of two hours were identified. Records of weekend use were discounted at this stage. Validation tests, based on the sequence of events, were made so as to discount 'spurious' data such as up and down landing calls registered by a single user. Tests were also run so as to reject records which gave evidence of major system faults.

Fully validated records were first analysed to determine whether response times showed any systematic dependence on height of floor above ground. As no consistent correlation was found all above-ground floors were subsequently treated as one homogeneous group.

All the a.m. and p.m. periods were then divided into four 30-minute periods. From each of these was extracted the numbers of calls and stops, and all the response times, for both above-ground and ground floors separately. Examination of these results revealed tht little would be lost, for design purpose, by taking only the whole two-hour values and corresponding statistics.

Tables 3 and 4 illustrate some of the case study results obtained from each site; demand/activity results from the p.m. period have been omitted from this outline treatment.

Table 3 — Demand/activity statistics: a.m.

	A	B	C	D	E	F	G	H
Event rates: above ground								
Sample mean: 2h	61	42	48	68	49	59	72	112
Worst-period value	75	53	45	73	42	74	77	135
Worst-case value	72	34	44	60	68	102	62	182
Call rates: above ground								
Simple mean: 2h	42	27	36	38	35	43	44	69
Worst-period value	59	30	35	37	30	57	43	90
Worst-case value	52	24	38	44	34	50	28	116
Stops above ground per ground floor departure								
Sample mean: 2h	2.0	1.6	1.5	1.7	1.6	1.7	2.1	3.1
Worst-period value	2.0	1.6	1.5	1.6	1.4	1.6	2.4	3.7
Worst-case value	2.8	2.3	1.8	2.0	1.9	2.7	2.4	4.8
Ratio of calls above ground to calls at ground								
Sample mean: 2h	3.6	1.8	3.5	3.0	3.4	3.7	4.5	1.9
Worst-period value	5.9	1.6	5.2	2.3	3.0	6.1	3.5	5.9
Worst-case value	5.2	1.7	6.3	2.9	1.7	1.9	1.9	4.5
Interval between ground floor departures (s)								
Sample mean: 2 h	119	134	114	90	116	103	105	102
Worst-period value	96	106	124	81	121	76	112	98
Worst-case value	138	180	150	68	109	95	138	95

Table 4 — Performance (response time) statistics

	A	B	C	D	E	F	G	H
A.M. values: above ground only								
Sample-mean: 2 h	34	44	40	31	35	33	42	53
Worst-period value	36	4842	31	36	38	46	57	
Worst-value	48	67	57	46	53	41	72	68
P.M. values (Ground only)								
Sample-mean: 2 h	27	36	28	5	33	26	14	19
Worst-period value	30	41	32	7	36	30	15	26
Worst-case value	40	75	47	43	54	52	31	47
(Above ground)								
Sample-mean: 2 h	34	41	39	27	45	31	37	52

Notes to Tables 3 & 4.
(1) Sample mean: 2 h = Mean, over the sample of (typically 10) days, of the two-hour values.
(2) Worst-period value = Sample-mean of the 30-minute values from that period with the highest relevant sample-mean response time, i.e. above-ground floors for a.m., ground-floor for p.m.
(3) Worst-case value = The value from the 30-minute period with the highest recorded response time mean.

5. POTENTIAL TRAFFIC DESIGN PROCEDURES

Response times obtained from this study were characteristically greater during the a.m. period than those of the p.m. period. Accordingly the search for an empirical approach to traffic design was made using a.m. data alone. A number of attempts at linear correlation between various system derived variables and response times were made. None gave a correlation coefficient of better than 0.48 and standard errors were all greater than around 20 s. A search was then made for a simple normalizing factor for the two-hour response times $T_1(120)$ for the above-ground calls during the a.m. period. Some of the potentially useful results are given in Table 5. The best of these shows a coefficient of variation amongst the eight buildings of the normalized variable of only 9%.

Table 5 — Normalized response time statistics a.m. values only (sample of eight buildings)

	$\dfrac{T_1(120)}{I(120)}$	$\dfrac{T_1(120)}{T_f}$	$\dfrac{T_1(120)}{T_f'+0.14B}$
Mean	0.36	0.44	0.32
Coefficient of variation	0.21	0.18	0.09

From this it might appear that the mean response time for above-ground calls $T_1(120)$ during the a.m. period may be predicted from a knowledge

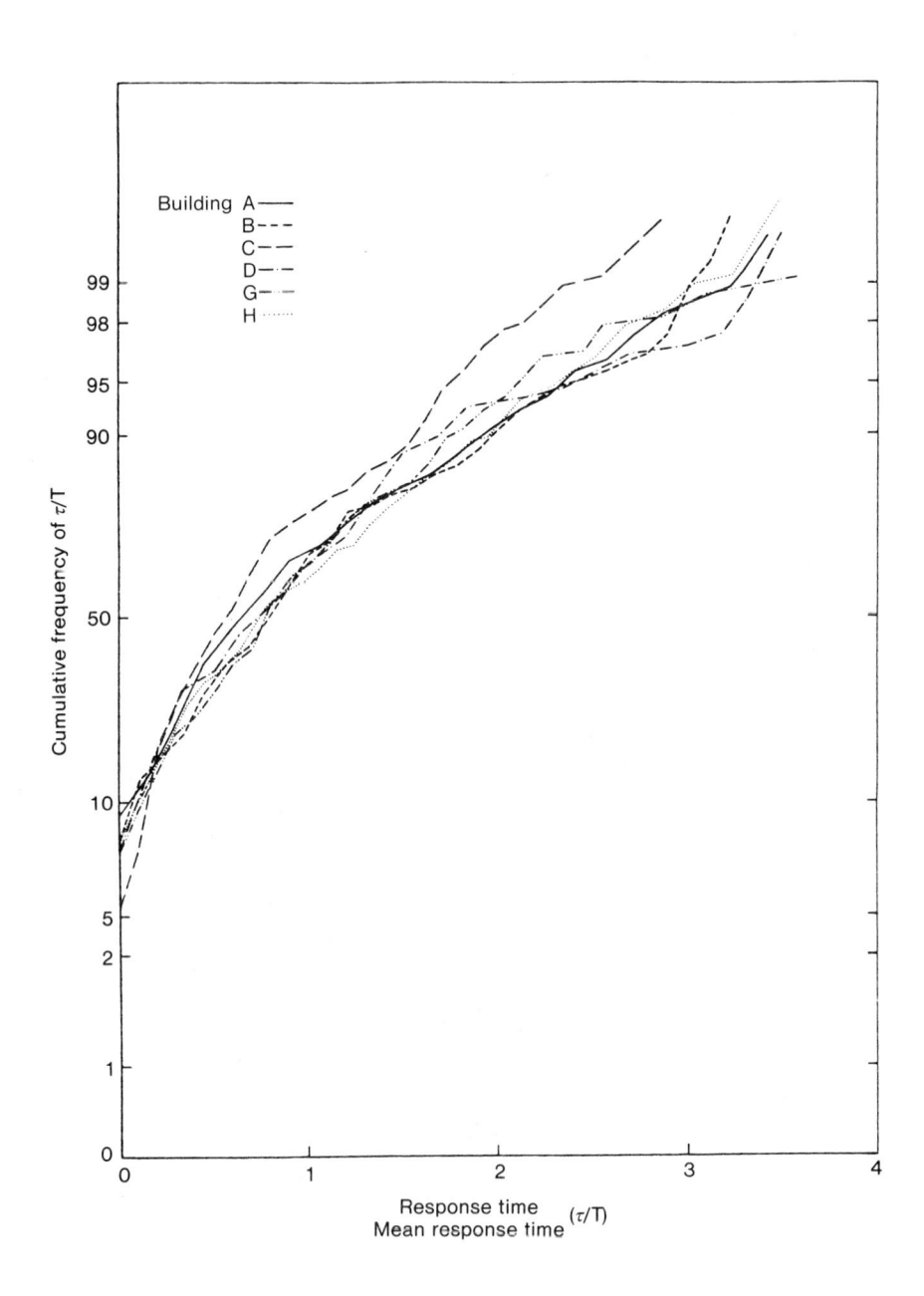

Fig. 1 — Cumulative frequency plots of the ratio τ/T: above-ground floors, 08.00–09.00 hours.

only of the number of bedspaces (B) and the sum of the flight times between terminal floors (T_f) using:

$$T_1(120) = 0.3\,(T_f + 0.14B)$$
$$= 0.3\,T_f + 0.04B \tag{1}$$

Finally it was found that the response time distributions from each building were very similar in shape (see Fig. 1). Calculations revealed a relatively constant *ratio*

$$\frac{\text{Response time exceeded on only 5\% of the calls}}{\text{Corresponding mean response time}} \tag{2}$$

Amongst the eight buildings this ratio had a mean value of 2.3 with a coefficient of variation of only 4%. This suggests as an alternative, a simple design formula for the prediction of the response time (to be exceeded on not more than 5% of the calls) as:

$$\tau(0.05) = 2.3\,T_1(120)$$
$$= 2.3\,(0.3T_f + 0.04B)$$
$$= 0.7T_f + 0.1B \tag{3}$$

6. ACKNOWLEDGEMENTS

The work described was carried out as part of the research programme of the Building Research Establishment of the Department of the Environment. This chapter is published by permission of the Director. The data logger was developed under BRE contract at UMIST under the supervision of Dr G. C. Barney.

7. REFERENCES

McKay, E. M. (1976) Proposals for a new approach to traffic design of passenger lift systems. *BRE Note N117/76*.

McKay, E. M. (1980a) A data logger for lift traffic engineering. *Elevator World*, **28** (1) 20–26.

McKay, E. M. (1980b) Lift systems in high rise flats: an exploratory study of their traffic performance. *Bldg & Environment*, **15** (1), 17–25.

Swindells, W. (1975) Software for computer control of lift systems. MPhil Thesis, UMIST.

6

Simulation and data logging

Eng. A. Lustig, S. Lustig Consulting Engineers, Tel Aviv, Israel

ABSTRACT

The simulation of lift systems has enabled theoretical studies of proposed and actual lift installation. The data logger enables the acitivity of an actual lift system to be monitored. This chapter shows how these two tools can be used together to verify and improve lift system design.

1. INTRODUCTION

The great advances in the electronics field and the development of microprocessors, have also entered the field of lifts, and together with the development of controls based mostly on pure microprocessors, they have contributed a great deal to the performance of traffic studies.

For the purpose of performing the different traffic studies, two stages can be distinguished:

(i) The first stage when the lift system has yet to be installed. At this stage it is possible to make several analyses on the way the lifts will behave when the building will be populated and will begin to operate.
(ii) The second stage when the lifts are already operating, the building has its own dynamics, and the initial problems have been solved.

This chapter will describe the most modern and advanced tools available to lift engineers for execution of traffic studies for each of the above stages, stimulation and data logging.

2. SIMULATION

The demand for performing simulations of lift systems has arisen owing to the advances in technology. This is reflected also by the development of advanced and sophisticated lift controls that have sometimes alerted lift engineers to the fact that the results received from using all the accepted formula that enable calculation of Highest Reversal Floor (H), Number of Stops (S), five minute. Handling Capacity (HC), Round Trip Time (RTT), etc. do not match reality in the building even if the results were obtained from traffic studies.

It is obvious to lift engineers that formulae whose results express averages cannot give solutions to those extreme cases such as very good situation, above average or below average situations.

With the realization that the accepted formulae are not reliable enough in many instances, and the beginning of computerization at companies, factories and institutions of higher learning have brought about the development of various simulation programs.

2.1 Lift Simulation Program (LSP)

The program developed is written in the Simula language (based on Algol) and is made up of several routines and sub-routines:

(i) Input routine where the characteristics of the builiding, its population, and the lifts are provided.
(ii) Simulation routine.
(iii) Output routine.

Figure 1 describes the main structure of the program.

Since the program should suit every type of lift, during development, great emphasis was put on making it as modular as possible and to permit the definition of every possible situation from the point of view of the building structure, behaviour of the population and the control of the lifts.

In the program, there are no basic assumptions that can not be changed, therefore every type of building or lift can be handled. Because of this flexibility, the program requires the definition of many parameters; as detailed in Table 1.

After defining all the parameters and the simulation program has been started, the program enables the user to obtain a very wide range of reports in tables and graphs.

3. THE DATA LOGGER

While performance of simulation in the various design stages is based on estimates, theories and professional experience, the operation of the data logger on the other hand, gives an exact and true picture of the real time behaviour of the lift systems in the building.

The data logger that will be described is the result of a number of years of experience, and was designed and built by using the most modern technologies from the point of view of equipment as well as program.

The system that was built is the most advanced stage in the development of a medium to collect information on lift behaviour. Its predecessor was the report that was filled in by observors stationed at the main lobby and in the cars. The role of these observers was to count the number of passengers

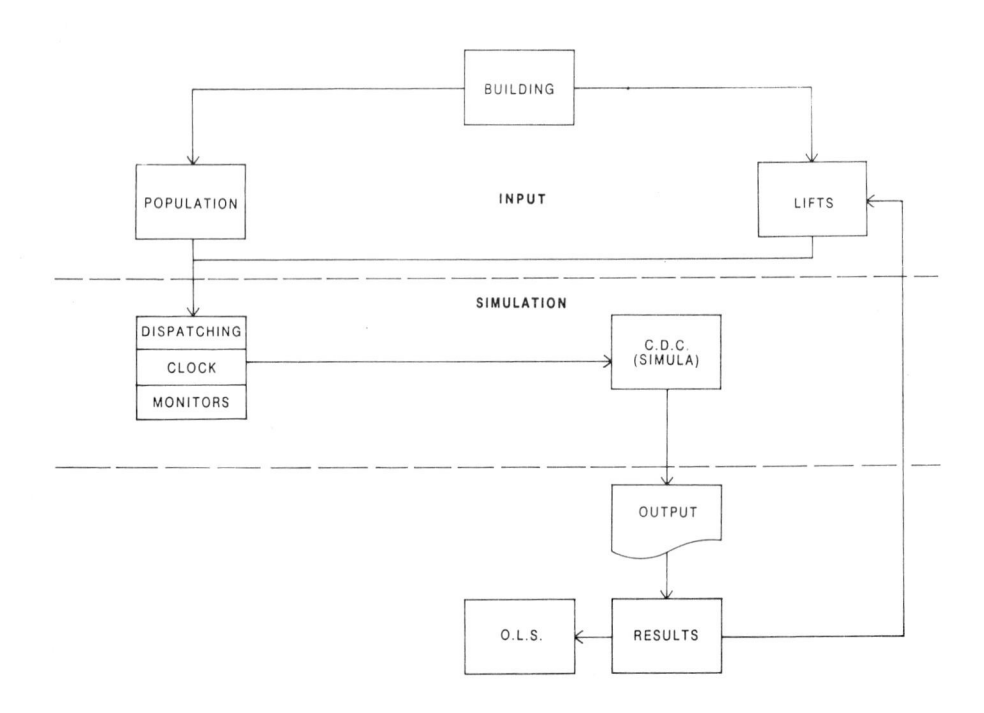

Fig. 1 — Structure of LSP.

Table 1 — Data required for simulation with LSP

Number of floors above M.L.	Total population
Lowest sub-floor	H.C. (%)
Main lobby	Dist. up
Alternate	Intra-up
Floor height	Down
No. of cars	From sub
Car capacity	To sub
Speed	Sub to 0
Acceleration and deceleration (s)	Upper floors dist.
Loading value (%)	Lower floors dist.
Loading interval (s)	Dispatching
Door width (m)	Simulation time (s)
Transfer time (s)	Begin time
Door opening time (s)	No. of monitors
Doors closing time (s)	Floors to monitor

Ch. 6]   **Simulation and data logging**   65

entering and leaving the lifts at the main lobby. With the aid of stopwatches, the observers also measured the time that elapsed for each lift to complete one round trip (RTT). The observers stationed inside the cars measured door opening and closing times, and counted the number of passengers leaving or entering at every floor.

At a later stage, a multi-channel registering device was added to the data logger (usually with 20 channels). This was built around a recorder using continuous paper and several pens that enabled collection of limited information, in quantity as well as quality.

The main limitations to these above processes: the manual and the recorder, were that only very limited information could be collected quantitatively. The main drawback was its inaccuracy during data analysis, and that this analysis was time consuming.

The system that was developed, of which a block diagram is shown in Fig. 2, can be connected to up to 234 sources of information (channels). The input voltage that it can handle is from 5 to 160 V (a.c. or d.c.).

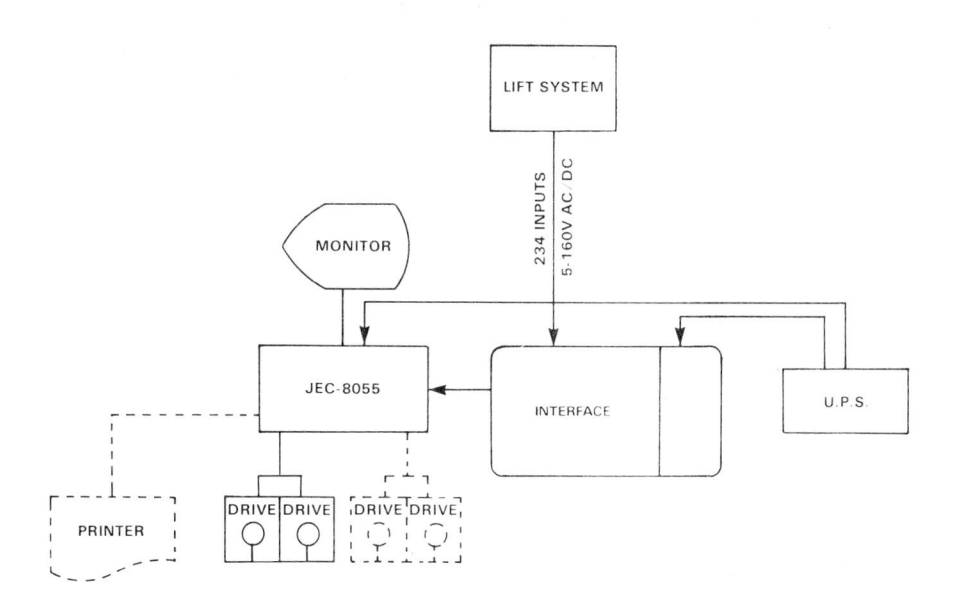

Fig. 2 — Block diagram of data logger.

Table 2 details the various sources of information to which it can be connected. The system is built to enable its easy mobility from building to building and during design special emphasis was placed on making the logger robust in order to withstand many moves. Easy identification of the various outgoing channels and the method of connection to the lift control panels enable connection of the system and its operation after a very short period of time (usually 3–4 hours).

Figures 3 and 4 describe the system when connected and operating in the

Table 2 — Sources of information

● Landing calls — up to 32 floors
● In every lift:
    M.G. (motor generator)
    direction of traffic (up/down)
    condition of car (moving, stopped)
    condition of doors (open, closed, opening, closing)
    loads (zero, full, overload)
    car position

Fig. 3 — Data logger installed in machine room.

machine-room. From these pictures, it can be seen that the method of connection of the system to the control pannels does not interfere with the continuous operation of the machine-room. As in the stimulation program, the operation of the data logger provides a very detailed output of the

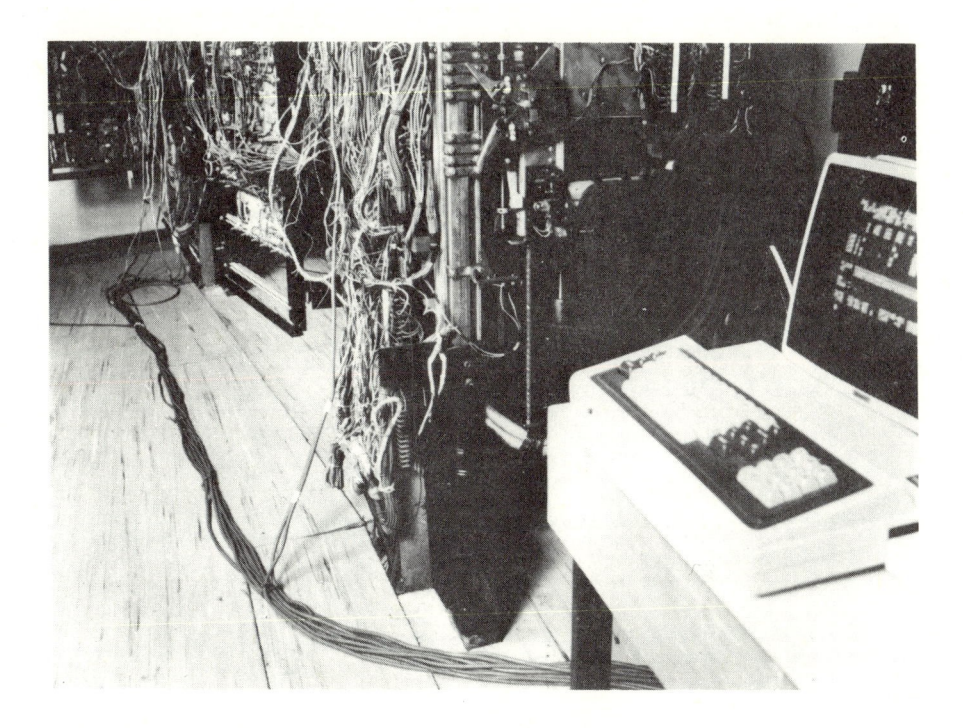

Fig. 4 — Connections to the data logger.

information that has been collected. This output includes information on landing calls (their direction, time of call registration and cancellation), door conditions, travel directions, number of stops, etc.). The type of information received is identical in principle to the reports and tables produced by the simulation program. Examples of these tables and graphs appear in Table 3 and Fig. 5.

4. SIMULATION AND DATA LOGGING

With the completion of the building and the lifts in operation, the lift engineer can check, if the theories and assumptions, that were the basis of the simulations performed during the various design steps, are correct and justified.

The examples given below were collected from a building that has 29 floors above the main lobby and three floors below it. In the building, three lifts were operating with the possibility of adding two lifts in shafts that were prepared during construction. Due to the bad service that the lift system was giving, the operation of the building was organized so that the lifts stopped only on the odd floors above the main lobby and at only one of the lower levels.

Table 3 — Example of tabulated data from logger.

Building	: Office building	23 Floors, 5 Cars.
Data date	: Wed 01/08/84	
File name	: ICO10884.DAT	
Time interval	: 0100	Start time : 0800 Finish time : 1600

Time	Car 1 Speed Starts Relevels	Car 2 Speed Starts Relevels	Car 3 Speed Starts Relevels	Car 4 Speed Starts Relevels
08:00–08:59	60 141 16	39 129 11	55 162 38	58 155 9
09:00–09:59	73 174 19	71 183 9	88 233 52	83 179 15
10:00–10:59	81 185 19	83 183 19	99 235 65	85 198 11
11:00–11:59	85 186 21	78 189 15	103 241 67	84 197 13
12:00–12:59	62 193 17	78 183 22	99 241 69	79 205 13
13:00–13:59	72 203 12	59 202 15	83 253 57	68 220 4
14:00–14:59	82 189 25	72 176 23	102 251 62	79 206 18
15:00–15:59	75 164 17	70 161 19	81 200 53	83 184 23
TOTAL	590 1435 146	550 1406 133	710 1816 463	619 1544 106
% SP OF ST	41.99	39.99	39.36	40.36
% RE OF ST	10.02	9.99	25.02	6.86

SPEED   Full speed in time interval
STARTS   Number of starts in time interval
RELEVELS   Number of relevels in time interval

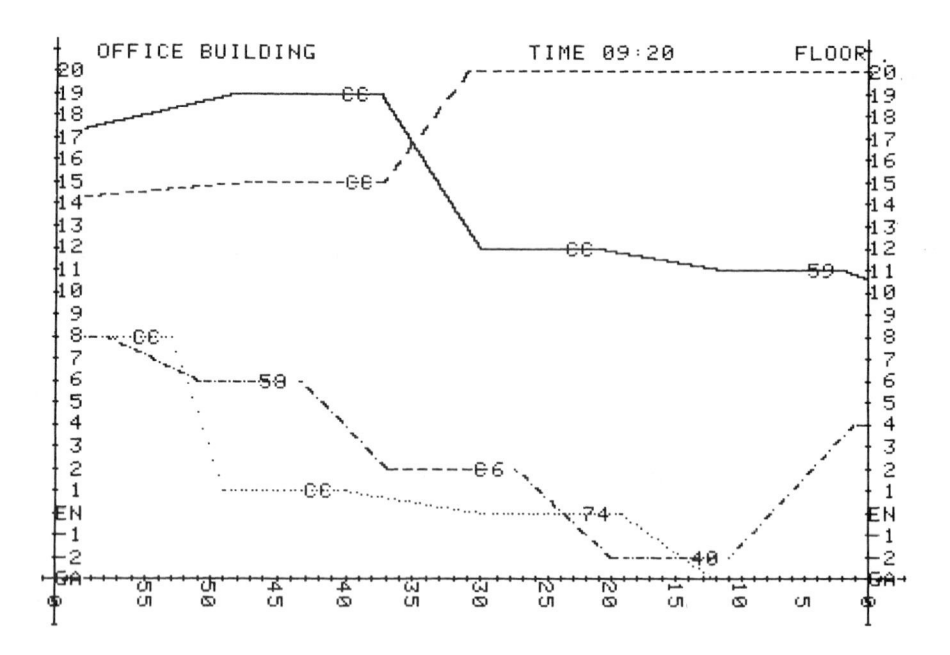

Fig. 5 — Example of graphical output from logger.

4.1 Comparison of Data Logger and Simulation Results

The data logger was attached to the different sources of information on the control panels. Table 4 details the distribution of waiting times as registered by the data logger. Looking at Table 4, it can be seen that there were a total

Table 4 — Distribution of waiting times

Haifa University, Eshkol Tower 11.03.81
Data Logger V.1 (Waiting Time Table)

No.	Floor	Direction	Calls	<30	31-60	61-90	>90	Longest wait (Sec)
1	1	P	39	16	7	8	8	240
2	4	Up	19	6	2	5	6	180
3	4	Down	9	3	3	1	2	240
4	7	Up	6	1	2	–	3	270
5	7	Down	22	5	3	7	7	425
6	11	Up	20	5	5	3	7	150
7	11	Down	16	9	3	1	3	180
8	13	Up	–	–	–	–	–	–
9	13	Down	12	5	2	2	3	210
10	17	Up	2	–	–	–	2	105
11	17	Down	14	3	2	3	6	210
12	19	Up	6	1	2	1	2	220
13	19	Down	17	5	3	2	7	225
14	21	Up	7	4	–	2	1	165
15	21	Down	12	3	1	2	6	180
16	25	Up	1	1	–	–	–	20
17	25	Down	13	6	1	3	3	195
18	27	Up	6	3	–	1	2	150
19	27	Down	8	2	2	1	3	190
20	29	Down	10	1	2	2	5	195
Total			239	79	40	44	76	

of 239 calls in the period of time to which this table relates, and their distribution is shown. The longest waiting time was 425 seconds (7th floor, down direction).

Table 5 details a small part of the information received from the simulation program for the same model. Note that passenger 76 entered the lift at 11:32:42 on the 6th floor with the intention of going down, and the waiting time was 425 seconds.

The exact result (425 seconds) is purely coincidental, a range such as 440 to 400 seconds, would be considered an identical result. It is also possible to compare the results of distribution of waiting time received from the data logger (Table 4) and the lift simulation program (Table 6) in Table 7.

The near identical results received enable the use of both tools available (the data logger and the lift simulation program) to be carried out with confidence in order to check the functioning of a lift system. It is possible to execute changes in lift operation (owing to changes taking place in the day to day functioning of the building) on the basis of data received from both tools.

Table 5 — Detail from Lift Simulation Program

Passenger Number	Entering Time	Waiting Time(sec)	From	To
175	11.36. 8	200	3	8
128	11.34.18	310	7	5
201	11.37. 7	141	3	8
167	11.35.56	213	7	4
172	11.36. 0	209	7	0
181	11.36.29	181	7	5
221	11.37.35	117	3	14
206	11.37.17	150	4	5
207	11.37.19	148	4	10
215	11.37.26	142	4	14
76	11.32.42	425	6	5
216	11.37.30	138	4	8
217	11.37.32	136	4	6
94	11.33. 8	400	6	5
238	11.38.12	97	4	14
169	11.35.57	233	6	−1

Haifa University, Eshkul Tower.   11.03.81.

Table 6 — System waiting time

Wating time (Sec)	Number of passengers	Percentage from total	Cumulative percentage	Remaining percentage
0	12	5.33	5.33	100.00
$0<T<=10$	20	8.99	14.22	85.78
$10<T<=20$	15	6.67	20.89	79.11
$20<T<=30$	21	9.33	30.22	69.88
$30<T<=40$	16	7.11	37.33	62.67
$40<T<=50$	15	6.67	44.00	56.00
$50<T<=60$	16	7.11	51.11	48.89
$60<T<=70$	18	8.00	59.11	40.89
$70<T<=80$	15	6.67	65.78	34.22
$80<T<=90$	10	4.44	70.22	29.78

Haifa University, Eshkol Tower.   11.03.81.

Table 7 — Comparison of waiting time distributions

Passenger waiting time (seconds)	Data logger (%)	Lift simulation (%)
<30	33.0	30.2
<60	49.7	51.1
<90	68.2	70.2
>90	31.8	29.8

4.2 Modernization

An additional area where the results received from using these two tools can be applied is modernization. There is a new and growing demand for the modernization of lift systems that are about twenty years old. Here it is possible to apply in the best and most efficient way the advantages of both the data logger and the lift simulation program.

First, a simulation should be performed on the system where the input data is based on information known about the building, its population, and its lifts. This is information collected from all the years of the building's operation.

The second stage is the operation of the data logger in the building. The information collected on the functioning of the lifts is analysed and the results received are fed into the simulation program.

The third stage begins where the simulation program is run with input data based on the information collected by the data logger. After running the simulation program and receiving its results, the results received can be compared from those obtained using the data logger.

The fourth and last stage is when simulations are performed after part of the input data relating to the lifts has been changed. This is usually improvements in the dispatching, the door operation, and shortening of the time required for a floor to floor jump.

The results received enable a check to be made on the improvement in the service to be obtained as a result of modernization, as well as if all the improvements and changes are indeed necessary and justified.

It is possible to repeat the above process (stage 4) several times until the best results for the building are reached.

5. SUMMARY

After several years' experience combining these two tools, their application has been proved and it is considered that simulation and data logging are very important tools for traffic studies and the modernization of lifts.

7

Practising lift simulations

Ir E. H. Allaart, Tebodin, The Hague, Netherlands

ABSTRACT

A lift consultant is called upon to design the installation for many unusual buildings. In these cases the use of standard calculations does not always lead to an authoritative design. The use of computer-aided design methods which simulate the specific characteristics of such buildings can then lead to a suitable lift configuration. This chapter will describe the use of lift simulations to design a building in the Netherlands.

1. INTRODUCTION

In this chapter the author's experience concerning computer simulations techniques, will be discussed. Simulation is a reliable tool to design the lift capacity in high buildings. Tebodin, is one of the largest firms to consulting engineers in the Netherlands. In a great number of projects simulations

Ch. 7]	**Practising lift simulations**	73

methods are used. It gives a better insight into the parameters of dynamic systems.

As an illustration a number of project types where simulation has been applied is given below:

(i) Simulations for harbour terminal systems, of which the EMO bulk terminal at Rotterdam was the most complex.
(ii) Another application field is the process industry, especially for start and stop procedures.
(iii) Difficult operations can be imitated in a model. Examples are:

> The positioning of a huge oil productioin platform on its location near Norway.
> The operation of the mat-laying vessel *Cardium*, part of the Dutch Delta Works.

(iv) The material flows inside the cargo handling building at Schiphol Airport have been simulated for KLM's Air Freight Centre.
(v) The logistic management of coal transport and distribution for five Dutch power plants has been simulated.

For all these purposes special simulation programs are written by experts. It takes at least one month, often three months, to achieve a program that can ensure getting reliable results. In this respect it is an advantage that for lift simulation a program was already available, which has been proved to be successful.

2. AMRO BANK PROJECT

In the period from 1981 to 1985 the author was involved in a project for AMRO, one of the largest banks in the Netherlands, concerning the design and engineering of the main office in Amsterdam. The building was to be used by about 3000 bank employees and would be about 100 metres high. It was the author's task to give advice about the transport facilities, which had to be installed in the building. The largest part of the work concerned the lifts.

At the beginning of the project the correct lift configuration was of great importance, as the lift installations were planned to be situated inside the concrete core. Any revision of the core dimensions in a later stage of the project would cause radical changes in the main design of the building. As a first approach to determine the lift capacity, calculations were carried out. Then the results were checked with the help of computer simulations. In order to prepare the input data, and afterwards, to evaluate the results of the simulation run the basic data of clients requirements should be determined. Figures 1 and 2 indicate the building form.

Fig. 1 — Floor layout.

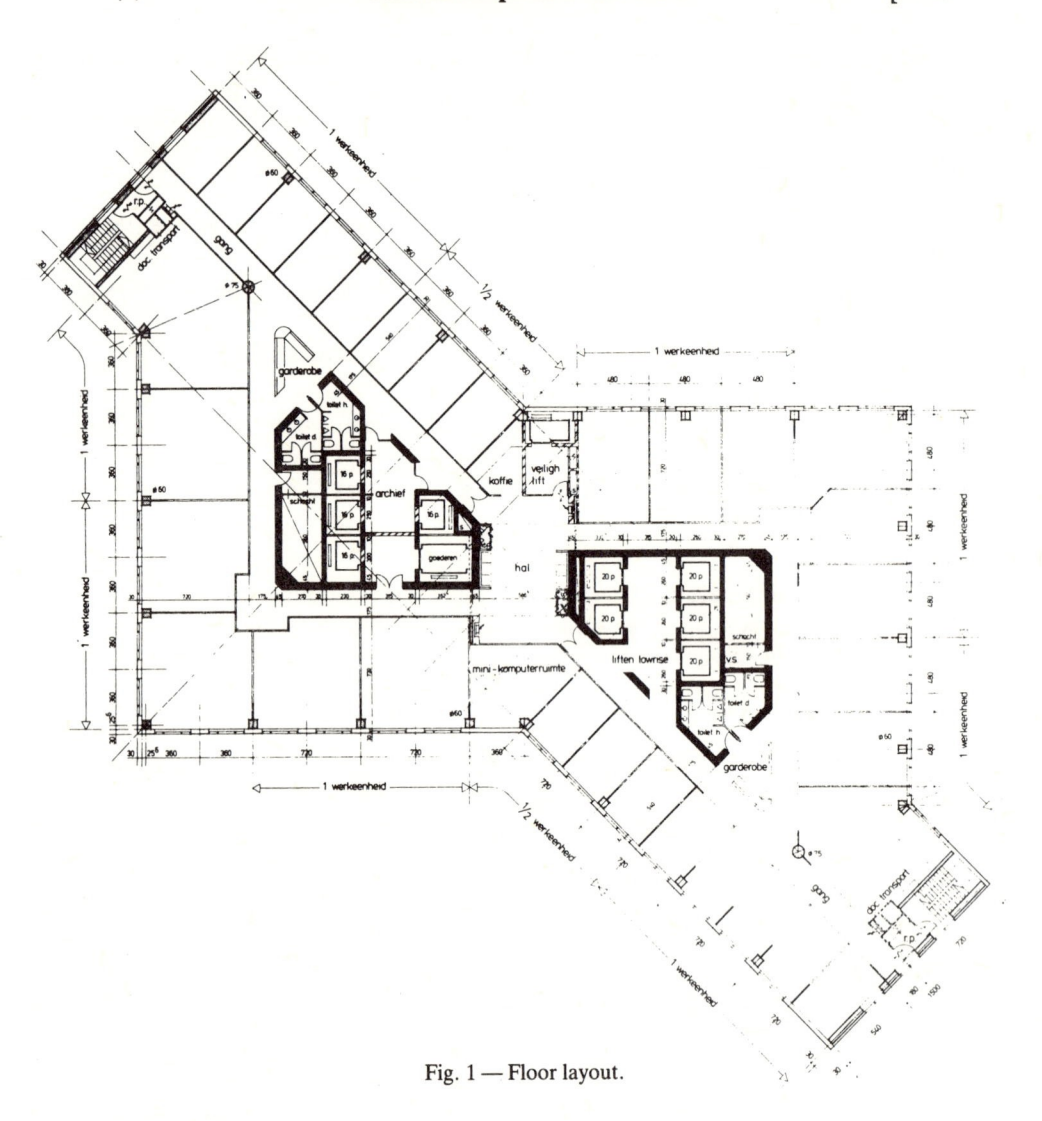

3. SIMULATION PROCEDURE
The most important data are:

    Number of persons in the building.
    Number of persons that will arrive during the monitoring-peak.
    Arrival pattern, maximum 5 minutes rate.
    Interfloor traffic.
    Lunch procedures.
    Number of persons on each floor.
    Special requirements, for example priorities to the conference floor, restaurant, etc.
    Accepted waiting times. Preferably defined as the average waiting time during the heaviest 5 minutes.

Fig. 2 — Landing levels.

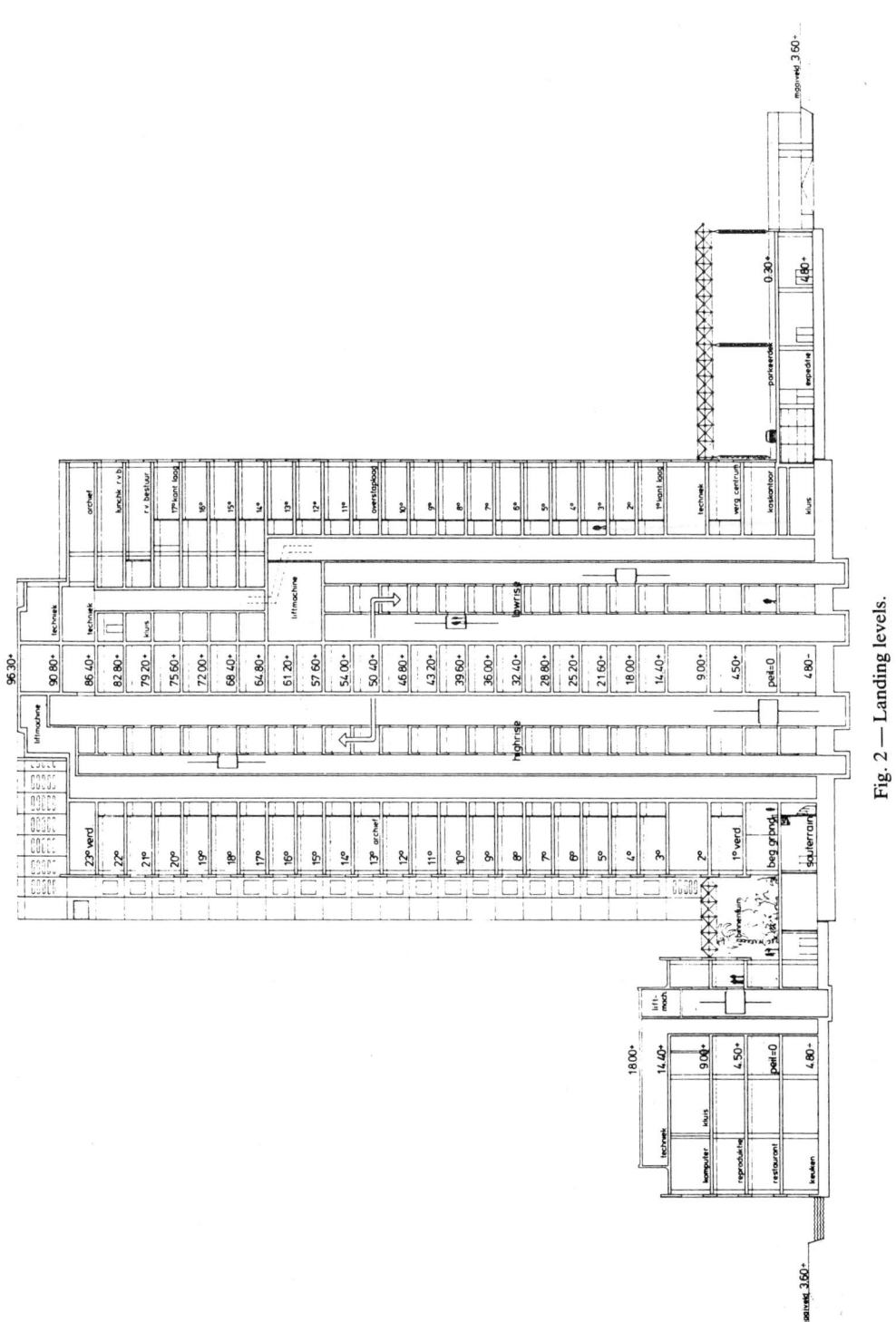

Other essential data are the results of the main building design:

> Available area for shafts and lobbies.
> Landing levels.
> Specific floors (restaurant, post office, technique etc.).

There are a number of other aspects which are of interest:

(i) A way to present the main results of each simulation run on one uniform sheet was developed.
(ii) The building was planned to be situated near the railway and underground stations; this concentrated extra attention to the arrival pattern.
(iii) The client required full lift service to both basement floors by the main lift groups. (Although the use of escalators was strongly advised the client insisted on lift service. Therefore the 'bunching' phenomenon could be observed in the results of the simulation runs.)

During the project as quite often happens the building design changed. One of the examples was the clients' requirement for a separate lift group to serve the upper three floors of the building. All these changes in the project design caused an extra number of computer runs. In each simulation run it is possible to distinguish three main parts of the procedure:

> Careful preparation of all input data.
> The computer run itself, taking usually not more than one hour.
> The evaluation of the output data.

The output data evaluation is the most important part of the work. When the results show for example a bad service to the basement an analysis of all data might be necessary to find out what was wrong.

The program used was written by Dos Santos and Barney. The program gives an extended set of output data, which is very useful for evaluation, but in a later stage, the main data and the results of the output data evaluation gave sufficient information. Table 1 shows the summary sheet containing all relevant data for one run. In this way it was easy to compare the several run results. Each sheet contains an input block an output block and and a short evaluation.

4. OTHER TRAFFIC REQUIREMENTS
4.1 Trains and Lifts
The building was planned to be situated a very short distance from railway and underground stations. This means that during the morning up-peak the arrival patten of lift users was expected to be different from the usual

Ch. 7] Practising lift simulations

RUNSHEET

client : AMRO
* run nr.   : 1
  order : 11516
* building part : HIGH RISE   project : Sloterdijk
* # x car cap. : 7 × 16 pers   date : 17 Jan '83.

INPUT

building part	HIGH RISE	arrival rate:
#lifts	: 7	main , -1 , -2 , total
car capacity	: 16 pers	781 230 159 1170
speed	: 4.5 m/s	interfloor traffic: 10 % of present pers./h
control system	: FS4	at transferfloor: 25 % of present pers./h
doors (mm)	: 1100 mm	arrival 14 % in heaviest 5 min.
doortimes (o/c)	: .3/3.0 s	remarks R.v B ⎫ no pop.
transfertimes	: 1.2 s	Lunch. ⎭
loading interval	: 10 s	

K20 : 106 [X]
K19 : 53 [X]
K18 : 53 [X]
K17 : 106 [X]
K16 : 1 [X]
K15 : 5 [X]
K14 : 1 [X]
K13 :   [X]
K12 :   [X]
K11 : 4 [X]
K10 : 1 [X]
K 9 : 106 [X]
T 2 : [X]
K 8
K 7
K 6
K 5
K 4
K 3
K 2
K 1
VC
T 1
+ 1
main : 731 [X]
-1 : 230 [X]
-2 : 159 [X]

OUTPUT

landing	: main , -1 , -2 , total building
av. waitingtime over heaviest 5 min.	: 30 25 25 31 sek
90 percentile	: 37 44 48 42
maximum waiting time	: 75 77 51 51
average R.T.T. over heaviest 5 min.	: 153
voyage time	: 100 57 100 95
car occupancy	: 72 %.

EVALUATION

remarks : No.

OVER ALL : Acceptable / Good.

Table 1 — Run sheet.

situation. Shortly after the arrival of one or more trains an increase of lift users could be expected. The arrival of passenger groups according to this pattern cannot be simulated within the existing program. To avoid results the arrival rate was therefore set 10 to 15% higher than usual.

4.2 Basement Service: Bunching

The building design had two basement floors for parking facilities. Although it was strongly advised not to use the main lift groups for full service to these floors, the client had insisted and turned down the suggested use of escalators. (Using escalators instead of lifts was much better from a point of view of traffic capacity and would have been cheaper.) As a result of advice in an early stage the client had agreed to research this option. This contributed to a good analysis of both options, showing the disadvantages of using lifts for basement service.

Instead of proper circulation through the building, the lifts had a tendency to move in phase. This phenomenon, called 'bunching', causes

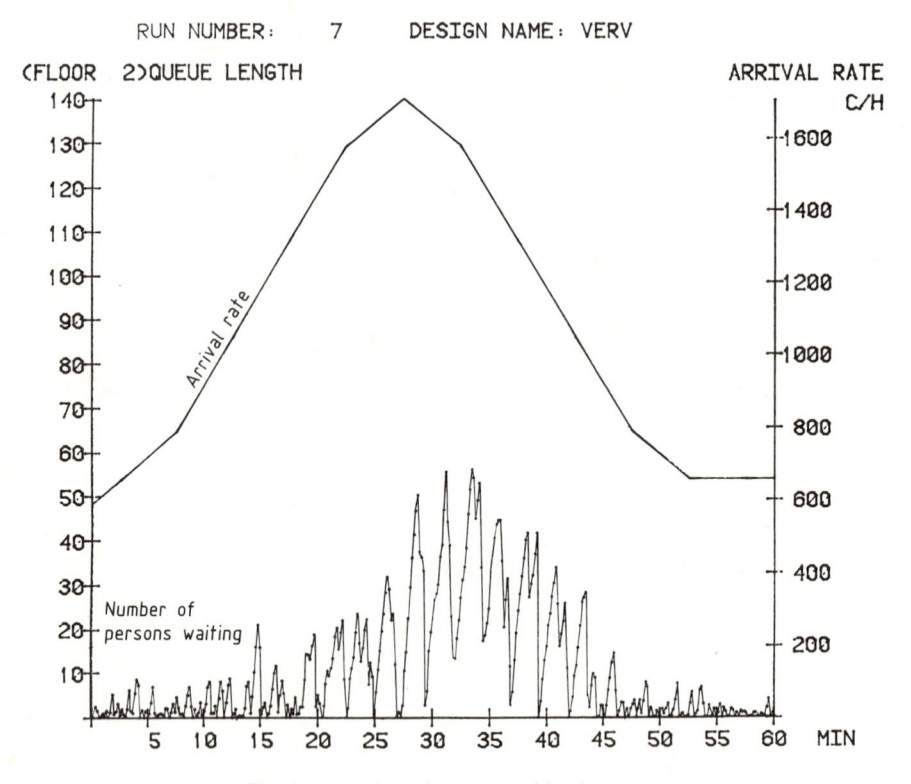

Fig. 3 — Number of persons waiting in a queue.

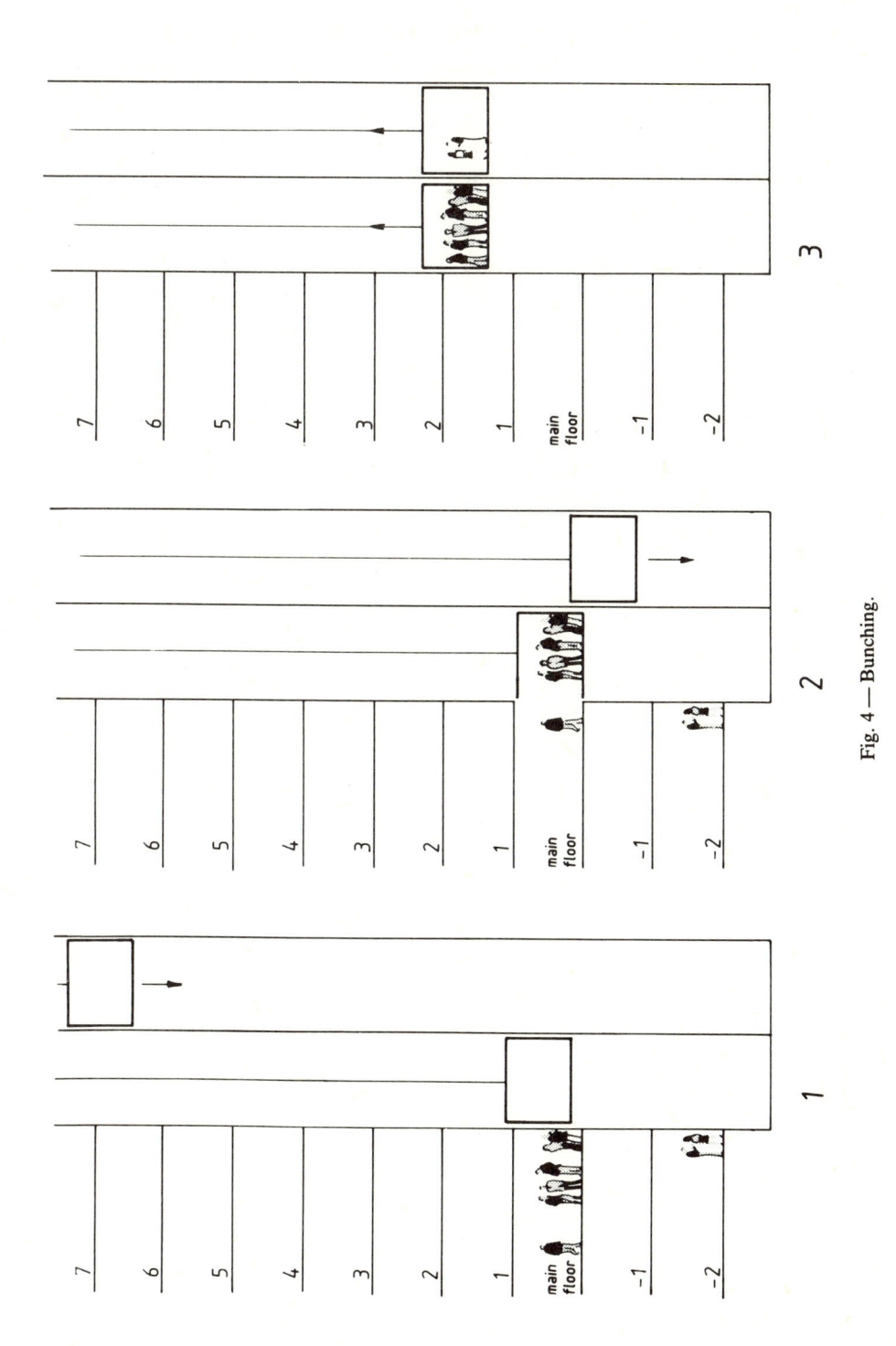

Ch. 7]   **Practising lift simulations**   79

Fig. 4 — Bunching.

long waiting times. For example, at a point of time (the 39th minute), three cars were at the main floor. Figure 3 shows what happens to the queue length in the same run. At the above-mentioned point of time the number of waiting persons is decreased with about three car loads, 48 persons, who entered. It can also be seen that the lift capacity in this run is insufficient as after the departure of a car not all the waiting persons have been able to enter the lifts.

Figure 4 gives an idea of how the basement service can lead to 'bunching'. The control system of the group should be designed to avoid this method of operation.

5. CONCLUSION

Computer simulation is recommended as it has been found to be a reliable and successful method. In the bank building project the required lift capacity was very important. In the case of existing installations, simulations can, if there are problems, indicate what might be wrong, and also can be used to try out suggested solutions.

It is always interesting to visit a building and see the installation in full operation. But what is the performance, the waiting times, etc., During peak demand? This can only be observed by the use of extensive measuring equipment and time-consuming traffic analysis. All these incur high costs.

What the simulation program does is to 'visit' all the buildings of interest, just as the time when the peak demand takes place. In this way all the necessary data is obtained in order to make a good lift design.

Part 3
Mechanical design

8

A new design of compensating cable

Mr Richard Laney and Mr William McCallum, Siecor Corporation, Rocky Mount, USA

ABSTRACT

Compensating cables are used in elevator systems to balance out hoist rope weight as the car moves up and down the shaft. There are a number of forms for compensating cables including, bare chain, plastic jacketed chain and chain with interwoven sash cord. The chapter will describe a new form of compensating cable composed of a core made from proof coil chain with its voids filled with a metallic bead-plastic mixture and covered with a black pvc jacket. The characteristics of this form of compensating cable will be described and experience in use will be indicated.

1. GENERAL PRINCIPLES

The basic elements of an elevator system are: a sheave, counterweight, an elevator car, a compensating cable or counterweight rope, a hoist rope and a travelling cable (Fig. 1).

All of these are assembled in an elevator well or shaft in a well known manner. As a general rule, the elevator car is connected to a counterweight by a hoist rope threaded over one or more sheaves or pulleys located in the upper reaches of the shaft. One end of the compensating cable is connected to the counterweight and the other to the bottom of the car, in some cases after having been threaded over a compensating cable sheave located in the bottom of the elevator well. In most instances, however, the compensating cable is left to hang free without being threaded over a sheave. The counterweight is essentially the same weight as the car and the weight of the

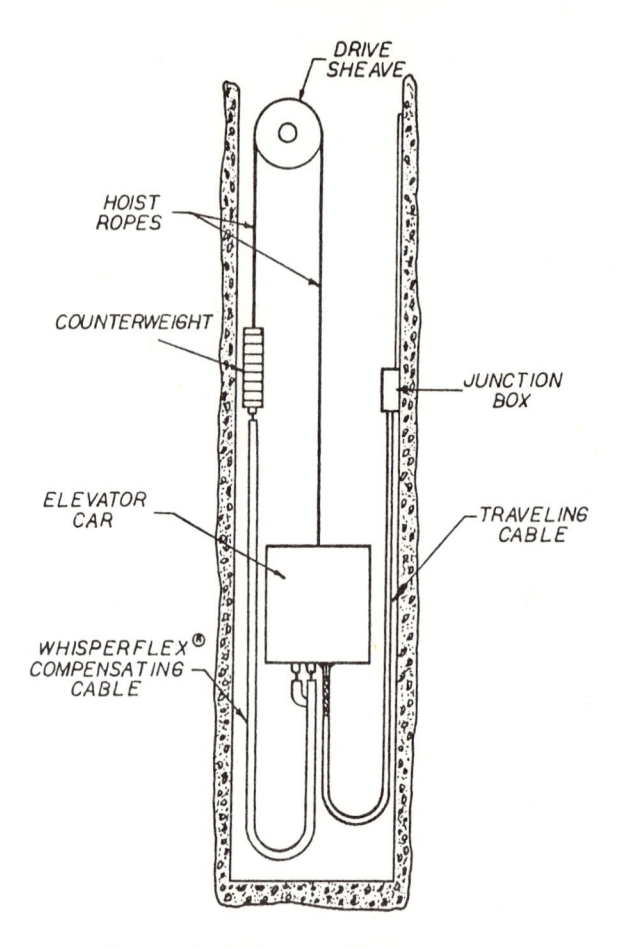

Fig. 1 — Basic elements of an elevator system.

hoist ropes essentially equal to the weight of the compensating cable. One end of the travelling cable is connected to the car bottom and the other into a junction box affixed to the elevator well sidewall. Signals are sent via the travelling cable to the drive mechanism causing the car to obey commands sent over it.

    A prime function of the compensating cable is to provide dynamic weight counterbalance to the weight of the hoist rope(s) as the car travels up and down in the elevator shaft so that the car is dynamically balanced at all times. If the compensating cable is omitted, the weight of the hoist rope(s) will be transferred from one side of the drive sheave to the other causing the load to vary as the car rises and descends in the elevator shaft. For example, when the car is in the lower position and the counterweight is in the upper position, the force required to begin lifting the car will be equal to the weight of the occupants plus the weight of the hoist rope(s). As the car ascends, the hoist rope(s) weight will transfer from the car side to the counterweight side of the drive sheave. For optimum performance, the aggregate weight of the hoist rope(s) should be essentially equal to the weight of the compensating cable at any given position of the car in the elevator shaft. In addition, the length of the hoist rope(s) between the car and sheave should be equal to the length of the compensating cable between the counterweight and the lowest portion of the compensating cable or, stated alternatively, the length of the hoist rope(s) from car to counterweight is essentially equal to the length of the compensating cable from car to counterweight. Because of safety factor reasons, there may be five or more hoist ropes and the aggregate weight of such hoist ropes should approximately equal the weight of compensating cable. This does not mean that if there are five compensating chains required, there must be five hoist ropes. There may be only one compensating chain and more than one hoist rope, so long as the length and weight requirements of the compensating chain are met.

Earlier compensating cables usually were nothing more than a link chain. Constant raising and lowering of the elevator car caused the chain also to be raised and lowered, rubbing one link against the other causing noise and abrasion (Fig. 2).

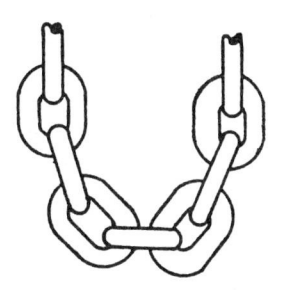

Fig. 2 — Detail of chain.

Link chains, when hung free in an elevator shaft (no bottom sheave), have a tendency to form a 'point' and not a loop, i.e., the side legs of the link chain tend to converge on a single link and form a point at the bottom of the 'loop' formed by the chain. Such a configuration causes one leg of the chain to rub against another during car movement causing noise and abrasion as the chain travels. More often than not, link chain type compensating cable would strike the sidewalls of the elevator shaft causing damage and additional noise. Noise was so much of a problem that some earlier compensating chain type cables either used a sash cord (a rope woven in the links of the chain) or employed a plastic coating over the link chain. These methods reduced the noise and damage to a degree.

2. DENSE AGGREGATE COMPENSATING CABLE

To overcome the inadequacies of other forms of compensating cable, Republic Wire, and Cable have developed a dense aggregate compensating cable known as Whisperflex.

Dense aggregate compensating cable when constructed in the cylindrical form, has a typical cross-section which comprises:

(a) A sheath of black polyvinyl chloride which provides a smooth abrasion resistant surface.
(b) A strength member which is either a chain or wire rope or similar material.
(c) The volume under the sheath not otherwise occupied by the strength member is ocupied by a mixture of metal beads and plastic which forms a continuous solid core. The metal particles can be ferrous and non-ferrous of any particle size and shape, preferably between 0.5 mm and 1.0 mm in diameter in an amount so that 50 to 75% of the volume is occupied by them. The balance of the volume is occupied by plastic which is the same material as listed above for the sheath.

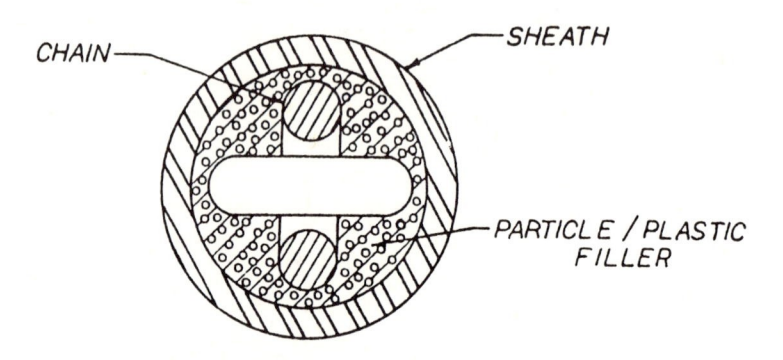

Fig. 3 — Cross-section of new compensating cable.

Dense aggregate compensating chain may also be produced as a flat compensating cable comprising:

(a) Multiple strength members each with their respective axis arranged in a line.
(b) A sheath of flexible polyvinyl chloride containing the strength members and a metallic particle and plastic mixture.

It should be noted that the outer surface of sheath is not necessarily undulating and may present an essentially circular cross-section as shown in Fig. 3 although it may be undulating if desired.

Metal particles result in a compensating cable having a greater weight per linear length than earlier compensating cables. For example, when earlier compensating cables are compared with dense aggregate compensating cable, it has been found that for a given equal length, an earlier compensating cable having links made of 9.5 mm diameter steel was equivalent to dense aggregate compensating cable made of steel links of only

6.4 mm in diameter. A more complete weight comparision is shown in Table 1. The balance of weight is comprised of the metallic bead/PVC core and PVC sheath.

Table 1 — Comparison of weight per unit length of various types of compensating cable.

Chain trade size mm	(inches)	Bare chain weight kg/m	(lb/ft)	Jacketed chain weight kg/m	(lb/ft)	Dense aggregate cable weight kg/m	(lb/ft)
3.2	(1/8)	—	—	—	—	1.49	(1.0)
4.8	(3/16)	0.60	(0.40)	0.91	(0.61)	2.24	(1.5)
6.4	(1/4)	1.07	(0.71)	1.45	(0.97)	2.98	(2.0)
7.9	(5/16)	1.61	(1.08)	2.03	(1.36)	3.73	(2.5)
9.5	(3/8)	2.33	(1.56)	2.82	(1.89)	5.22	(3.5)

As mentioned earlier, most elevator systems do not employ a bottom sheave, especially when the system is installed in a well or shaft. Systems installed in non-shaft or well situations where the compensating cable is left to hang free would be subject to wind or other forces usually employing a sheave or similar guidance system. Earlier chain type compensating members, have a tendency to form a point at this location where two links are joined at the bottom of the loop. This arises out of the relatively limber nature of the chain and the restricted lateral space in the elevator shaft. The closer together the legs of compensating member are to one another, the more likely a compensating chain type member would exhibit this 'point' tendency. It is at this point, which is a dynamic one as the car moves up and down, where one link strikes another, giving rise to undesirable noise and abrasion and a tendency for one leg of the chain to strike the car. The links of the chain of dense aggregate construction stay fully extended because of filler material, contrary to earlier compensating cables that permitted the link chain to shrink in length as a result of one link sliding within the link to which it is connected. A fully extended link chain results in an evenly distributed weight, eliminates noise and abrasion of one link on another, preserves the cylindrical surface of sheath, and avoids the problem of sheath cracking.

3. INSTALLATION

Installation procedures are basically the same as with bare chain. Once the plastic composition is removed to expose one chain link, the link can be attached to a U-bolt in the same manner as bare chain (Fig. 4). An 'S' hook passed through the chain 0.6 m to 1.0 m from the car end U-bolt should be attached close to, but not onto, the U-bolt. This will form a safety loop of cable that will be available should be compensating cable catch on an obstruction. The 'S' hook will yield and the extra length of cable should free the compensating cable from the obstruction.

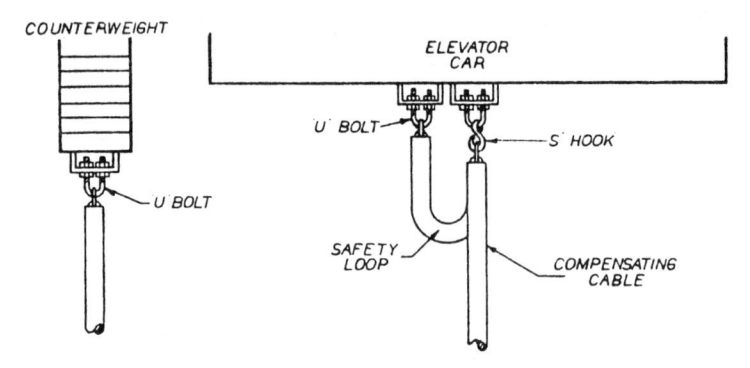

Fig. 4 — Installation details.

Multiple lengths of dense aggregate compensating cable can be attached to the counterweight and elevator car so long as the cables are spaced evenly about the centreline of the counterweight and elevator car. Uneven spacing and unequal loading of compensating cable on the counterweight and elevator car could cause poor performance and are not recommended.

4. ECONOMICS AND FUTURE OUTLOOK

The installed cost of dense aggregate compensating cable is comparable to the available alternatives (raw chain covered with a plastic jacket or raw chain interwoven with a sash cord). Obviously the cost differential would be the cost of the raw chain material to achieve a certain weight versus the extra materials and manufacturing cost, to make the dense aggregate compensating cable.

The true advantage of dense aggregate cable lies not only in cost but also in improved product performance. By varying the amount of metallic bead/plastic mixture in the core, a wider range of cables is available for use in all compensating cable applications. The bare chain is limited to the weight of the individual chain links for its linear weight per foot. Dense aggregate compensating cable is available in 0.75 kg/m increments from 1.5 kg/m to 6.0 kg/m which makes it possible to more closely approximate the weight of the hoist ropes. Each cable is designed with a 4:1 safety factor considered for each cable size.

The future outlook for this new type of compensating cable is very exciting. The product will continue to be used as a direct replacement for conventional compensating chains utilizing the free loop hanging method, but continued growth is seen in the market as industry standardizes on this cable.

The more exciting possibility is the utilization of dense aggregate cable in jobs which would normally use compensating ropes and a sheave arrangement. These would typically be high speed high rise installations (Fig. 5). The economical advantages and improved product performance of dense

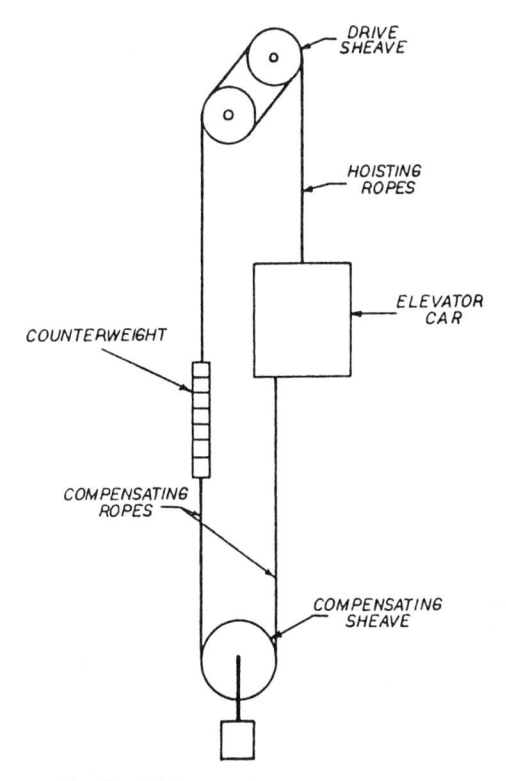

Fig. 5 — High speed, high rise arrangement.

aggregate compensating cable become immediately obvious in this application. Not only would the initial investment of wire ropes and sheaves be decreased but the ongoing maintenance and replacement cost of the ropes would be virtually eliminated. The installed cost of dense aggregate compensating cable could be as little as half the cost of the compensating ropes and sheave arrangement.

5. EXPERIENCE

The dense aggregate compensating cable manufactured by Republic Wire and Cable as Whisperflex† is available in seven sizes (1.5 kg/m×0.75 kg/m to 6.0 kg/m). It has been in service since 1983 extensively in installations with speeds up to 1.75 m/s as a direct replacement for bare chain, chain with sash cord and jacketed chain. Some installations have been made in high rise 3.5 m/s systems. Whisperflex compensating cable has a higher weight per linear length of a given chain size, is smaller in diameter for a given weight per linear length, and has less lateral cable sway. It virtually eliminates the problems of noise, abrasion and chain wear. In the future developments for high speed, high rise applications it should provide substantial savings to the elevator contractor whilst maintaining a quality installation.

9

A means of minimizing the effect of steel member torsion in elevator travelling cable

Mr Alfred Garshick, BIW Cable Systems, Boston, USA

ABSTRACT

The dynamic torsional characteristics of elevator travelling cable can be variable due to the shift in load on it when the car moves from its upper to lower limit. This chapter reports on the characteristics of the supporting steel member of conventional elevator cable and compares the behaviour to a proposed design that provides greater torsion stability.

1. INTRODUCTION

As an elevator car travels from one extreme position at the top of the hoistway to the bottom of the hoistway, the travelling cable weight load is transferred from one end to the other. The dynamics of transferring this load can influence the behaviour of the catenary and cause it to twist and oscillate under the car, while it is in motion. The tracking of this is important to the proper performance of the elevator, particularly, when multiple cables are hung together.

The ideal condition is that, which maintains the cable catenary in one plane. If there is any movement of the catenaries, they should be kept small and in the same direction. Factors which influence this behaviour are cable type, size, configuration and characteristics of the central steel supporting member.

This chapter deals only with the steel core characteristics as they relate to the finished control cable twist tendencies.

2. CURRENT PRACTICE

The steel supporting members that are used in travelling cable are typically of galvanised steel wire rope as shown in Table 1.

Table 1 — Typical steel ropes

Diameter	3/32-inch	5/32-inch
Stranding	7 × 19	7 × 19
Weight/1000 ft	16 lbs	45 lbs
Breaking strength	1050 lbs	2800 lbs

The choice in the selection of a particular rope depends on the maximum length of travelling cable that will be supported in use. Because of the inherent twist characteristics of conventional steel rope, the ratio of the breaking strength of the rope to the actual load on the cable must be kept at a high value. Current practice is to use a ratio of 5:1 or greater.

3. TEST PROGRAMME

3.1 Standard Strength Members

Short lengths of wire ropes made by different manufacturers were evaluated in the 'as received' condition and pre-conditioned as shown in Table 2.

Table 2 — Ropes tested

Diameter	3/32-inch	5/32-inch
Breaking strength	1050 lbs	2800 lbs
As received	Tested	Tested
Prestressed 500 lbs	Tested	—
Prestressed 600 lbs	Tested	—
Prestressed 1000 lbs	—	Tested
Prestressed 1650 lbs	—	Tested

Convenient 23 ft lengths of the wire ropes were suspended vertically with a platform to accept weights attached to the lower end. The platform weight of 21 lbs was taken as the starting point. With the lower end free to rotate as weights were added, the rotation was recorded for each increment of weight.

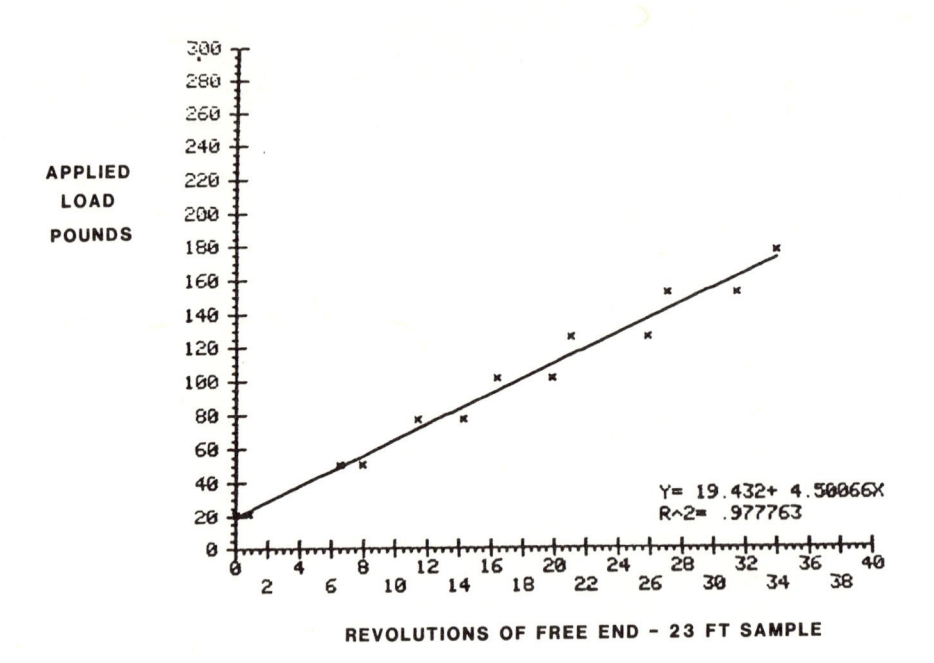

Fig. 1 — 3/32-inch steel rope as purchased.

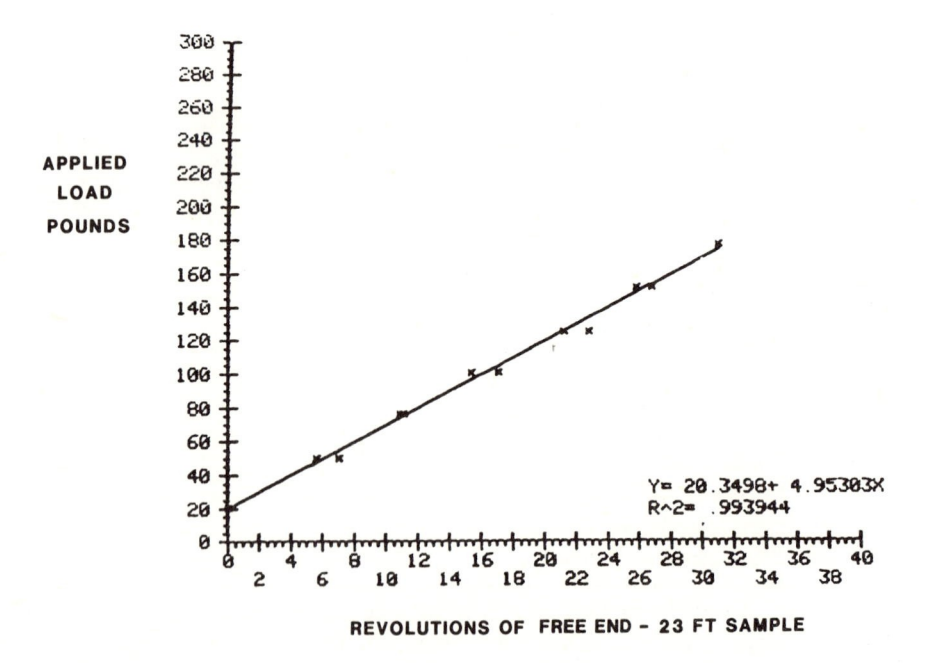

Fig. 2 — 3/32-inch steel rope — 500 lbs prestress.

Figures 1, 2 and 3 are for the 3/32-inch diameter conventional rope.

On the assumption that the steel supports a 150 ft long control cable which weighs 1 lb/ft, the twist for the 3/32-inch rope is shown in Table 3.

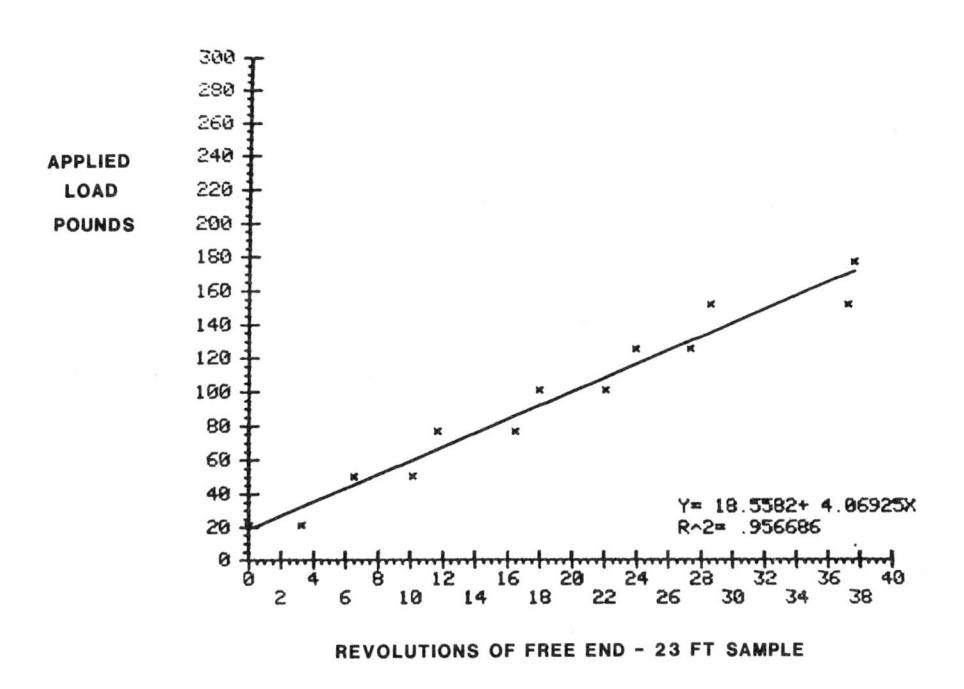

Fig. 3 — 3/32-inch steel rope — 600 pound prestress.

Table 3 — Turns in a 150 ft long cable

Diameter	3/32-inch	5/32-inch
As received	209	59
500 lbs prestress	183	
600 lbs prestress	209	
1000 lbs prestress		65
1650 lbs prestress		62

Similar data assumptions for the 5/32-inch diameter rope, as shown in Figs 4, 5 and 6 can be also seen in Table 3.

Note, however, that the loading of 5/32 inch rope for a 150 ft length is:

$$\frac{150}{2800} \times 100 = 5.35\%$$

and a twist tendency of 59 to 65 turns for the length. The loading for the 3/32 inch rope for a 150 ft length is:

$$\frac{150}{1050} \times 100 = 14.3\%$$

and a twist tendency of 183 to 209 turns/ft. One method of limiting the twist due to the rope behaviour would be to use larger size ropes.

This simple test with one end free top rotate clearly demonstrates that the twist is high when such wire ropes are loaded. Prestress in excess of 50% of the breaking strength had essentially no effect on this characteristic.

The equation of the regression line for each set of data is shown in the figure for that test. The slope of this line is expressed as the ratio of applied load to angle of rotation.

$$m = \frac{Y}{X} = \frac{\text{Load}}{\text{Rotation}} \tag{1}$$

This slope is a convenient index to indicate the rotation tendency due to the applied load.

This problem of the inherent twist of wire ropes has been overcome in the oceanographic, towing and oil-well logging industries by the use of torque balanced layers of steel wires. A new type of steel supporting member was designed using this principle of torque balance. Figure 7(a) and (b) show the old and new designs.

3.2 New Design Strength Members

The new strength member was tested in the same fashion as the standard ropes. The angular rotation test result is shown in Fig. 8. It is readily noted that the twist due to applied load has been reduced by a factor of about 700. The new strength member exhibits no significant rotation under load.

Employing similar assumptions to those applied to the standard cores to support 150 ft of cable weighing 1 lb/ft, the rotation would be 0.08 turn.

3.3 Performance in Control Cable with New Steel Core

To further evaluate the comparative behaviour, samples of multi-conductor elevator control cables of standard construction were made using the new steel core, as well as conventional core. These cables were the same in all respects except for the steel core.

Lengths of both constructions were tested using 23-ft sections in the same manner as were the steel ropes. In order to simulate actual installation, the cables were pre-loaded with weights equivalent to a total support length of 600 ft and then removed. Six hundred feet was selected as the test length to

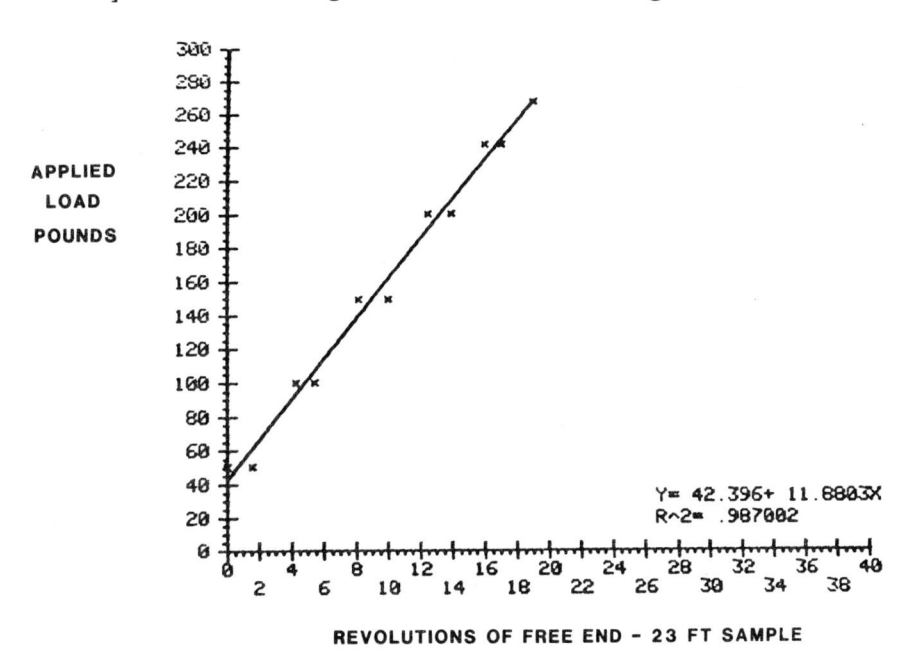

Fig. 4 — 5/32-inch steel rope as purchased.

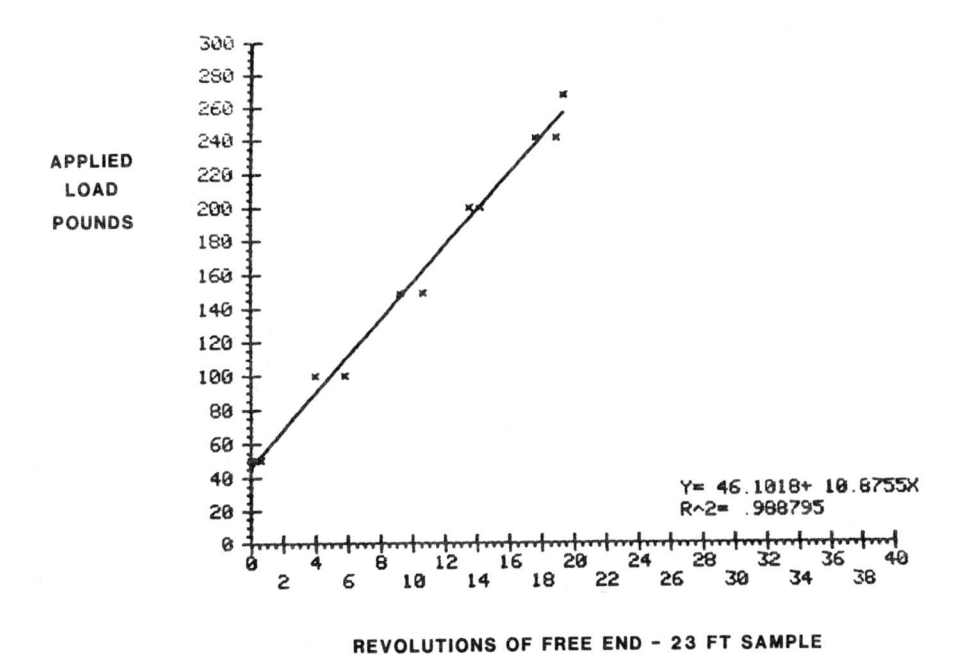

Fig. 5 — 5/32-inch steel rope — 1000 lbs prestress.

96   Mechanical design   [Pt. 3

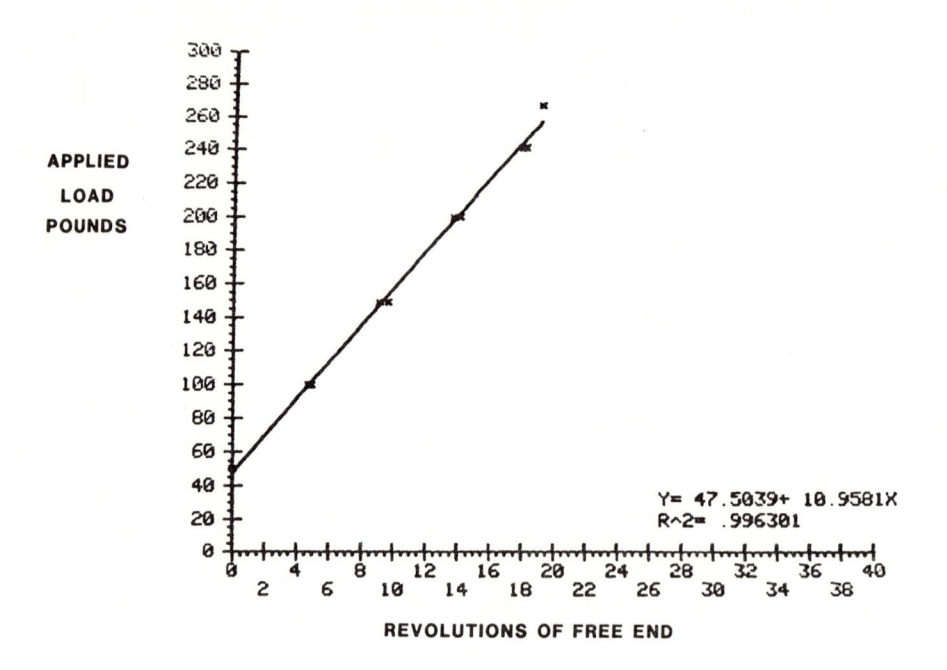

Fig. 6 — 5/32-inch steel rope — 1650 lbs prestress.

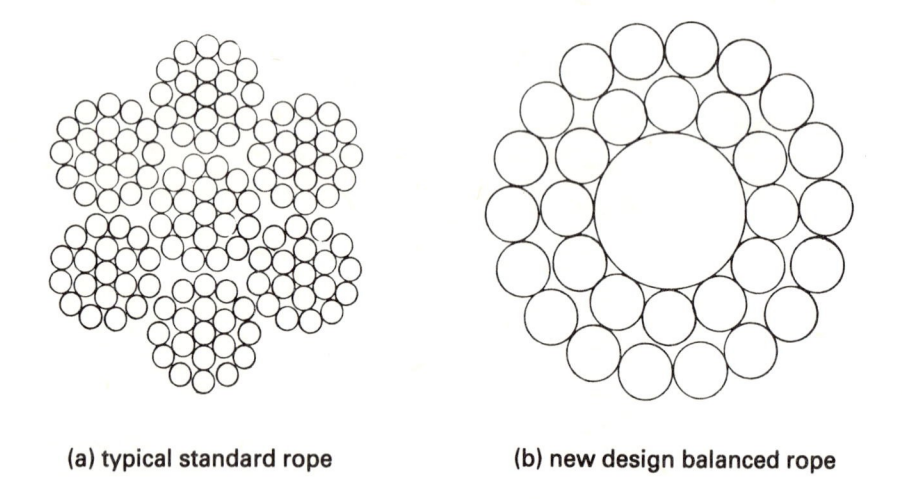

(a) typical standard rope   (b) new design balanced rope

Fig. 7 — Wire ropes.

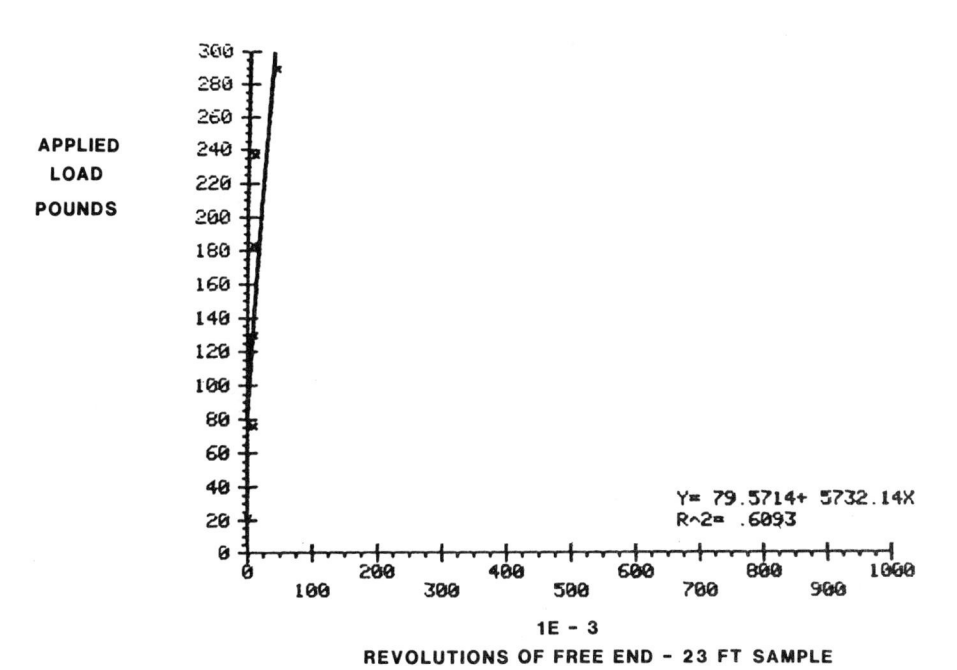

Fig. 8 — 5/32-inch steel rope — balanced design.

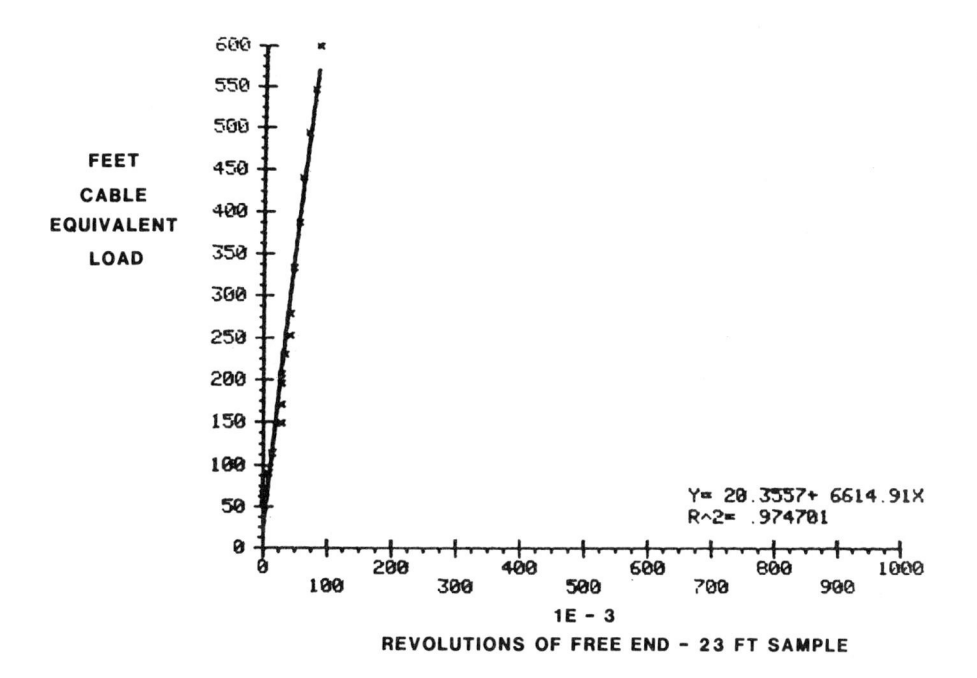

Fig. 9 — Multicore control cable — standard steel core.

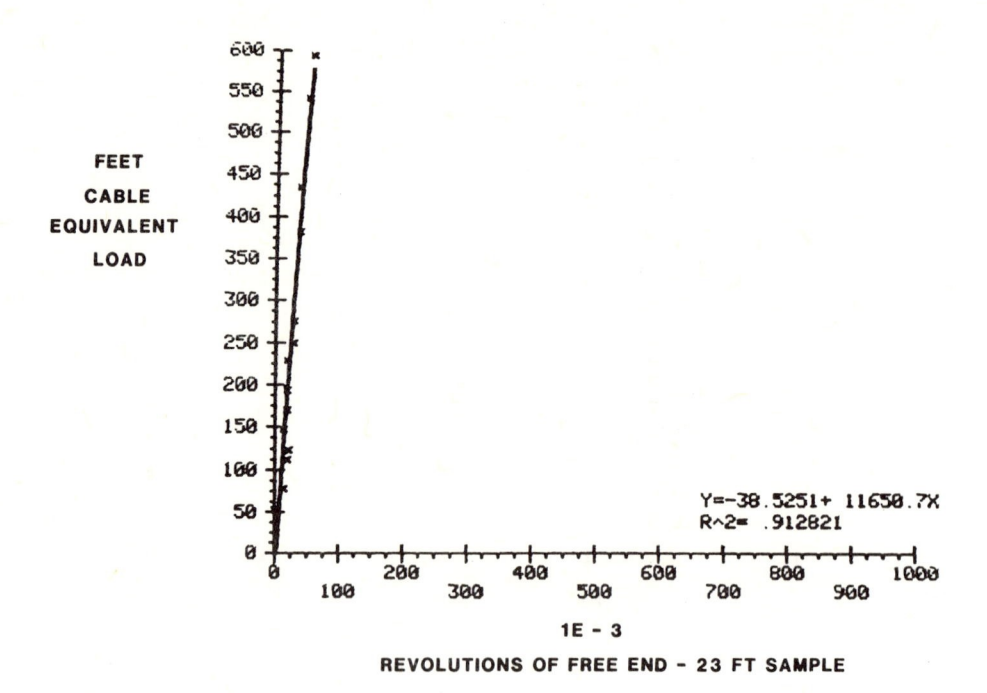

[Graph: Y-axis "FEET CABLE EQUIVALENT LOAD" from 0 to 600; X-axis "REVOLUTIONS OF FREE END - 23 FT SAMPLE" (1E-3) from 0 to 1000. Fit line: Y = -38.5251 + 11650.7X, R^2 = .912821]

Fig. 10 — Multicore control cable — balanced steel core.

Table 4 — Summary of rope types

Description	Slope = Load/Twist
3/32-inch rope as purchased	4.5
3/32-inch rope prestressed 500 lbs	4.9
3/32-inch rope prestressed 600 lbs	4.1
5/32-inch rope as purchased	11.8
5/32-inch rope prestressed 1000 lbs	10.9
5/32-inch rope prestressed 1650 lbs	10.9
5/32-inch new design steel rope	5732
Multi-core control cable Standard steel core	6615
Multi-core control cable Balanced steel core	11 651

simulate a very tall building that would use the particular cable tested. Weights were then added in increments and the rotation of the free end was recorded.

Figures 9 and 10 show that the rotation has been reduced through the use of the new steel supporting member by 74%.

A summary of the line slopes (Table 4) indicates the significant improvement in rotation offered by this new steel rope construction.

The inverse of this slope expresses the twist tendency due to mechanical loading. In the 5/32-inch steel rope this has been improved by a factor of approximately 500. The influence of this new rope in the multipcore elevator control cable has reduced the twist tendency by about half.

4. CONCLUSION

A means has been demonstrated to reduce the twisting effect of steel rope with changing load on travelling cable through the use of a newly designed steel strength member. The inherent stability of this core will permit other design features, that affect twist to be considered independently.

10

The problem of application of plastic lining in friction drives

Dr Wojciech Cholewa and Dr Józef Hansel, University of Mining and Metallurgy, Cracow, Poland

ABSTRACT

The chapter presents the problem of application of plastic lining on sheaves and driving drums of rope transport equipment. The advantages of the application of linings such as a favourable influence on the durability of ropes, the possibility of an increase in hoisting capacity caused by a higher friction coefficient as well as the problem of a partial compensation of the non-uniform load distribution onto individual ropes of multi-rope drives are discussed.

The effect on the rope durability is shown based on the analysis of stresses in rope wires. The results of the investiagation of the influence of linings on the rope life are described. The influence of several factors on the friction coefficient of linings based on the selected results of the investigations is discussed. The effect of the lining on load compensation in ropes of multi-rope drives is discussed using equations that indicate a favourable influence of the lining elasticity in reducing the non-uniformity of load.

1. INTRODUCTION

For many years, the lining of the driving wheel grooves, has been applied to rope transport friction drives. The use of lining material with a shear modulus much lower than that of the rope wire increases the durability of ropes, gives the possibility of increasing the load due to the higher friction between the rope and the lining and also produces a self-acting equalizing of the rope loads in multi-rope installations. The particular character of operation of the drive-wheel lining imposes the following lining requirements:

Resistance to contact pressures.
Value of the friction factor.
Fatigue life.
Wear resistance.
Resistance to rope lubricants.
Resistance to ageing.
Stability of mechanical properties for different temperature and humidity.

In some cases of use there are also requirements regarding:

Inflammability.
Formation of electric charge.
Gas evolution during a thermal decomposition.

This chapter relates to the problem of the favourable influence of the plastic lining on the durability of wire ropes and gives test results for the determination of the operational adaptability of the three types of material used for the manufacture of the lining which is used in Polish rope transport installations.

2. EFFECT OF LINING ON THE LIFE OF ROPES

The durability of wire ropes is a function of many variables which create difficulties in an empirical determination of the relation between the durability and those variables. Investigators working on that problem give various relations, often complicated ones, in which there are many factors determined by laboratory tests. Most of those relations can be reduced to the following general form:

$$N \sim \frac{c}{\sigma^n} \qquad (1)$$

where c is the factor of proportionality and σ is reduced stress in rope wires.

The value of the reduced stresses is the component of the stresses resulting from the longitudinal loads and from the running of the rope over pulleys and rollers. Based on the relation given by Wyss (1956) and making some simplifying assumptions, the formula is obtained:

$$\sigma = (\sigma_r - \sigma_z) + 0.35p \qquad (2)$$

where σ_r is the tensile stress, σ_z is the bending stress and p is pressure on the bending arc.

The actual pressures occurring during bending depend on the rope load, wheel groove material, rope material, rope construction, wheel diameter, angle of lap and the geometric parameters of the groove. The determination of exact values of actual pressures between the groove material and the wires is a very complicated problem. The formulae given by the investigators of that problem are based on the Hertz formula while for practical calculation many simplifying assumptions are made.

Without closer investigation of the pressures it can be said, that an application of the plastic lining reduces the pressures between the rope wires and the groove material by 4 to 12 times. That reduction of pressure increases the life of ropes. Owing to a simultaneous effect of other stresses and factors on the wear process of a rope it is impossible to determine the direct relation between these parameters. It is obvious, that the application of elastic lining will increase the life of ropes, but the effect will be smaller than the reduction of the pressures between the rope and the lining (see Equation (2)). The quantitative results of tests resolving the influence of the elastic lining on the durability of ropes differ considerably. This is due to the use of fatigue testing machines of different design, different rope constructions and the lining material with different properties and also owing to the individual testing conditions. The tests carried out by Benoit (1915) revealed an increase in durability of the 5×7 construction ordinary-lay ropes with organic cores on leather lining by 1.6 to 5 times. Dukelski (1916) recorded a 1.5 to 2.5 times increase in durability of the 6×19 construction ordinary-lay rope with a steel core, and Müller (1962) obtained an increase of 2.2 and 3.5 times for a rope of the same construction with an organic core on the polyamide lining which is similar to the data stated by BASF (1972). For similar material, Chen and Gage (1981) obtained the values of 1.3 to 4.5 for 6×25 construction ropes with steel cores. Hansel and Stachurski (1971) present the values of 1.3 to 2.6 for polyamide lining and 2.2 to 5.3 for vinyl-rubber lining and different rope constructions.

3. OPERATIONAL SUITABILITY OF LINING MATERIAL

The use of plastics as the lining material increases the life of ropes and at the same time the load can be increased because of the higher friction factor obtained by a proper choice of the physical and chemical properties of the lining material. It is also necessary to ensure the high durability of the lining to make it competitive with the conventional material. Investigations of the drive-wheel lining have for a long time been carried out in many research centres which deal with rope friction drives. In the first period when the flexible lining was introduced the conventional materials like wood and leather were of interest. The research into the suitability of plastics for the drive-wheel lining put stress on the determination of friction factors for different values of pressure and surface condition of the cooperating elements.

The problems of the choice of plastic lining for friction drives have been dealt with for many years by the Interbranch Laboratory for Testing Wire Ropes and Rope Transport Equipment in Cracow.

Figure 1 presents diagrams of test stands for determining the properties characterizing the suitability of materials for rope wheel lining. The a, b, c stands are used for abrasion tests and friction factor determination. The resistance to wear according to the PN/C-89081 standard is tested at the stand d. The e, f, g stands are used for tests on resistance to changing pressures. Later some tests (University of Mining and Metallurgy, Cracow, 1983) are presented for the operational suitability of three kinds of plastics proposed as a lining for rope friction drive wheels. The plastic lining produced on the base of butadiene acrylonitrile rubber and polyvinyl chloride is denoted by R-7, a similar composition lining with addition of polyurethan rubber is denoted by N-1, and the polyurethane plastic by PU-1.

3.1 Tests of Friction Factor

The measurements of the friction factor between a rope and a straight lining element were carried out at the stand shown in Fig. 1(a). The tests were made at the relative speed between the rope and the lining of 1.7 mm/s, thus the results are the measure of the frictional contact during the elastic slip of the rope on the friction wheel. The plastics tested worked with the 18 mm T 6 × 19 +A_0 right-hand Langs lay ropes. One was dry and the other lubricated with the LWKP grease. The tests were made in different temperatures using the KTK-800 climatic chamber for this purpose.

To determine the kinetic friction factor, the mean value of the three measurements in the same conditions was taken. These results were plotted as the relation between the kinetic friction factor of the plastics tested and the mean pressures for various temperatures (Fig. 2), and also as the relation between the friction factor and the temperatures for various pressures (Fig. 3). The mean values of the pressures were calculated from the formula:

$$P_0 = \frac{N}{dl} \tag{3}$$

where N is the pressure of the rope on the sample and l is the length of the sample.

In the case of a rope operated on an arc lining element the measurements were made at the stand shown in Fig 1(c). The measurements were taken after the rope reached a speed of 0.5 m/s and the wheel has been loaded with the braking moment $P_2 . R_2$ for about 5 second. The static and kinetic friction factors were calculated from the data taken by the recorder using the following formula:

$$\mu = A \ln\left(1 + B\frac{p_2}{p_1}\right) \tag{4}$$

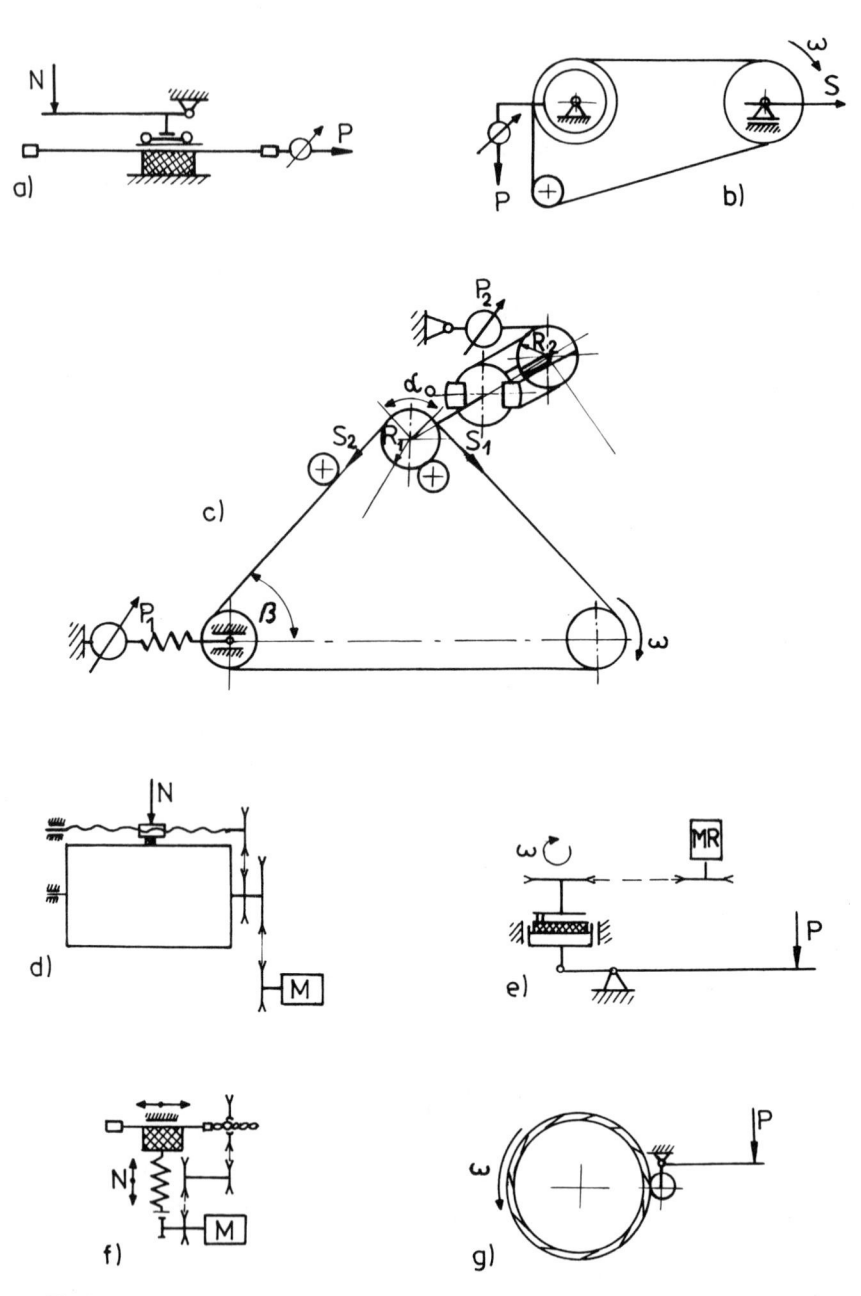

Fig. 1 — Diagrams of test stands. *Note:* a, b, c are used for abrasion tests and friction factor of determination; d is used for resistance to wear; e, f, g are used for tests on resistance to changing pressure.

Ch. 10]   **The problem of application of plastic lining**   105

where:

$$A = \frac{1}{\alpha_0} \quad \text{and} \quad (1 + \cos\beta)\frac{R_2}{R_1}$$

The quantities contained in the formula are indicated in Fig. 1(c). Some results obtained by calculation are presented in Figure 4.

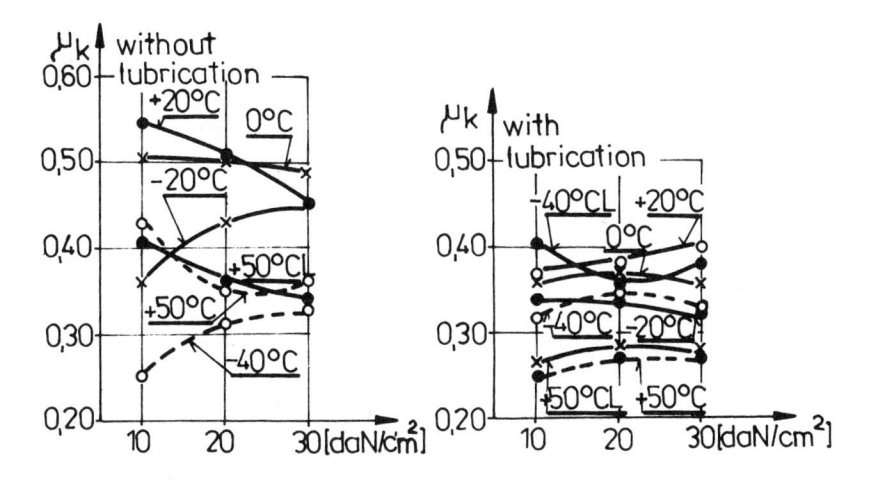

Fig. 2 — Relation between kinetic friction factor and pressure for R–7 plastic.

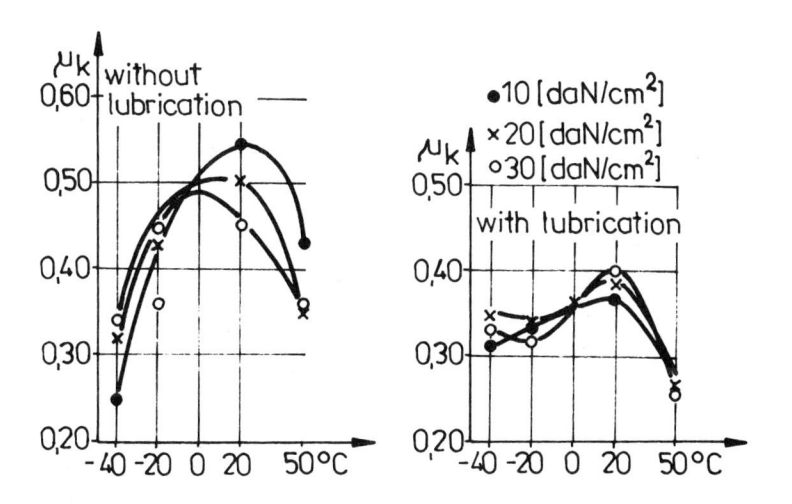

Fig. 3 — Relation between kinetic friction factor and temperature for R–7 plastic.

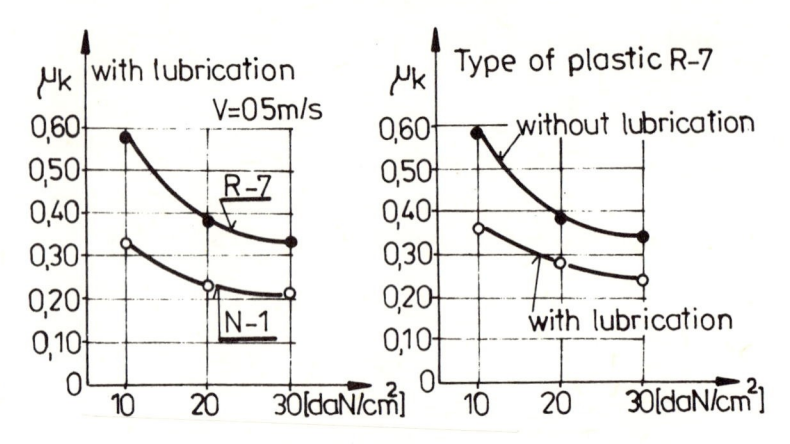

Fig. 4 — Relation between static and kinetic friction factor and pressure for N–1 and R–7 plastics.

3.2 Tests of Wear Resistance during Rope Slip
Wear resistance tests were carried out on the stand shown in Fig. 1(c), for two slip speeds (rope speeds) 0.5 and 1 m/s, during time intervals of 10 and 15 s. After the slip the radial loss of the lining material was measured in the middle of the lap angle, by directly taking the difference of the groove depths before and after the slip. A measure of unit wear was assumed as the loss of the lining material, in the radial direction, in relation to 1 m of rope displacement. The results obtained are listed in Table 1.

3.3 Tests of Resistance to Altering Pressures
Tests were made on the stand shown diagrammatically in Fig. 1(e), for mean pressures of 15, 20, 25 and 30 daN/cn^2. For each of these values two samples taken from each plastic were tested. In certain time intervals the sinking of a ball in the lining and the temperature on the lining sample surface were measured.

The tests were continued until the ball sank deep enough so that it could not roll any more.

The ambient temperature during the tests were 20°C and 50°C. In the preliminary tests a high dependency of results on the hardness of plastic has been found. For this reason the samples made from plastics with great differences of hardness were tested.

Some test results are shown in Fig. 5.

4. CONCLUSIONS
(i) The kinetic factors of friction of the plastics tested shown differences depending on the testing conditions. For a straight section of the lining made from the N–1 plastic, the temperature 20°C and the pressure of 20 daN/cm^2 the friction factor for a dry rope is 0.63 and for a lubricated rope 0.43. Similarly for the PU–1 plastic 0.30 and 0.28.

Ch. 10]   **The problem of application of plastic lining**   107

Table 1 — Results of tests of wear resistance.

Plastic	Pressure (daN/cm^2)	Slip speed (m/s)	Time of slip (s)	Loss of lining (mm)	Wear (mm/m)
R–7	30	0.5	10	—*	—
			15	—	—
		1.0	10	0.4	0.05
			15	0.6	0.04
	40	0.5	10	0.7	0.14
			15	1.2	0.16
		1.0	10	1.6	0.16
			15	2.6	0.17
N–1	30	0.5	10	1.0	0.20
			15	1.8	0.24
		1.0	10	2.5	0.25
			15	4.0	0.27
	40	0.5	10	5.5	1.10
			15	10.0	1.33
		1.0	10	10.8	1.08
		15	15	18.0	1.20

*0 unmeasurable wear

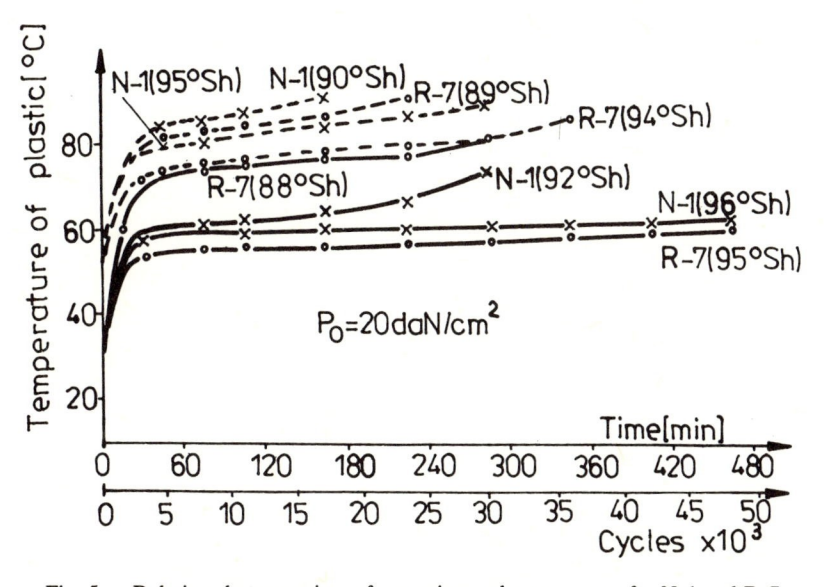

Fig. 5 — Relations between time of operation and temperature for N–1 and R–7 plastics during fatigue testing.

(ii) The lubrication of ropes reduces the friction factor for N–1 and R–7 plastics by 15 to 40% depending on the surface condition and pressure. The PU–1 plastic does not obey this rule.
(iii) The values of the static friction factor are about 15 to 50% higher than those of the kinetic factor. For pressures of the order of 10 daN/cm^2 the differences are 15 to 20%, but for pressures of 30 daN/cm^2 they amount to 30 to 50%.

REFERENCES

BASF (1972). Werkstofblatt 3142. 1.
Benoit, G. (1915). *Die Drahtseilfrage*. Friedrich Gutsch, Karlsruhe-Berlin.
Chen, J. and Gage, P. (1981). Improved Wire Rope Endurance Life with Nylon Sheaves. 13th Offshore Technology Conference, 4–7 May, 1981, pp. 443–447.
Dukelsky, A. I. (1966). Vynoslivost kanatov pri robotie na futerovanyh blokah. *Stalnyie Kanaty*, no. 3, Izd. Tehnika, Kiev.
Hansel, J. and Stachurski, J. (1971). Choice of lining material for headframe wheels and their effect of lining on the fatigue life of ropes. *Przeglad Mechaniczny*, No. 13, pp. 464–467.
Müller, H. (1962). Das Verhalten der Drahtseile bei Wechselbeanspruchungen. *Deutsche Hebe- und Fördertechnik*, No. 2, S.27–30.
University of Mining and Metallurgy, Cracow (1983). Investigation of operational properties of plastics used for drive wheel lining. Research work of SLBLSiUTL.
Wyss, T. (1966). *Stahldrahtseile*. Springer-Verlag, Berlin.

Part 4
Control systems

11

Use of proportional valves for hydraulic elevators

Mr. Eduard Hadorn, Beringer Hydraulik, Neuheim, Switzerland

ABSTRACT

In this chapter the principal construction of a new lift valve system, the electronic control board and the advantages of this system will be described.

The new valve functions without electronic equipment and is designed for travelling speeds up to 0.65 m/s. The integrated hydraulic control permits load-independent travelling characteristics in upward and downward directions.

1. INTRODUCTION

During the last two decades, hydraulic elevators have enjoyed an enormous boom. And there is still no change in the growing trend towards hydraulic drives. One of the main reasons for this trend is definitely the advanced development of the hydraulic valves which opens further possibilities for the use of the hydraulic elevator.

2. THE REQUIREMENT FOR AN IDEAL ELEVATOR DRIVEN BY A HYDRAULIC SYSTEM

A number of requirements can be stated:

(i) Ideal acceleration up to the maximum of travel speed; possibly long travelling time with maximum speed; optimal deceleration with a soft stop to follow.)
(ii) Exact stop in floor level.
(iii) Constant travelling times.

(iv) Ideal travelling comfort for speeds up to 1 m/s as to use hydraulic system also when higher speeds together with excellent performances are required.
(v) Load and temperature independent (i.e. these requirements have to be met by empty or loaded cabins and have to cover a wide range of temperature).
(vi) Little use of energy.
(vii) Low noise emission.
(viii) Low maintenance requirements.
(ix) Good rate of costs versus efficiency.

To meet these requirements a system will be described which consists of the mechanical valve, type LRV and the electronic control board, type ELRV as shown in Fig. 1. Worldwide more than 20 000 electronically controlled valves have been installed by Beringer.

3. DESCRIPTION

3.1 Valve Type LRV (Fig. 2)

The elevator valve is presently offered in three sizes for flow rates up to 1000 l/min. Whilst travelling up, it will be seen (Fig. 3) that an electrically controlled proportional relief valve will control the opening of the by-pass valve according to its regulation voltage. The pressure produced by moving the by-pass valve allows the oil to flow into the cylinder through a check valve. The increasing voltage at the proportional solenoid causes the by-pass valve to be slowly closed against the force of a spring, effecting a smooth acceleration of the elevator.

The down travel functions in a similar, but separated system. The oil flow during upward travel flows towards the cylinder and, in downward travel away from the cylinder, passing the flowmeter integrated in the valve. As a function of the pressure drop, the baffle plate will be axially displaced. This movement is converted into an electrical d.c. voltage signal by a potentiometer, and this signal is supplied as the existing value to the electronic control board. The output signal is proportional to the flow rate which, directly determines the lift travel speed.

An emergency lowering valve with deadman's handle permits a slow lowering of the cabin (e.g. in the event of power failure).

3.2 The Electronic Control Card

The electronic control card is the brain of the lift control valve system and receives the same information signals as for all standard lifts.

The card is of the European standard size and is plugged into the power module or directly into the 19-inch rack. By means of five potentiometers the total travel curve in upward and downward directions can be 'programmed'. The travel curve which is set once at an optimum, will be constantly compared with the actual value during the total travel. In case of deviations, the proportional solenoids, which are controlled through an

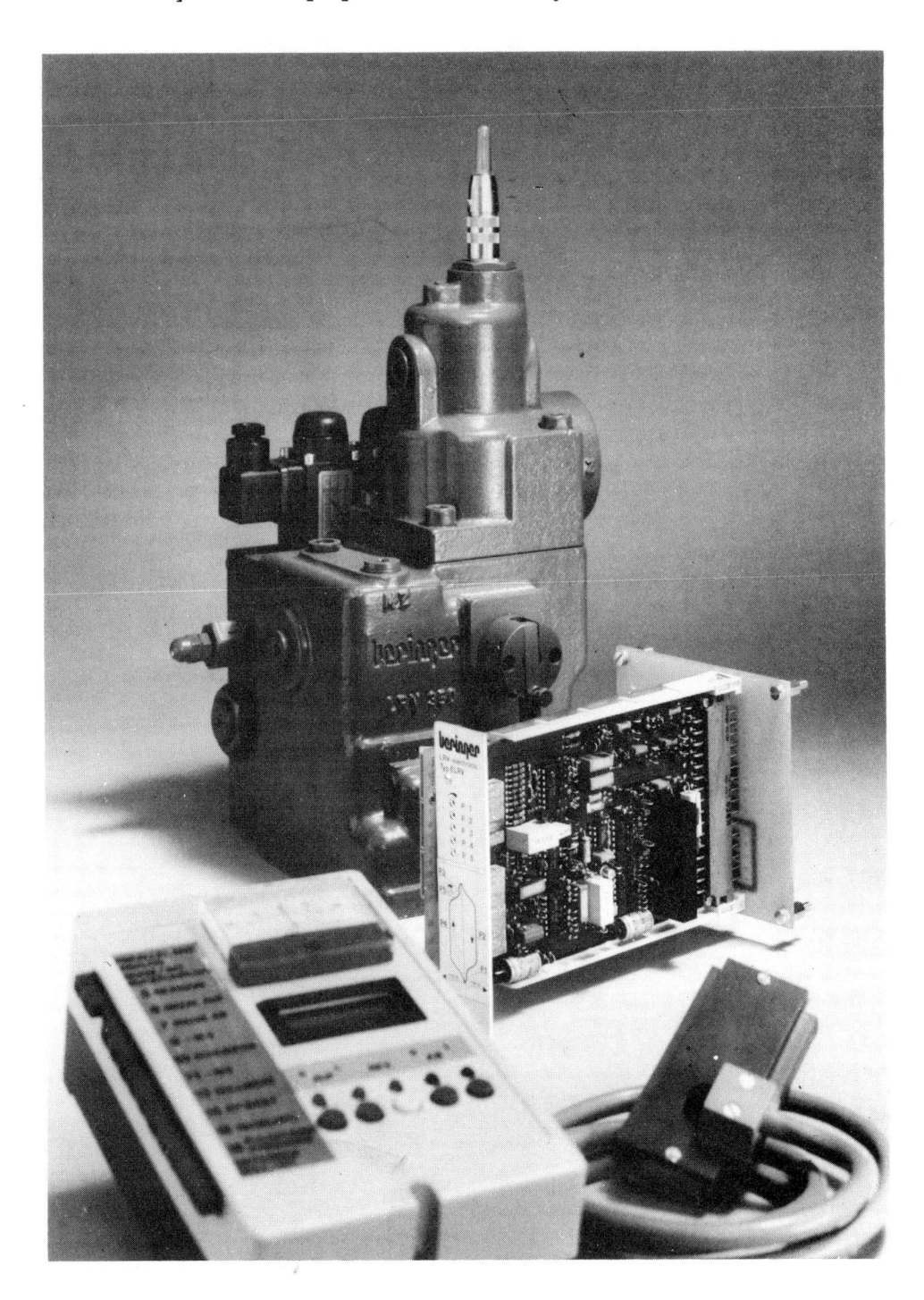

Ch. 11] **Use of proportional valves for hydraulic elevators**

Fig. 1 — Complete system of electronically controlled valves.

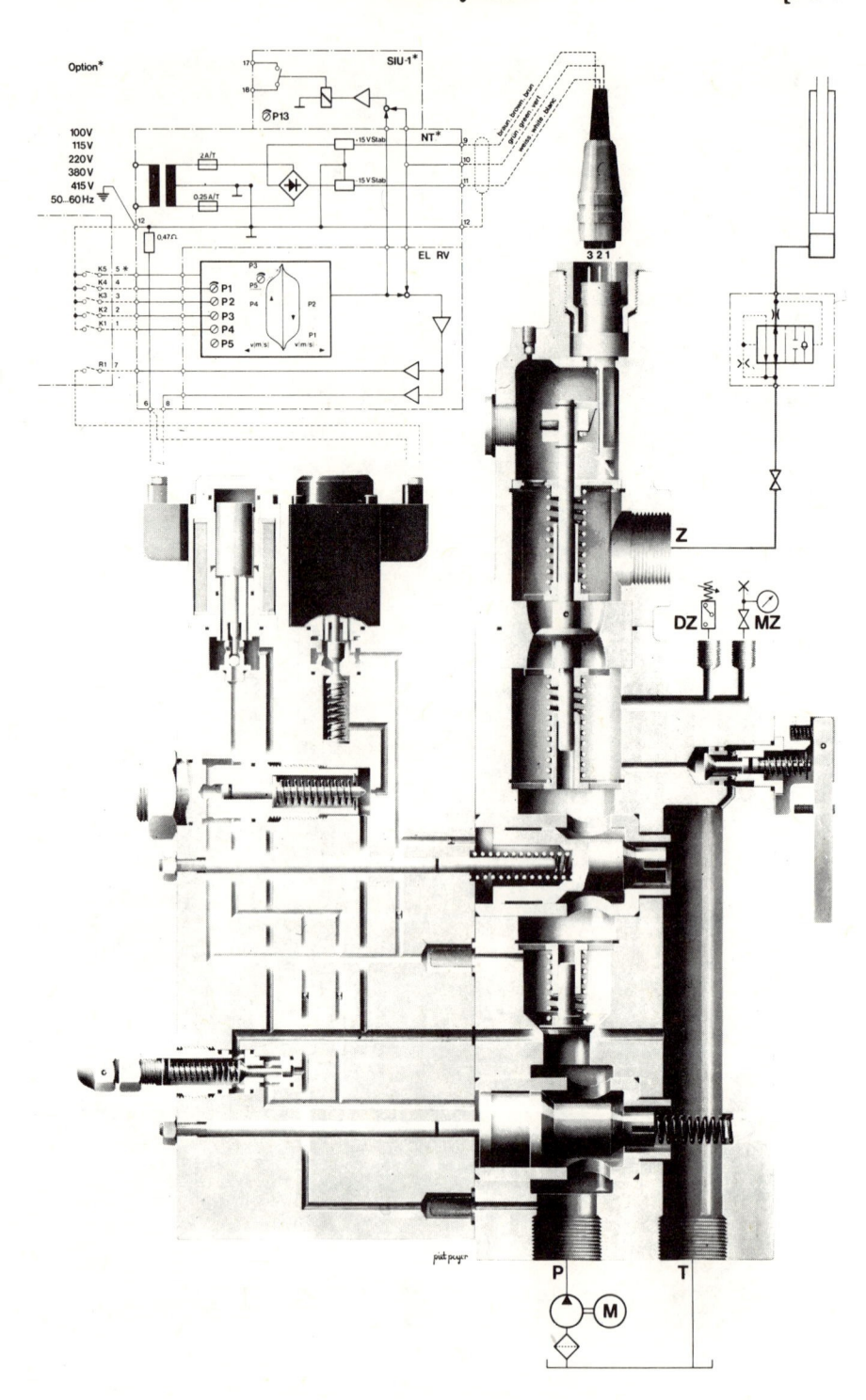

Fig. 2 — Details of valve type LRV.

Ch. 11]   **Use of proportional valves for hydraulic elevators**   115

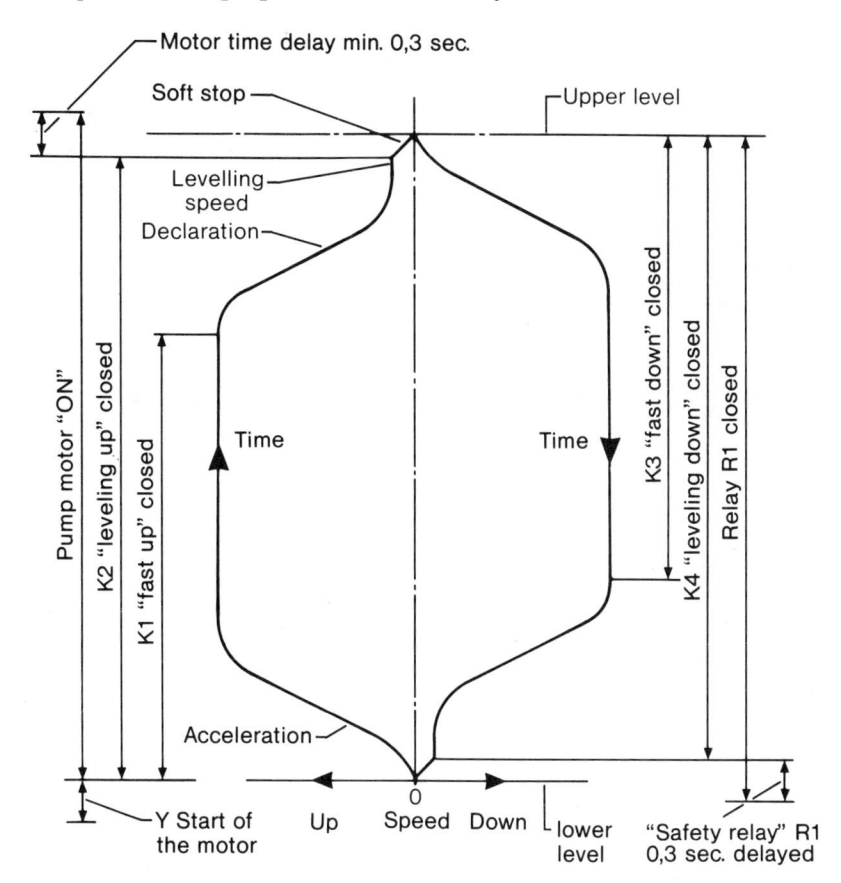

Fig. 3 — Control sequences.

amplifier, will correct the by-pass or the lowering valve in such a way that the actual value corresponds to the preset value. In this closed-circuit system the effect of the varying oil temperature and lift car loads are so negligible, that they can be practically ignored.

3.3 Adjustment and Test

The ability to preset an ideal travel curve in the form of d.c. voltages on the electronic control card, when the car is stopped, makes the taking into service and the checking of the lift system very simple. The requirements are either a simple voltmeter or a test box, which has been developed especially for the electronically controlled valve system.

4. FINAL SUMMARY

The last years have shown that hydraulic elevators are an ideal field for the application of electronically controlled proportiona valves. The author is convinced that the higher demands for hydraulic elevators in the coming years requires an increasing use of proportional valves.

12

A new elevator controlled by inverter

Mr E. Watanabe, Mr H. Kamaike, Mr S. Suzuki, Mr T. Ishii and Mr S. Yokota,
Mitsubishi Electric Corporation, Inazawa, Japan

ABSTRACT

Today, elevators are in the process of major change. An important technological development is inverter (VVVF — variable voltage, variable frequency) control which is replacing traditional Ward-Leonard or thyristor-Leonard for high speed elevator and primary voltage control of induction motors for medium or low speed elevators. Another important change is tht use of the helical reduction gear for high speed AC elevators, which may become a competitor of the highly satisfactory 100-year application of the d.c. gearless machine.

Mitsubishi's new elevators have achieved improvements such as increased energy saving and power supply reduction. Compactness, light-weight design, high reliability and safety, and simplified installation and maintenance.

This chapter describes Mitsubishi inverter control and helical reducer in comparison with conventional techniques.

1. INTRODUCTION

Today, elevators are on the verge of major change owing to technological developments. The most epoch-making changes of these is the VVVF (variable voltage, variable frequency) control system for elevators covering all speed ranges from low speed to very high speed and helical gear reducer for high speed a.c. elevators. In recent years, a number of technological developments have been made. One is in motor control which shows that an induction motor can be driven at the same performance level as a d.c. motor by controlling frequency and voltage, backed by the hardware technology such as power devices and microcomputers which can put the theory into practice at an economic cost. Other developments are CAD, have enabled the mass production of silent helical gears. Thus this new technology has overtaken the 100-years tradition of high-speed elevators using the d.c. gearless system.

2. STRUCTURE OF THE VVVF CONTROLLED ELEVATORS

2.1 General Comparison

Table 1 shows a comparison of the VVVF controlled elevators. Low and medium speed elevators are equipped with proven worm reduction gear. A diode bridge circuit is incorporated as the converter to obtain direct current from an alternating current power source. High speed and very high speed

Table 1 — Comparison of the VVVF controlled elevators.

	Low and medium speed elevator	High speed elevator	Very high speed elevator
Speed range m/s	0.75–1.75	2–4	5–6
Motor capacity (kW)	3.7–18.5	18–45	35–65
Driving device	Worm gear	Helical gear	Gearless
Converter	Diode without Pulse Amplitude Modulation	Thyristor with Pulse Amplitude Modulation	Thyristor with Pulse Amplitude Modulation

elevators employ more efficient helical reduction gear or no reduction gear since the energy level in operation is relatively high. Also, a thyristor converter is used for regeneration to the power supply. For the elevators ranging from 2–4 m/s of operating speed, the drive unit built with an induction motor which runs at high speed coupled with a high-efficiency reduction gear is an ideal device for small and lightweight design of the elevator facility. For very high speed elevators which range at 5 m/s and over, however, it is better not to use a reduction gear for design efficiency because the reduction ratio is small.

2.2 VVVF Control Device for High Speed and Very High Speed Elevators

Figure 1 shows the configuration of VVVF control device for high speed and very high speed elevators.

The thyristor converter turns a three-phase alternating current into a direct current. The speed of the induction motor is detected by the pulse generator and fed back to the regulator as the speed feedback signal. The regulator compares the speed feedback signal and speed command signal

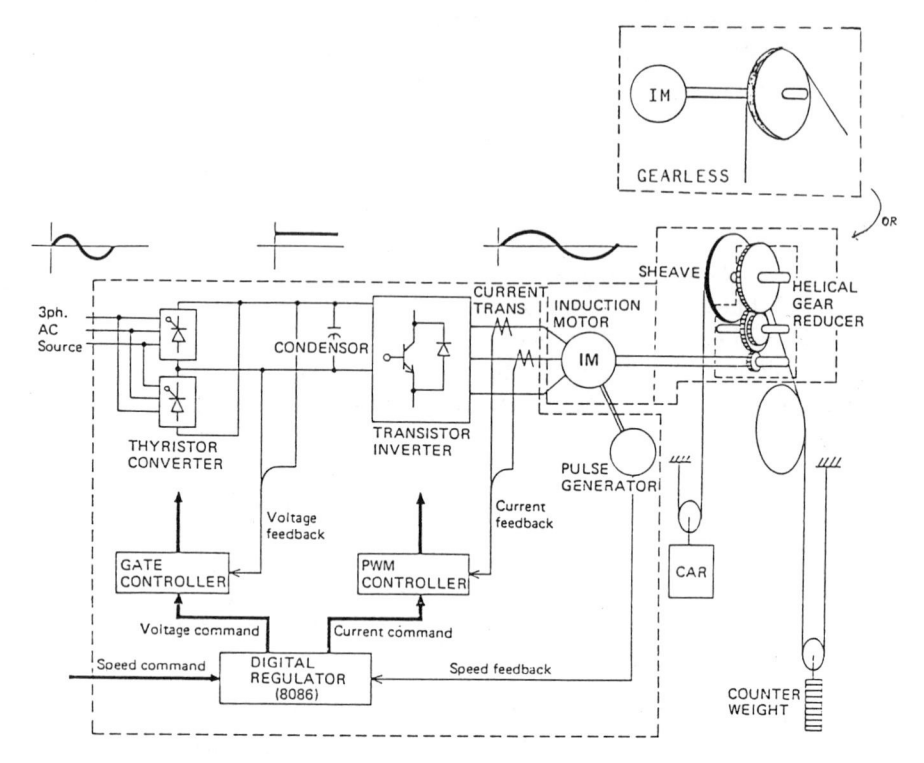

Fig. 1 — VVVF system configuration for high and very high speed elevators.

and outputs the voltage command signal and current command signal. A gate control device meters the output voltage of the thyristor converter corresponding to the voltage reference signal. This voltage control is used for the output PAM (Pulse Amplitude Modulation) of the transistor inverter. The thyristor converter also allows the regenerated power produced by the motor to be returned to the power supply. This action makes it possible to reduce power consumption. The PWM (Pulse Width Modulation) control device alters the output current of the transistor inverter corresponding to the current reference signal. The transistor inverter performs PWM to ensure the current supplied to the motor is approximately sinusoidal in order to reduce the torque ripple produced by the motor.

2.3 VVVF Control Device for Low and Medium Speed Elevators

Figure 2 shows the configuration of a VVVF control system for medium and low speed elevators. A three-phase alternating current is turned to a direct current through the diode converter, and then smoothed by the capacitor. The smoothed current is again turned into three-phase alternating current of variable voltage frequency by the inverter and supplied to the induction

Ch. 12] A new elevator controlled by inverter 119

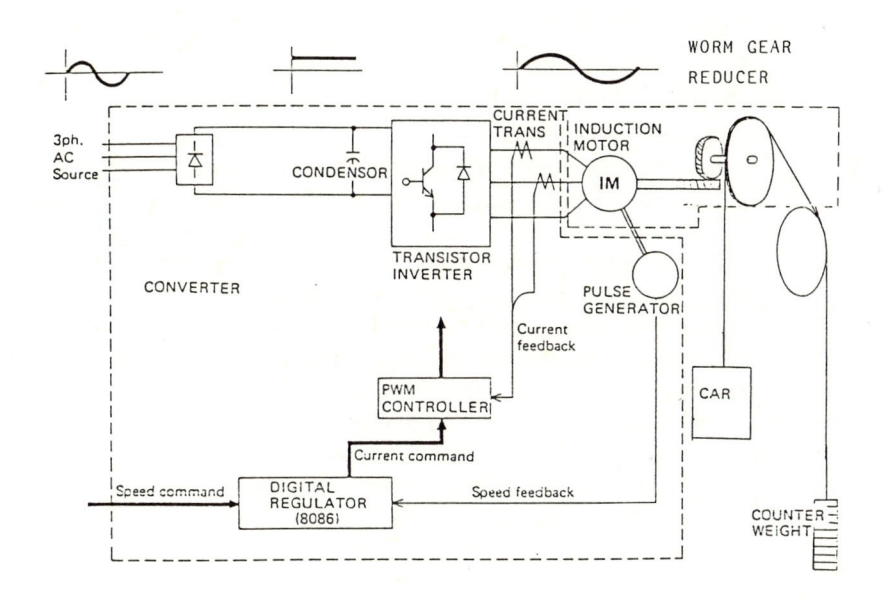

Fig. 2 — VVVF system configuration for low and medium speed elevator.

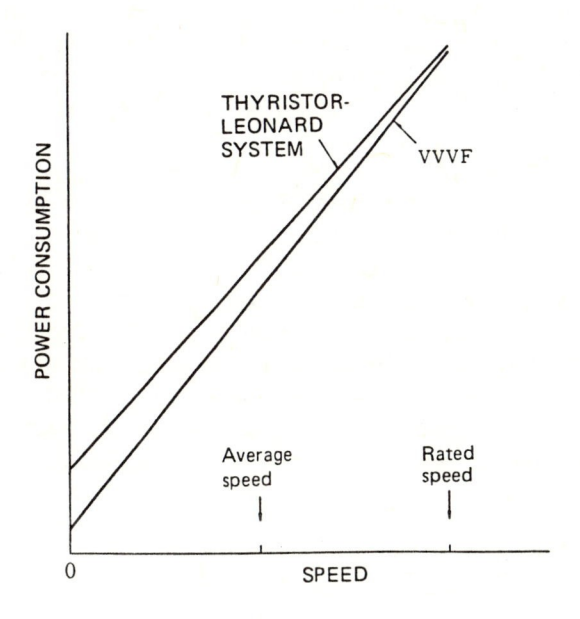

Fig. 3 — Comparison in power consumption.

motor. The inverter is comprised of transistors and outputs alternating current of variable voltage and frequency by PWM.

3. ADVANTAGES

3.1 Power Consumption

(A) *Power consumption of high and very high speed elevators*

The energy savings made by the VVVF control is not so large when compared with low speed elevators since high and super-high speed elevators are already equipped with a high-efficiency thyristor–Leonard control system. Figure 3 shows a comparison between the VVVF control and the thyristor–Leonard control on power consumption at various speeds. The figure indicates that the energy saving effect is more conspicuous in the low speed range than during rated speed operation. Generally, elevators are controlled in various operation modes, such as, partial speed operation and rated speed operation. According to the statistics of the distribution of operation modes, the average operating speed is about half of the rated speed. At this speed, the VVVF system consumes less energy than the thyristor–Leonard system by 5–10%.

(B) *Power consumption of medium and low speed elevators*

Figure 4 shows the relationship of torque, loss and speed of induction motor. The VVVF control is designed to operate always with the least loss of motor speed. Therefore it controls voltage and frequency supplied to the motor corresponding to the actual speed and required torgue of the motor. On the other hand, the conventional control system which has been used for medium and low speed elevators only controls the voltage which corresponds to required torque. This system suffers a large loss in the low speed range of the motor.

Figure 5 shows a comparison between a conventional system (primary voltage control) and VVVF control system on instant power consumption corresponding to car speed. As described above, the conventional control system suffers a large loss, and thus consumes a larger power in the low speed range during acceleration.

In the conventional control system, the decelerating elevator provides d.c. braking by the direct current flow in the motor coil and it consumes relatively less power. This is because it is sufficient to supply power to the motor to generate a d.c. field to provide the mechanical energy to be consumed within the motor. During rated load operation, power consumption increases in the latter half of deceleration, which is because the car is controlled to approach a floor at slow speed which makes operating efficiency low. The VVVF control, shown in Fig. 5 consumes power almost in proportion to the mechanical output of the motor during acceleration. During deceleration and rated load operation, regenerated power is supplied from the motor to the d.c. side of the inverter. As shown in Fig. 5 the VVVF control system requires half or less power, when compared with the conventional control system.

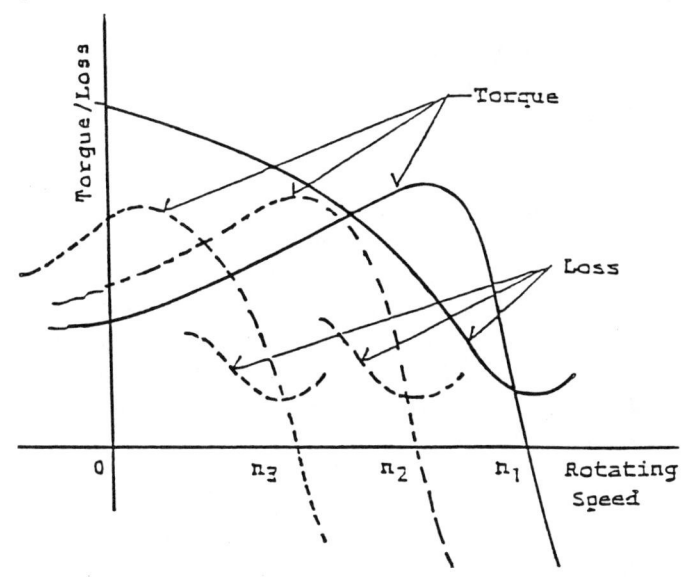

Fig. 4 — Relationship between motor torque/loss and rotating speed.

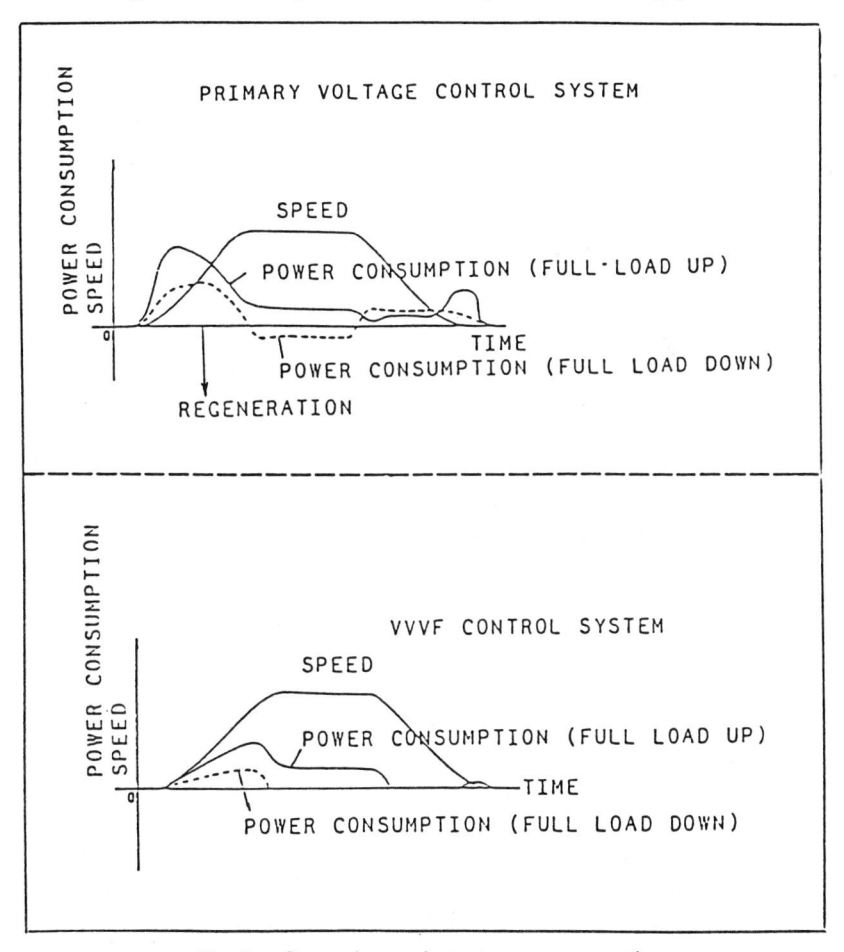

Fig. 5 — Comparison on instant power consumption.

3.2 Power Supply Capacity
(A) *Input waveform*
The VVVF device for elevators presents a rectifier load, when viewed from the power source side, and has a large capacitor on the d.c. side. Figure 6 shows the current and voltage waveforms at the input side of this rectifier. Figure 6(a) gives the equivalent circuit including power source when the VVVF device is regarded as rectifier load. In the diagram, Z represents power source impedance, E_D represents voltage on DC side and I_D represents direct input current supplied to the inverter. The graph in Fig. 6(b) shows the phase voltage of the power source and the graph in Fig. 6(c) shows the relationship of rectified waveform of interphase voltage and d.c. voltage E_D. From this graph, it will be seen that current can flow from the power source to the rectifier when the rectified waveform of interphase voltage is higher than the d.c. voltage E_D. The graph (Fig. 6(d)) shows the current flow in the S-phase and Fig. 6(e) is the interphase voltage waveform at the input end of the rectifier, whose peak value is clipped by the constant voltage E_D on the d.c. side. This is a distorted waveform which contains low-order (especially 5th and 7th order) harmonic current. The shape of this distorted wave is determined by the current on the d.c. side, power source impedance and the value of the smoothing capacitor, and can be obtained by relatively simple calculation. The relative harmonic content and power factor are easily obtained from the waveform. Figure 7 provides an example of them. The vertical axes indicate the ratio of the power factor and the 5th and 7th harmonic content with respect to the fundamental wave, and the horizontal axis indicates the direct current. A different power source impedance is used for this calculation as the parameter. The figure shows also that the relative harmonic content decreases as the direct current and power source impedance increase, which means the power factor is also improved.

(B) *Reduction on power supply capacity of building*
Figure 8 shows a comparison of power factor between the VVVF system for high and super-high speed elevators and the thyristor–Leonard system against varying speed (rated 4 m/s 1.150 kg). In the thyristor–Leonard system, the power factor increases as elevator speed increases, but in the VVVF system, the power factor does not decrease so much as the elevator speed decreases. Therefore, the VVVF system shows much improvement in the power factor in the low speed range although it shows almost the same level as the thyristor–Leonard system does around the rated speed. This is because the d.c. voltage in the thyristor speed, which makes the phase angle of thyristor large for low speed operation. Conversely, the d.c. voltage in the VVVF system need not to be decreased so much for low speed operation, and accordingly, the phase angle of thyristor can be small. The average operating speed of the elevator is about half of the rated speed. Therefore, the power factor of the VVVF system is much better than thyristor–Leonard system when compared at the average speed as shown in Figure 8.

The VVVF system for low and medium speed elevators achieves even more improvement in power factor since its converter is comprised of

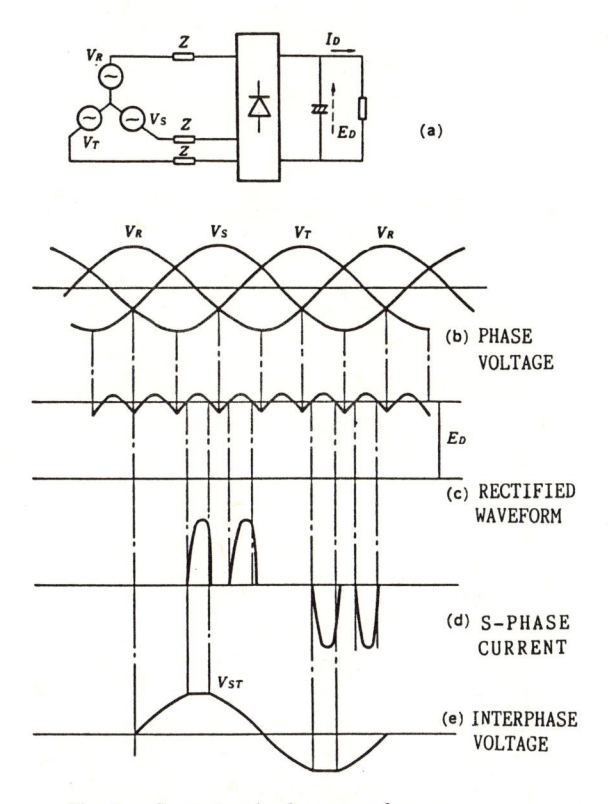

Fig. 6 — Current and voltage waveform.

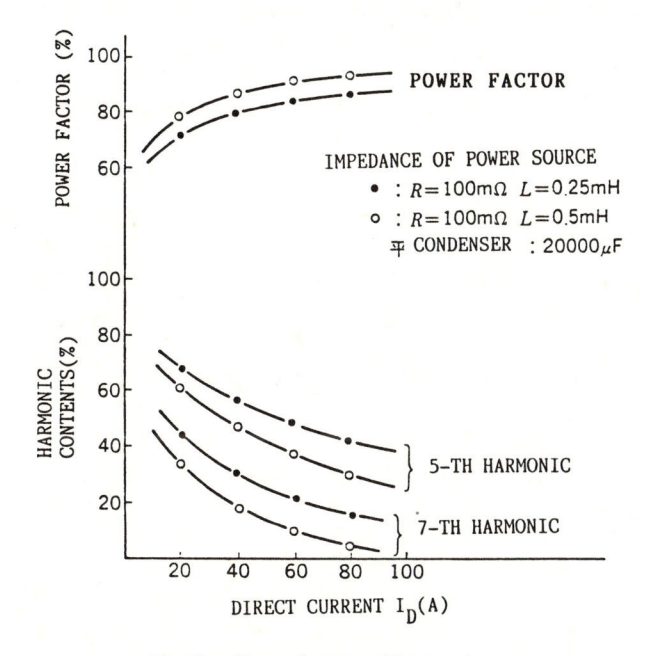

Fig. 7 — Power factor and harmonic contents.

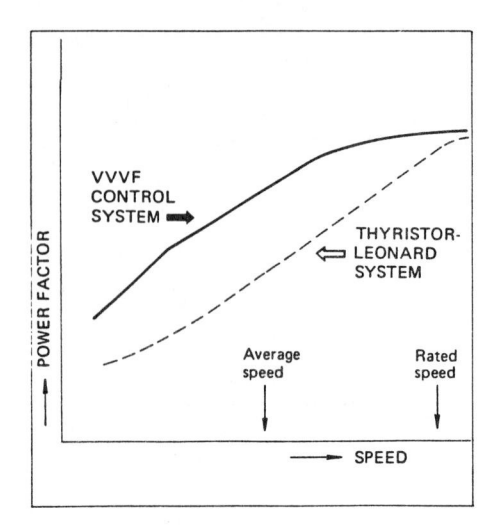

Fig. 8 — Comparison in power factor.

diodes, which eliminates the need for phase control operation. Thus, the elevators controlled by the VVVF system can reduce the demand on power supply as a result of improved efficiency and power factors. Reduction in power supply capacity achieved by the high and super-high speed elevators is 20–30%, and it reaches about 50% by the low and medium speed elevators, when compared with standard elevators.

3.3 Reduction on Bearing Load of the Elevator Machine Room
Figure 9 depicts a comparison, in outline dimensions, between the new driving device, which uses a helical reduction gear and a conventional d.c. gearless driving device. The new driving device is built more compactly, the

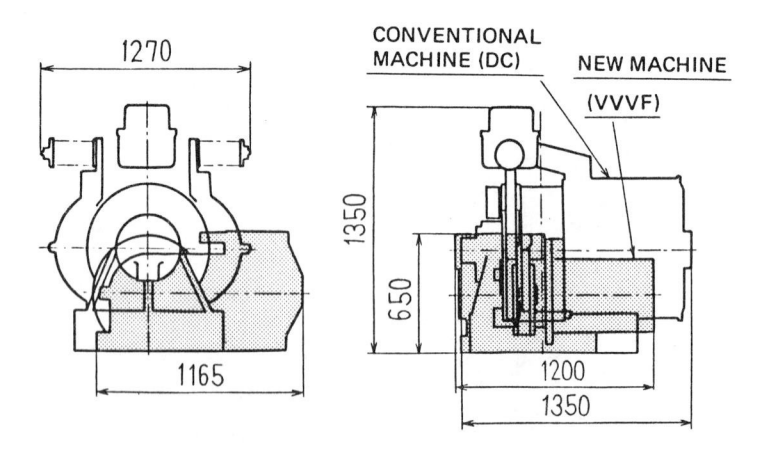

Fig. 9 — Comparison of traction machine (typical example).

reduction in size by volume being 50%. In the case of an elevator with a rated speed of 4 m/s and capacity of 1600 kg, the bearing load of the elevator machine room has been reduced by approximately 1 tonne.

4. CONCLUSION

This completely new elevator system is more advanced in ride comfort, safety and reliability compared with conventional elevator systems. Th Mitsubishi high speed VVVF system elevator is already in service at Gotanda NN Building, Japan which was opened in March 1984, and a number of other places, and is acquiring a favourable reputation. The ARK (Akasaka-Roppongi Area Redevelopment plan) Office Tower, Japan is to install 12 VVVF elevators including four 5 m/s and four 4 m/s. Furthermore, several hundred medium and low speed VVVF elevators are in service.

13

Variable voltage and frequency elevator drive

Dr Jalal T. Salihi, United States Elevator Coporation, Spring Valley, USA.

ABSTRACT

The first VVF elevator drive system was developed in the United States and put into service at the beginning of 1983. Since then, more than a thousand have been built and put into service in the United States. Other countries, in particulare Japan, have active VVF elevator development programmes and several elevators of these types have been installed.

Owing to the inherent advantages of the VVF drive system and anticipated cost reduction of power electronics, it is expected that the use of these systems will continue to increase and eventually dominate the elevator market.

This chapter includes a brief description of the basic components of a VVF elevator drive system, and a brief discussion of the principle of operation. Also, several configurations of the VVF elevator drive system are discussed and compared with respect to cost efficiency and other operational features.

1. BASIC COMPONENTS

This chapter describes the basic components of a four-quadrant variable volage and frequenty (VVF) elevator drive and gives a brief explanation of the theory of operation. Depending on the method of handling the regenerative energy, at least three versions of the drive system can be used for elevators. The choice is dictated by the elevator requirements and economic considerations.

Figure 1 depicts the basic components of a VVF drive system consisting of two main sections: a transistor inverter and a d.c. supply.

The transistor inverter operates from a d.c. source applied to its input (referred to as d.c. link), and produces a variable voltage and frequency three-phase sinusoidal output. The output drives the three-phase induction motor. The inverter–motor combination provides four-quadrant operation, such that during overhauling, the induction motor becomes a generator, converting the mechanical energy to a three-phase supply at the motor terminals. The inverter works backwards and converts the three-phase supply to a d.c. supply appearing at the input of the inverter. As shown in Fig. 1, the polarity of the d.c. voltage at the input of the inverter remains the same during motoring and regeneration, while the direction of current flow changes. The top arrow pointing to the right is the direction of current during motoring. The bottom arrow pointing to the left is the direction of current during regeneration.

The second section is a means for providing the d.c. supply to the input of the inverter during motoring and means for handling the reverse current at the input of the inverter during overhauling. The left portion of Fig. 1 shows three versions.

(i) The power to the d.c. link during motoring is supplied by a rectifier. During overhauling, the d.c. link voltage rises and blocks the rectifier and the regenerative energy associated with reverse current is dissipated in a resistance. The transistor in series with the resistance regulates the current flow. It acts as a switching regulator and turns on whenever the d.c. link voltage exceeds a set limit.

(ii) A bi-directional controlled rectifier is used to supply the power to the d.c. link during motoring and return the regenerative power back to the 60 Hz line by providing a path for the reverse current.

(iii) The d.c. link consists of an energy strage element such as a lead acid battery. A battery of sufficient energy capacity can provide the instantaneous motoring power and absorb the regenerative power, and can practically isolate the impact of the elevator on the 60 Hz mains. A relatively small battery charger is required to keep the batteries charged. Care must be taken to prevent the battery from overcharging during sustained elevator overhauling periods (e.g. evening down peaks). A resistance in series with a transistor is connected across the battery. The transistor behaves like a switching regulator and turns on whenever the battery voltage exceeds a set limit, thus preventing overcharging.

2. OPERATION OF INDUCTION MOTOR FROM A VARIABLE FREQUENCY AND VOLTAGE SOURCE

Figure 2 depicts the torque-speed characteristics of a four-pole, three-phase induction motor operating from a fixed frequency of 60 Hz and a fixed voltage of 220 V. The characteristic is exaggerated for the purpose of explanation. The portion of the curve above the x-axis corresponds to motoring and the bottom portion corresponds to overhauling (regeneration). The motor

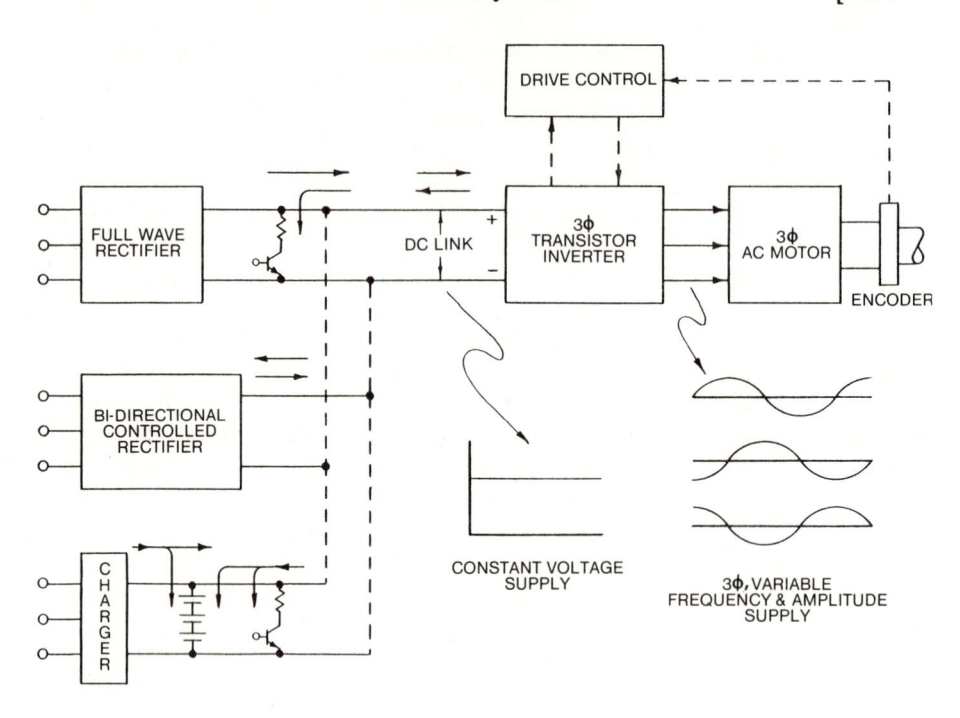

Fig. 1 — Basic VVF drive system.

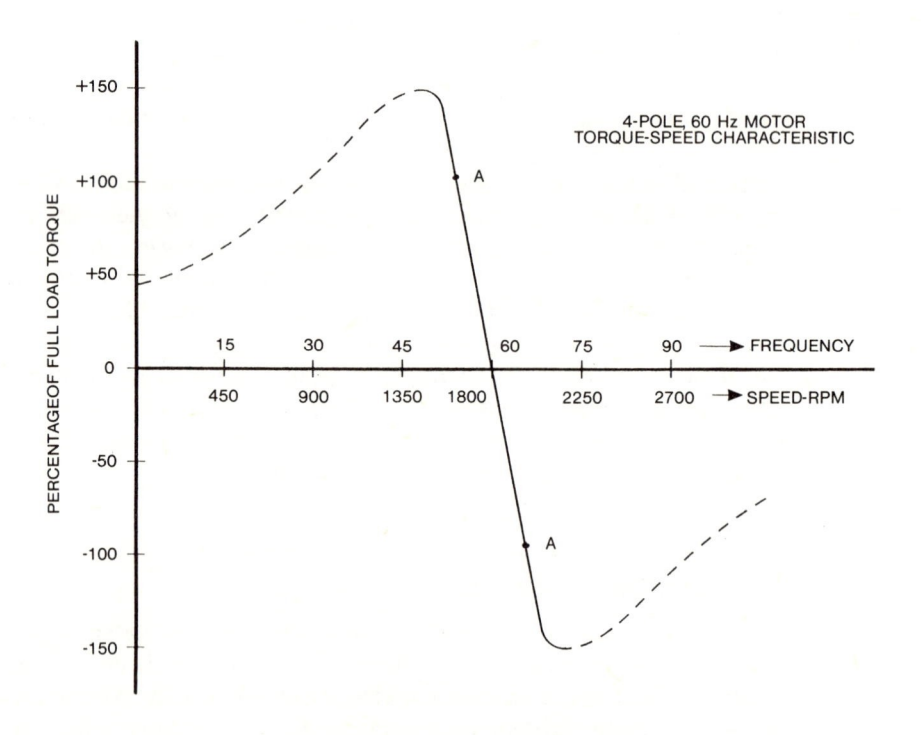

Fig. 2 — Torque-speed characteristics.

Ch. 13] Variable voltage and frequency elevator drive

normally operates within the linear range of the characteristic designated by the line AA between the knees of the characteristic.

At no load, the motor runs at a synchronous speed of 1800 rpm corresponding to a synchronous frequency (F) of 60 Hz. As the load increases, the motor speed and the equivalent mechanical frequency (FM) (or rotor frequency), decreases. The relationship between motoring torque and slip frequency (FSL)

$$FSL = F - FM = 60 - FM \qquad (1)$$

is approximately linea. Similarly, during overhauling, the motor becomes a generator, thus providing a negative braking torque. As overhauling torque increases, the motor speed increases above 1800 rpm and the equivalent mechanical frequency increases above 60 Hz. As in the case of motoring, the relationship between the braking torque and slip frequency is linear and is of the same slope. The only difference is in the polarity of slip frequency which is positive during motoring and negative during overhauling.

For example, in a well-designed 30 hp, 4-pole commecial induction motor of NEMA-B design, the slip frequency is about +1 Hz, during motoring at full load, and −1 Hz during regeneration at full load.

Figure 3 shows the effect of supply frequency on motor operation. The

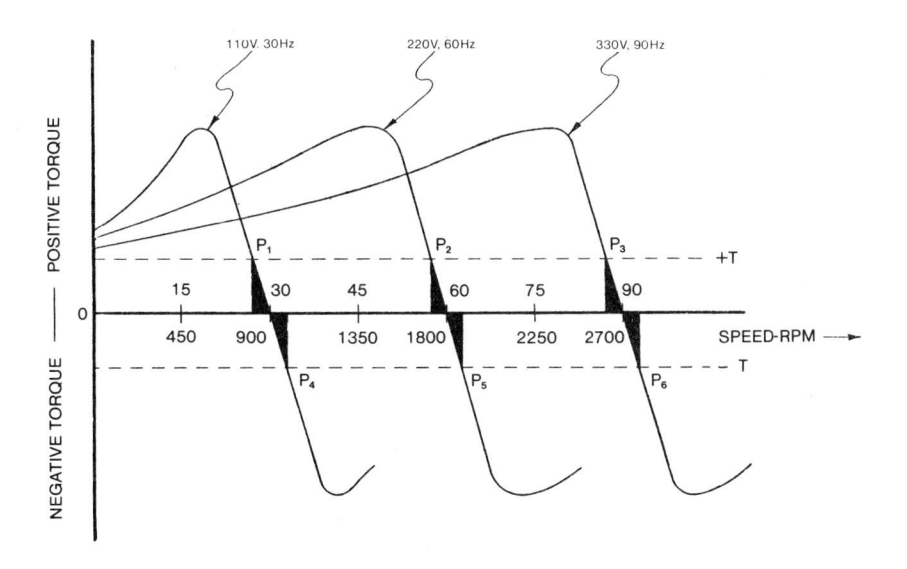

Fig. 3 — Effect of supply frequency.

three curves depict the characteristics of the motor of Fig. 2 when operating at three different frequencies of 30, 60 and 90 Hz corresponding to the three voltages of 110, 220 and 330 V, respectively.

Note that the ratio of voltage to frequency is the same, meaning that the flux for the three cases is equal. This makes the linear portions of the three curves approximately parallel to each other. Consequently, the relationship between torque and slip frequency is the same in all three cases and independent of motor speed.

The above relationship is further clarified by the triangles of Fig. 3 which shows that a constant motoring torque $(+T)$ can be produced by the same slip frequency at three speeds (Points P1, P2, P3). Similarly, a negative motoring torque of equal value $(-T)$ can be produced at an equal slip of opposite polarity at three different speeds.

3. CONTROLLED SLIP MODE OF OPERATION IN INDUCTION MOTOR

The above example illustrates the concept of slip control in induction motor which can be generalized as follows:

In an induction motor operating from a three-phase source of smoothly variable voltage and frequency, the torque is proportional to slip frequency and is independent of motor speed provided that the flux (which is determined by the ratio of voltage to frequency), is kept constant. This is true during motoring where both the slip frequency and torque are positive and during regeneration where both the slip frequency and torque are negative.

Figure 4 is a schematic diagram showing how the slip control mode of

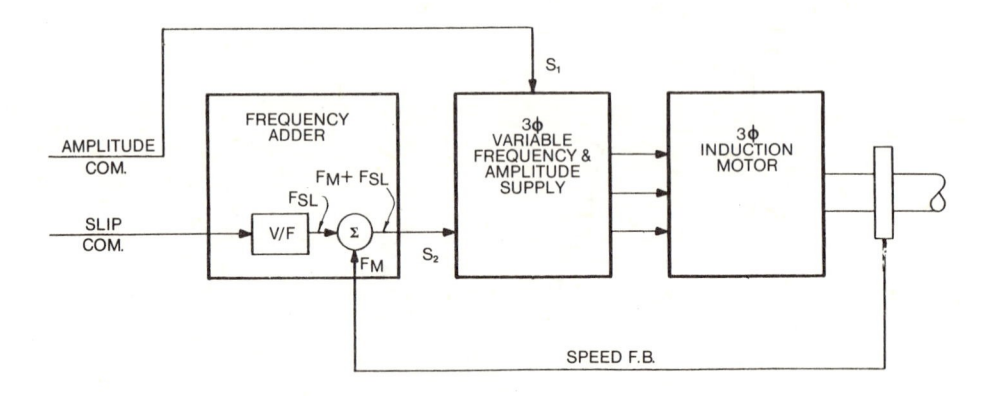

Fig. 4 — Schematic diagram of slip control mode.

operation can be implemented. The middle block produces a three-phase, smoothly variable and frequency supply feeding an induction motor. The input (S1) designated by 'amplitude control' is a d.c. signal controlling the amplitude of the three-phase output. The other input signal (S2) is a frequency signal controlling the frequency of the output. The block designated (frequency adder) produces the signal S2 which is the sum of the

Ch. 13]   **Variable voltage and frequency elevator drive**   131

mechanical frequency (FM) and the slip frequency (FSL). The slip frequency is produced by a voltage to frequency converter and is determined by a d.c. voltage designated by the 'slip command' applied to the converter. The mechanical frequency is derived from a digital tachometer mounted on the motor shaft.

The overall function of the control scheme is to provide means for dictating the slip frequency of the motor by a d.c. voltage command designated 'slip command' and dictating the amplitude of the motor voltage by a d.c. command designated 'amplitude command'. The frequency applied to the motor (F=FSL+FM) is the sum of the slip frequency and the mechanical frequency. In a practical system, the slip command signal is bipolar and the frequency adder can either add or subtract frequency depending on the polarity of slip command and the direction of motor rotation. Four-quadrant motor operation can thus be realized.

One of the main advantages of slip control is to make the behaviour of an induction motor similar to a separately-excited d.c. motor. This can be seen from a comparison of Figs 5(a) and 5(b). Figure 5(a) is applicable to d.c.

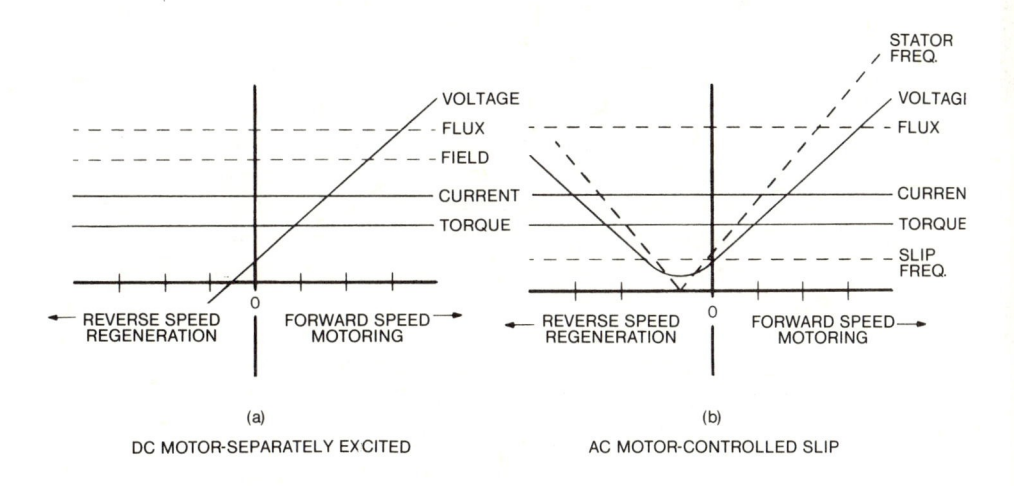

Fig. 5 — Motor control.

motor and shows the variation of motor current and terminal voltage with speed for a constant torque produced at a fixed field current (or flux). The characteristics cover both the motoring and regeneration region (right and left of the y-axis, respectively). Figure 5(b) shows the corresponding characteristics for a controlled slip induction motor, where the slip frequency and flux are kept constant. One of the main differences between the two motors is that in the case of a d.c. motor, the polarity of the terminal voltage changes in the reverse direction. This corresponds to a reversal in phase sequence of the three-phase supply feeding the induction motor.

4. CURRENT OR TORQUE CONTROL

Keeping the voltage to frequency ratio constant at the motor terminal is a crude method of keeping the flux constant. The flux remains approximately constant at high and low speeds and begins to change as the motor speed is reduced where the effect of primary resistance and reactance voltage drops becomes appreciable. This method is suitable for most practical applications where precise torque control at low speed is not required.

In elevator applications, precise torque control down to zero speed is required. Referring to the equivalent motor circuit of Fig. 6, this can be

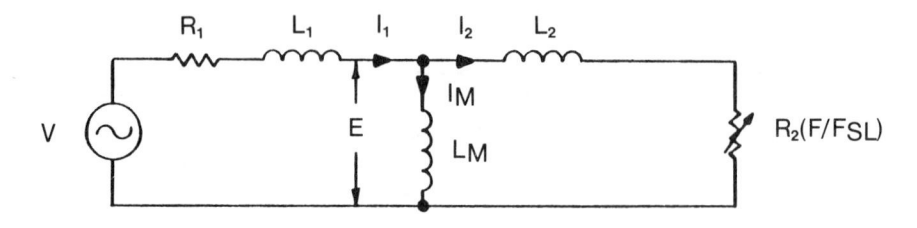

Fig. 6 — Equivalent motor circuit

achieved by compensating for the primary voltage drop across R1, L1 such that the voltage to frequency ratio across the excitation inductance LM remains constant at all speeds (that is the excitation current IM and associated air gap flux remains constant) The Appendix gives the analytical basis for the torque control scheme meeting the above requirement. The control scheme is shown in Fig. 7, and consists of slip control and torque control.

The slip control scheme consists of a modification of Fig. 4 to act as a current source instead of a voltage source. This is achieved by incorporating a three-phase current regulator and associated motor current feedback where the amplitude of the motor current is dictated by the amplitude command.

The torque control section provides both the amplitude and slip command to the modified slip control scheme. It correlates the amplitude and slip command based on the results of the Appendix, such that the input to the torque control block designated 'torque command' dictates the magntide and polarity of torque independent of motor speed. the signals A, B, C simulate the currents I2, IM, I1 which are related by Equation (4) of the Appendix. The gains K1, K2 make the slip command proportional to current I2 (simulated by Signal A).

5. SPEED CONTROL

The block designated 'speed control' in Fig. 7 compares the speed command with the actual motor speed, thus providing a speed error. The error is amplified and applied to the torque control block described above. The

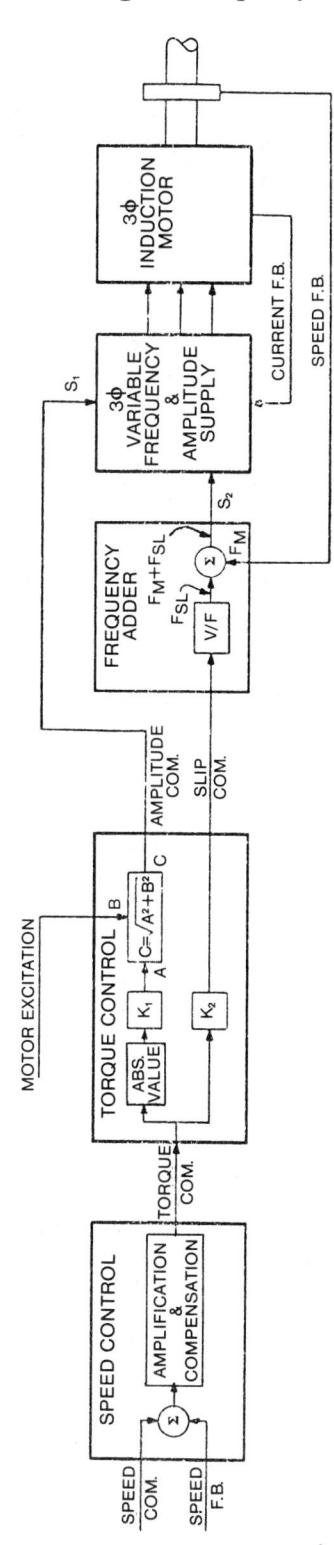

Ch. 13]   **Variable voltage and frequency elevator drive**   133

Fig. 7 — Speed control.

torque produced acts to minimize the speed error and make the motor follow the speed command.

In conclusion, Fig. 7 depicts the basic components of a four-quadrant VVF drive system suitable for elevator application. It will make the elevator follow a desired elevator speed trajectory accurately and automatically compensates for load variations.

APPENDIX: ANALYSIS OF TORQUE CURRENT CONTROL SCHEME

The equivalent circuit of the motor is shown in Fig. 6, where:

V	= Applied voltage	$R1, L1$	= Primary resistance and inductance
E	= Counter EMF voltage	$R2, L2$	= Secondary resistance and inductance
$I1$	= Primary current	F	= Input (synchronous) frequency
$I2$	= Secondary current	FM	= Rotor (mechanical) frequency
IM	= Excitation current	FSL	= Slip frequency
LM	= Excitation inductance	$F=$	= $FM + FSL$

Referring to the motor equivalent circuit:

$$\text{Flux} = K1 \times \frac{E}{F},$$

$$E = I2 \times R2 \times \frac{F}{FSL}$$

(Neglecting voltage drop across L2)

$$\therefore \text{Flux } (0) = K2 \times \frac{I2}{FSL} \tag{2}$$

$$\text{Torque } (T) = K3 \times 0 \times I2 \tag{3}$$

If the slip frequency is made proportional to current (I), then ($FSL/I2$) is a constant, then equations (1) and (3) gives:

$$\text{Flux } (0) = \text{Constant}$$

(E/F) = Constant
Torque (T) = Constant × $I2$

Also, since (E/F) is constant, then the magnetizing current (IM) is constant. Therefore:

$$I1 = \sqrt{(I2)^2 + (IM)^2}$$
$$= \sqrt{(I2)^2 + K} \tag{4}$$

14

Microprocessors in elevator controls

Dr-Ing. Joris Schroeder, Schindler Management AG, Lucerne, Switzerland

ABSTRACT

There are very few discoveries and technological developments that cause drastic changes to our society, the introduction however, of the microprocessor was definitely one of them! The elevator industry stated to apply microprocessors in 1980. Since then, a new generation of the most sophisticated controls has become available representing new and highly advanced technology. The latest microprocessor controllers offer digital drive controls, predictive dispatch algorithms, system diagnostic capabilities and extremely 'user-friendly' man/machine communications.

1. INTRODUCTION

Microprocessors have made computer techniques available for industrial control application. Today, a microprocessor-based controller is usually less costly than previous generations of relay or parallel logic solid-state controls — and much more 'intelligent', compact, reliable and durable.

Elevator controls are a particularly attractive applicaiton for microprocessors not only because they can handle large amounts of data quickly, but because the system becomes programmable. Adapting the controller for

Ch. 14] Microprocessors in elevator controls 137

a new application no longer requires new wiring diagrams, possibly additional components and different electronic assemblies. The new function can be achieved by new software, programmed into EPROMs (erasable Programmable Read Only memories), without any changes to the circuits. Fast microprocessors, EPROM memories and RAMs (Random Access Memories) are the key elements in any microprocessor control system. Large scale integration of these elements has permitted drastic reduction in physical size, (Fig. 1) and recent developments of market prices permit

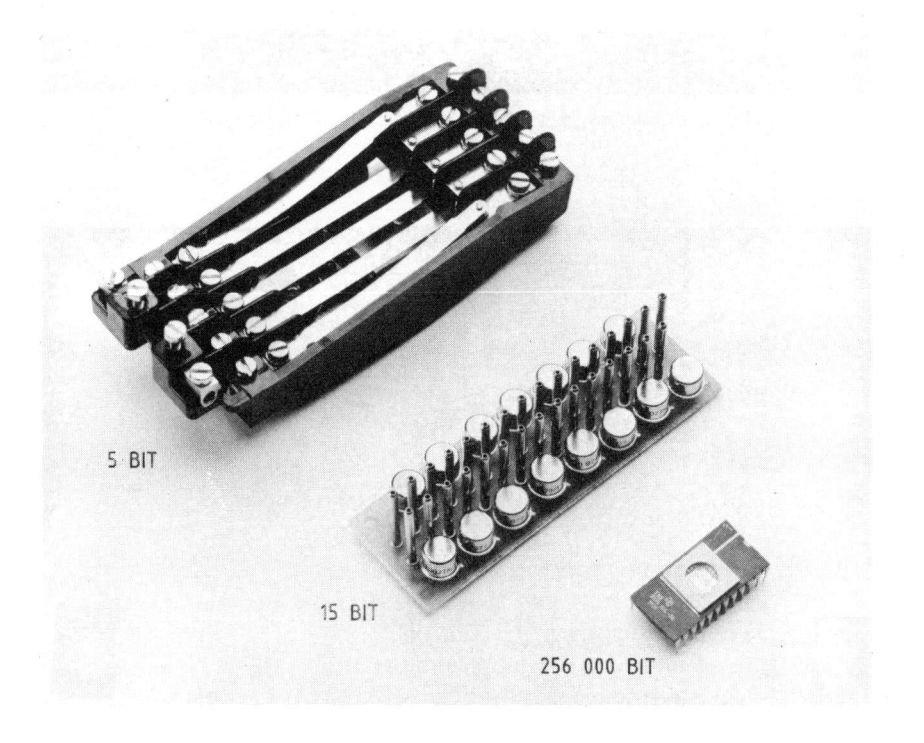

Fig. 1 — Reduction in physical size.

designers to apply the latest versions of these components rather generously.

Owing to the small size of the components used, microprocessor-based controls are more compact than previous generations of elevator controls as shown in Fig. 2, and they are less costly. In fact, the cost per function or feature has dramatically reduced.

Microprocessor elevator controls have permitted many departures from previously used techniques. Innovation has been particularly impressive in the following areas:

Fig. 2 — Microbased controller.

 (i) Positioning controls
 (ii) Drive controls
(iii) Supervisory or group controls.
(iv) Indicator and communication systems.
 (v) Diagnostic techniques.
(vi) General man/machine interface.

2. POSITIONING CONTROLS

The mechanical floor controller or selector of the relay control era has now been replaced by tachometers (Fig. 3) and solid-state memories. Tachometers are capable of sensing the car position within ±1/2 mm. Similarly, floor locations are stored in memory at a resolution of ±1/2 mm. This permits the car to stop more accurately. Also, floor locations can be monitored, and if the building height should shrink because of floor loadings or other reasons, memories can be altered automatically to reflect the changes.

Computerized selectors are usually programmed automatically after the equipment has been installed in a building: Car-mounted secondary position sensors detect the building floors, and the corresponding tachometer count

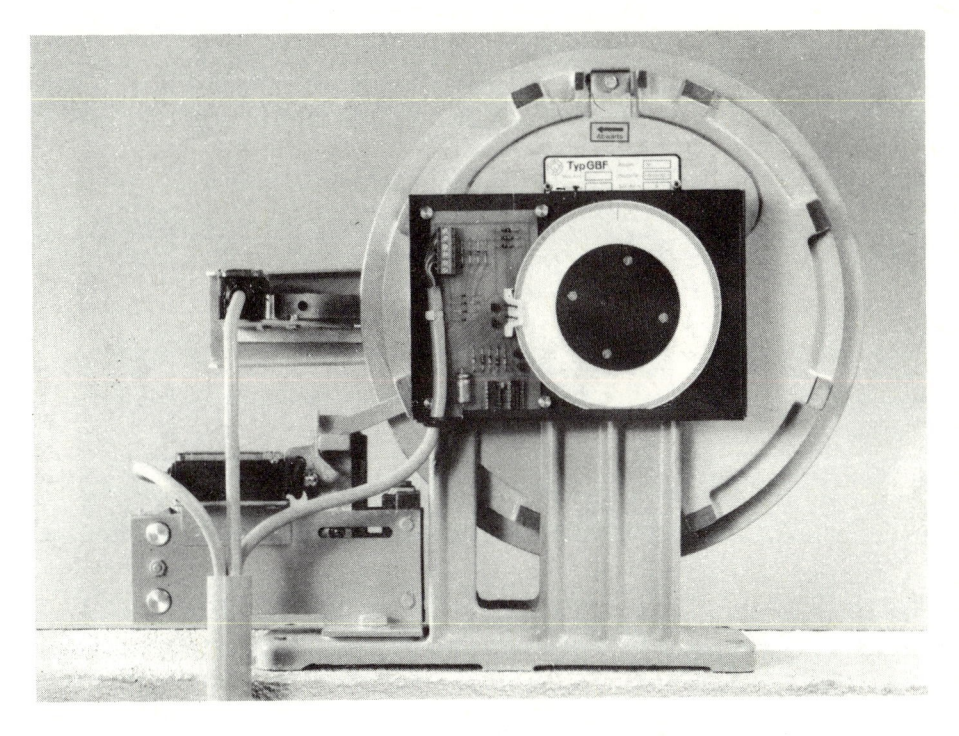

Fig. 3 — Tachometer

is stored to define the floor location. This procedure, performed while the car is slowly travelling over its full-rise, eliminates the need for costly mechanical adjustments.

3. DRIVE CONTROLS

The use of microprocessors has permitted the design of low-cost closed-loop ('feed-back') drive controls. Today, these controls are applied even to small geared machines with a.c. induction motors.

All microprocessor-based drive controls are digital instead of analogue and provide drift-free operation over years of use, eliminating the need for fine-tuning adjustments after the drive has been put into operation.

Microprocessor drive controls can be programmed for specific limits of acceleration and jerk allowing modification to change the degree of riding comfort. Digital techniques and feedback concepts permit the drive control system to follow the reference speed-pattern precisely as indicated in Fig 4, thus minimizing the travel time floor-to-floor while maximizing passenger

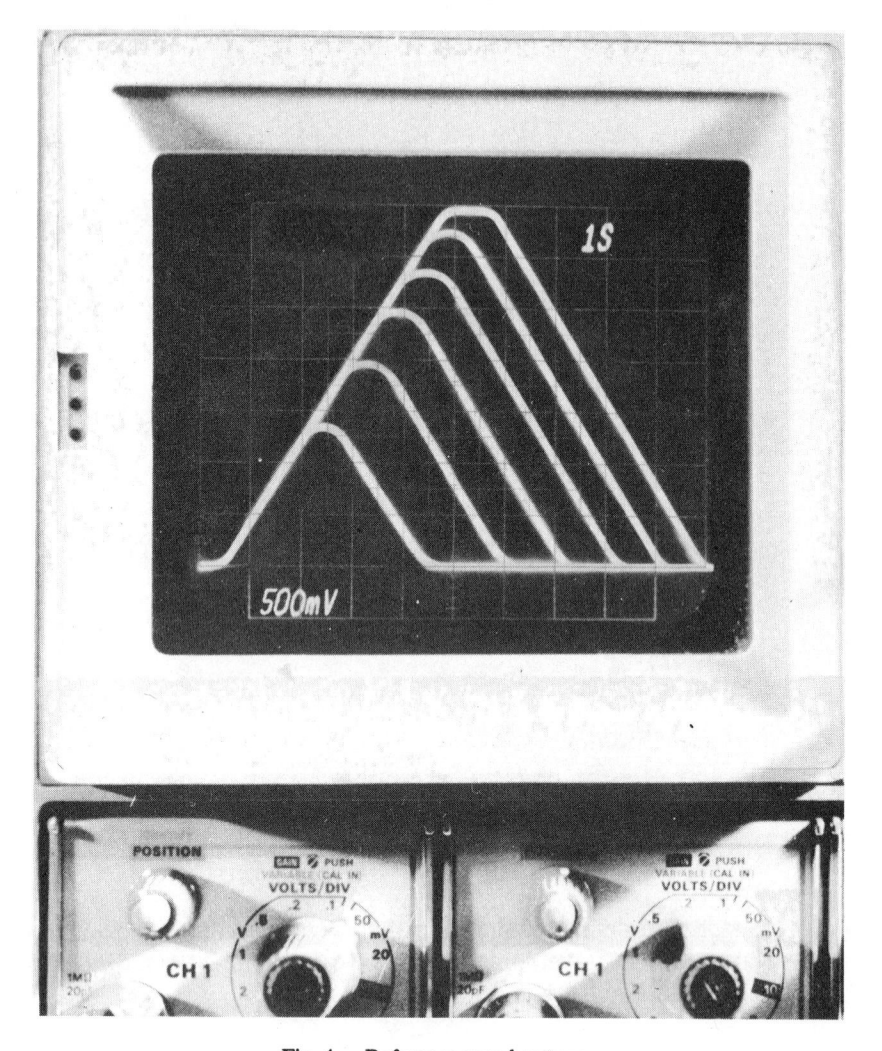

Fig. 4 — Reference speed pattern.

comfort. As a rule, 16-bit processors are used for drive controls. Smaller (8-bit) microprocessors do not provide adequate speed and resolution for drive control purposes.

4. SUPERVISORY OR GROUP CONTROLS

Older supervisory system assign cars to hall calls when the car arrives at slow-down distance from the floor. Owing to the very limited time available to make the stopping dicision only very few group operation parameters can be considered in the decision making, such as car loading and hall call registration time. Normally, the first car reaching the slow-down point will answer the call. This algorithm is sometimes referred to as the 'first come, first served' concept.

Microprocessor dispatchers usually apply a totally different technique. Assignments of hall calls to cars are made long before the cars reach their respective slow-down distance and are based on service factors reflecting the suitability of the car to serve the hall call efficiently (i.e. with minimum delays in the system). At least one system continuously simulates group performance one full round-trip ahead of each car. This system also has the objective of minimizing total destination time for all passengers instead of merely minimizing floor waiting times. Also, it applies FIFO (first in, first out) assignment principles during down peak thus totally eliminating 'forgotten man' situation by preventive action instead of acting after a hall call has been timed out.

Generally speaking, microprocessor-based dispatchers should be predictive instead of reactive. The arithmetic capability of the processor and its speed provide the means for this approach.

5. INDICATOR AND COMMUNICATION SYSTEMS

Any microprocessor system can operate a TV monitor to display information. This permits indicating elevator group operation in graphic form which is more legible than traditional multi-light panels. Also, more information can be shown in less space. One TV can be switched to indicate all elevators in a building, as illustrated in Fig. 5, or 'zoom in' on one specific group for additional detail such as hall calls, car calls, car loadings, traffic program in effect, etc. This ability is demonstrated in Fig. 6.

The TV or monitor screen can also be used for security purposes by installing a video camera in the cars. If a passenger presses the alarm button, the video picture of the car interior automatically appears on the TV screen and the intercom system is activated. Obviously, the video connection can also be established manually via a computer-type keyboard.

6. DIAGNOSTIC TECHNIQUES

Microprocessors inherently have the capability of recongizing errors occurring within the processor system. By appropriate programming, this capability can be extended to the processor's periphery, i.e. the elevator's machine, safety circuits, doors, operating fixtures etc. In addition, all events detected by the processor can be stored in a special, protected memory. The event log thus maintained can be displayed on the TV screen or printed out on a printer. The printed output provides an excellent daily record of unusual events in the elevator system. Event logs and print-outs are requested by the user through a computer-type keyboard located near the TV or monitor display.

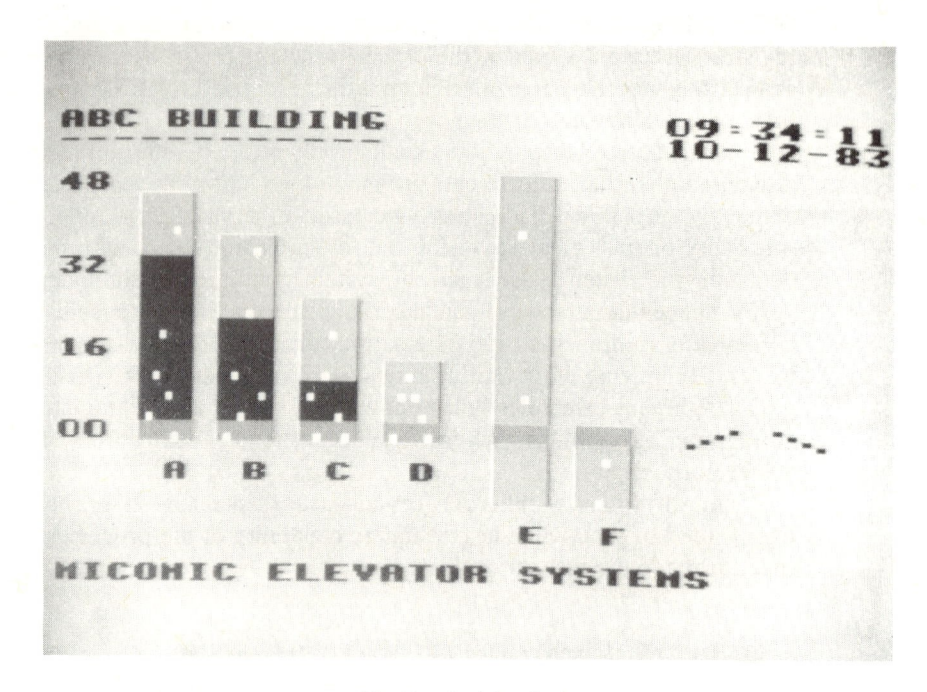

Fig. 5 — Activity display.

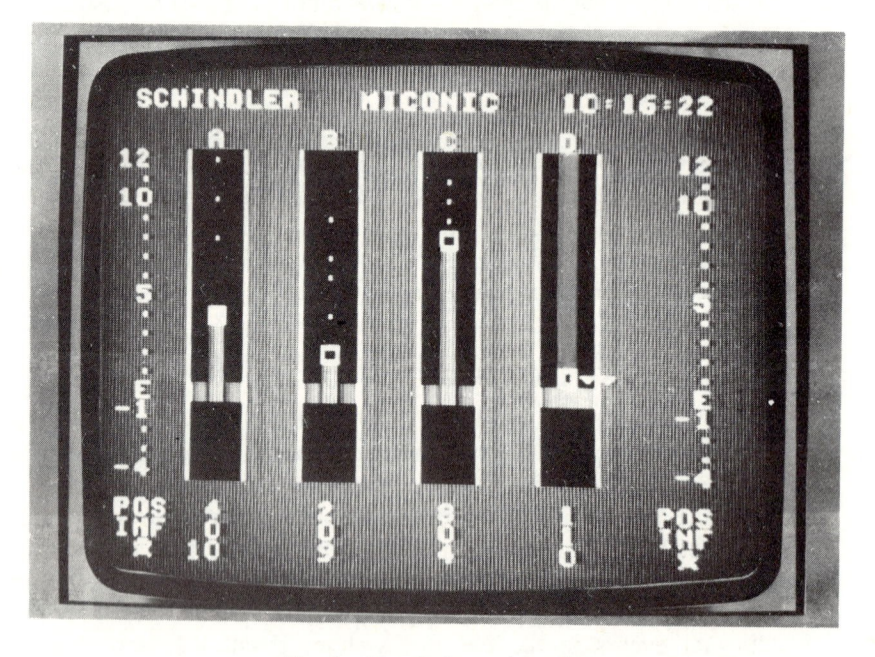

Fig. 6 — Detail of an activity display.

The data logging capability can also be used to store a listing of 'near failures' of the elevator system for the service man (Fig. 7). By examining the listing in the machine room, the mechanic will recognize components requiring action by him. The diagnostic capability of the system thus permits truly preventive maintenance of the elevator equipment. Microprocessor equipment can be serviced more efficiently and more effectively.

7. GENERAL MAN/MACHINE INTERFACE

Microprocessors are 'smarter' than conventional controls and their intelligence can be utilized to communicate with the user via TV screen (output) and keyboard (input). Communications with the building engineer and the service mechanic have been described in Section 6.

Similarly, the field engineer starting up and adjusting the elevator is communicating with the equipment. In fact, he is guided through the procedures by information displayed on the TV screen and follows a step-by-step sequence. In the course of this activity, the TV becomes instrumentation providing the functions of a voltmeter, an ammeter and even an oscilloscope (Fig. 8). Figure 9 shows a field engineer using a portable keyboard and monitor to set up an elevator. The type of instruction received is depicted in Fig 10.

8. CONCLUSION

The use of microprocessors in elevator controls has not only permitted the design of better controls but has also offered the capability to provide a greater amount of 'user-friendliness'. This is true as much for the passenger, as for the owner, the field engineer and the service mechanic.

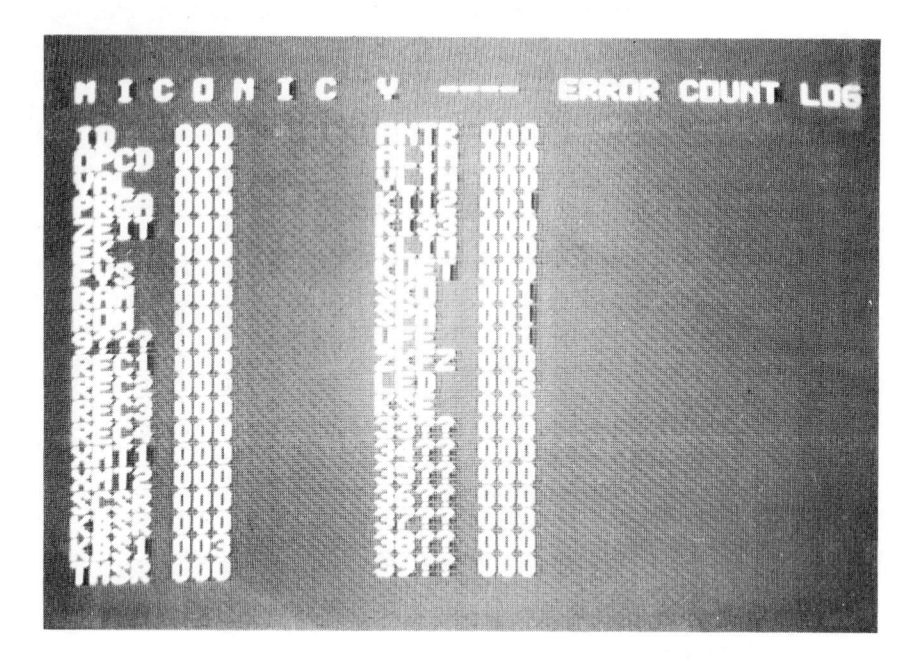

Fig. 7 — Logged data.

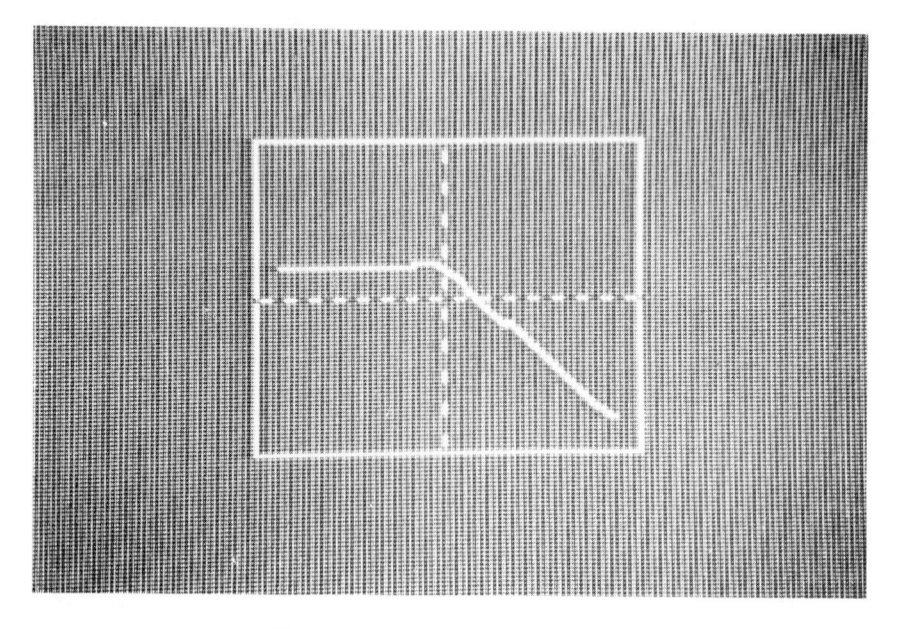

Fig. 8 — The display as an instrument.

Fig. 9 — Portable keyboard and monitor.

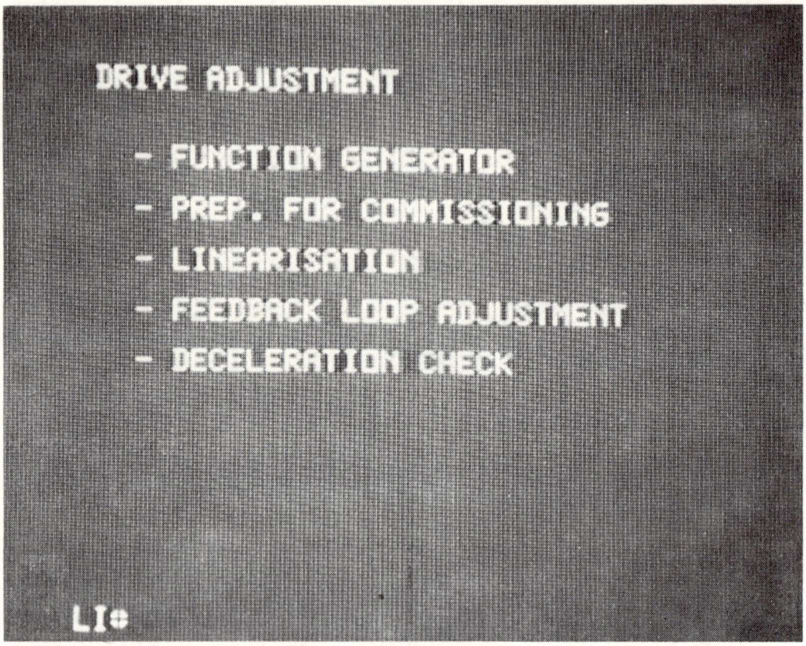

Fig. 10 — Typical instruction.

Part 5
Elevatoring buildings

15

Top/down sky lobby lift design

Mr James W. Fortune, Lerch, Bates and Associates Inc., Littleton, USA

ABSTRACT

This chapter will review the pros and cons of sky lobby designs for large office buildings (those in excess of 60 storeys) with double deck or single deck shuttle lifts feeding sky lobbies. The impact of zones of local elevators when serving the lower and upper portions of the buildings off sky lobby(s) will be discussed. This simultaneous feeding of local zones up and down from the sky lobby is called the top/down approach to sky lobby lift designs.

1. INTRODUCTION

In Japan, because of the extreme costs of land and labour, it is common practice to start constructing an office building by tunnel excavating the subterranean parking and mechanical spaces while at the same time, the above grade superstructure is also being constructed. This construction method is called the Top/Down approach to building construction. A similar method of servicing the local zones of elevators from an upper floor sky lobby(s) is called the Top/Down approach to sky lobby lift design and is the subject of this chapter.

2. REVIEW OF CONCEPTS AND TERMINOLOGIES

The lift designs for a major office building that would be large enough and sufficiently tall to warrant the use of sky lobbies, should meet the general design criteria during a 5 minute, morning, up-peak traffic condition given in Table 1.

Table 1 — Design criteria

Parameter	Sky lobby shuttles	Local lifts
Average interval	28–30 seconds	25–30 seconds
Group handling capacity	15–25% of combined local zone populations moved	12–15% of zone population moved
Transit time to destination	60–90 seconds* — en route from main lobby to sky lobby	60 seconds† — from sky lobby to local destination floor

* Calculated at half the average interval plus the total time on the lift.
† Calculated at quarter of the round trip time plus half of the average interval.

General rules of thumb for elevator designs are:

(i) No more than eight lifts (four opposite four) should be used in any one local group. Larger units can be considered (10–12 lifts/group) for sky lobby shuttles and local lifts, if a prediction type dispatch algorithm is provided.

(ii) No more than four local elevator groups should be provided for each building segment. Each single deck elevator local zone should serve no more than 15–16 upper level stops while each double deck local zone should be limited to no more than 18–20 upper level stops.

(iii) The physical lift size should be at least 3000 lb and not larger than 4000 lb for local group lifts (single deck or double deck). Sky lobby shuttles are typically rated at 5000–10 000 lb for single decks and 7000 (3500/3500 lb) to 10 000 lb (5000/5000 lb) for double decks. The sky lobby shuttles should be sized large enough to handle the required number of zone tenants *with one car removed from the group*. This is done to ensure sufficient group handling capacity, even though the intervals and waiting times may become protracted, if one lift is shut down or due to a malfunction or evacuation requirement.

3. A MODEL FOR A TOP/DOWN SKY LOBBY

By the time a building height extends beyond 60 storeys, the practical height limit for four groups of single deck elevators with 15 stops/group has been reached (about 80 storeys for four zones of pure double deck lifts). Beyond this limit, the application of sky lobbies should be considered. Otherwise, the local lift shafts take up too much of building core and the speeds required to overcome the express distances from the ground floor become impractical. Historically, the elevatoring solution for building higher than 60 storeys has been to add one or more sky lobbies. The local zones are then being dispatched up from them to feed the upper levels. With this approach, the single sky lobby would be utilized for a building with between 60 and 120 storeys while two sky lobbies would be required for a 120 to 180 storey project. In reality, the two super-tall buildings in the United States that utilize two sky lobbies are the twin, 110 storey World Trade Central Towers, located in New York City, and the 110 storey Sears Tower located in Chicago. The reason these buildings employ multiple sky lobbies is because the floor sizes and populations are so large that the local zone stops are required to be less than the ideal 15 floors each. If the Top/Down sky lobby approach were substituted, the number of sky lobbies could be reduced from two to a single lobby located at about two-thirds of the way up the building rather than at the one-third points. A single sky lobby can then be made to serve up to a 180 storey building if single deck local zones are utilized or possibly up to a 240 storey building (Fig. 1) if pure double deck local cars are provided. Obviously, these design projections are based upon idealized floor plates and population parameters and actual conditions would probably be less floors per zone because of the required slenderness to height building ratios. Thus it can be seen that the Top/Down approach saves considerable area due to the reduced number of sky lobbies required for a very tall building.

Sky lobby space is very expensive because it usually entails the design of special two-floor height structures, requires large holding areas and evacuation stairs, and is normally lost space from a revenue producing standpoint. Obviously, anything that can be done to reduce the number of sky lobbies in a project contributes to the overall efficiency.

4. SHUTTLE SERVICE

Traditionally, two elevatoring methods have been employed for shuttling building occupants to and from a sky lobby.

(i) Large single deck elevators, sometimes with front and rear doors to facilitate rapid passenger movements in and out.
(ii) Double deck lifts, where the upper and lower decks are uniformly loaded and discharged from dual loading lobbies and dual sky lobbies.)

The time factors associated with loading and unloading a double deck lift are substantially less than a comparable single deck scheme. Thus most large projects utilizing sky lobbies, that will be constructed in the future, will undoubtedly choose to utilize double deck shuttles, as they are much more efficient. Coupling the double deck sky lobby shuttles with a Top/Down approach to local lift service, means that the upper floor passengers simply have to identify whether their work station is located up or down from the upper sky lobby. If it were down, they would load and unload onto the bottom deck of the sky lobby shuttle and conversely, if it is up they would utilize the upper deck of the sky lobby shuttle.

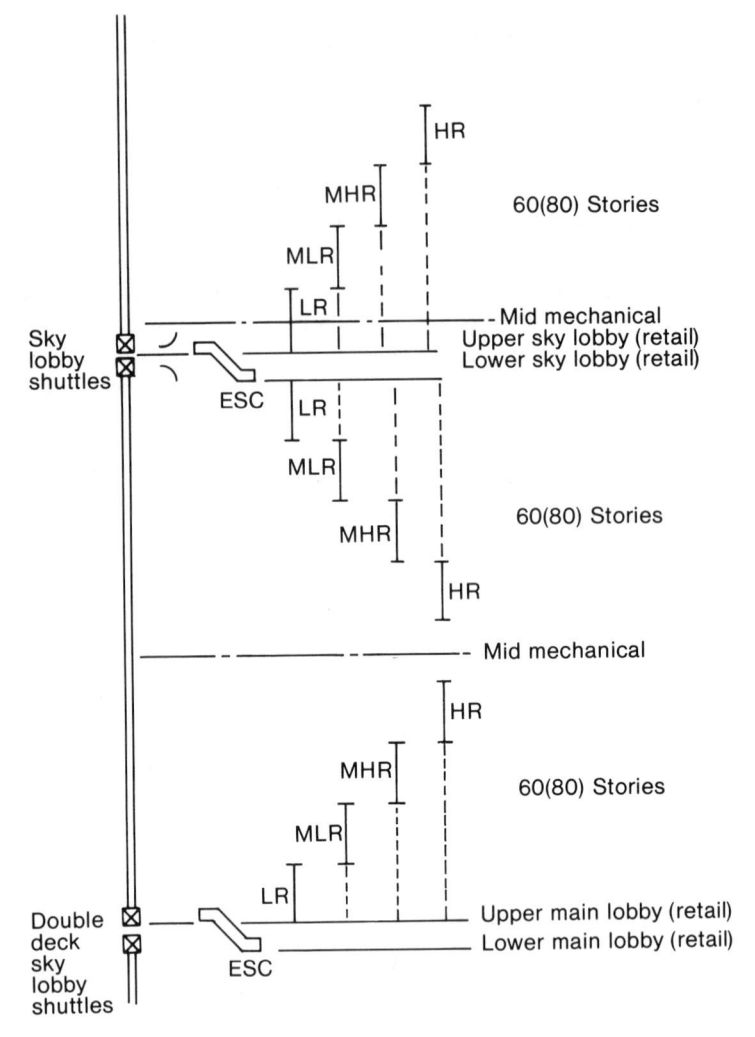

Fig. 1 — A 180 (240)-storey Top/Down sick lobby design with double deck shuttles and single deck locals.

5. CONCLUSIONS

The advantages of the Top/Down sky lobby designs are:

(i) Lift core savings — fewer sky lobby shuttles should be required.
(ii) Increased building efficiency — less floor areas devoted to sky lobby functions required.
(iii) Double deck shuttles and even double deck locals can be effectively utilized due to the split sky lobby floors.
(iv) A very large project requiring the use of triple deck, quadruple deck or quintuple deck sky lobby shuttles could be accommodated much more readily at a single sky lobby than at multiple sky lobbies.

The disadvantages of the Top/Down sky lobby designs are:

(i) It may be psychologically difficult to condition people to think in terms of first going up to go down and vice versa during the different parts of a business day.

(ii) To fight a fire in the building might be difficult if the staging area is at the sky lobby and the fire is located below this area.

In summary, the physical and cost saving advantages of the Top/Down sky lobby approach so far outweigh any psychological disadvantage of the counterflow traffic, that it is reasonable to guess that the next 'Worlds' Tallest Building' will be constructed utilizing these techniques.

16

Bush House: lifts of the world

Mr Michael Godwin, Lift Design Partnership, London UK

ABSTRACT

The BBC have been in residence at Bush House for over 50 years and during that time have built 40 studios within the complex. These are served by two four-car groups and two three-car groups of lifts.

As the World Service is operational 24 hours a day there is continuous passenger and trolley traffic flowing not only between the different wings but also of course within each block. The 1959 modernization of the lifts by Otis Elevator comprised making them UMV and incorporating 6 programme 'Automatic' with touch buttons on each group. The winding machines and hoist motors were retained and by 1980 were 60 years old. Owing to the excessive milage and heavy demand breakdowns were frequent and consequently the quality of lift service deteriorated to the point where the BBC operational planning had to allow 35 minutes for Announcers to get to studios. In essence the problem therefore was how to transform the lifts so that people with the news could get from, say, Centre Block to a studio in South East Wing via the lift system in the shortest possible time.

The design solutions chosen for this difficult lift problem are the subject of this chapter.

1. INTRODUCTION

The centre block of the Bush House complex is now 60 years old. Built on a grand scale, it was conceived by anglophile Irvin T. Bush as the then equivalent of a world trade centre. Whilst its design was appropriate 50 years ago for operaing as prestige headquarters offices for major British industrial companies, that era has long passed and since 1932 it has increasingly become the home of the BBC External Services, for which it is considerably less appropriate. The large lobby areas of travertine marble and the grand staircases cannot be efficiently utilized and thus studios and working areas covering 11 levels, occupy significantly less of the gross area than a normal office building.

The adjacent north-east and south-east wings being similar buildings, though on a perhaps less grand scale, and also occupied by the External Services of the BBC. It is the nature of broadcasting that there is always tremendous activity and consequent movement of people within and around the complex with newsreaders and announcers particularly relying on excellent lift performance and reliability to get them to their destination lift performance and reliability to get them to their destination on time. The heavy ciruclation of people and materials between floors and buildings demands the maximum performance and reliability from what today would be considered an inadequately sized lift installation, bearing in mind that new broadcasting centres tend to be designed horizontally with each level being self-contained.

In 1979 LDP was commissioned to propose a long-term solution to the circulation problem when related to the installed lift capacity at Bush House. The existing lifts had already been modernized twice since their original installation by Waygood Otis Ltd. In 1959 Otis installed the best equipment available, namely 62 UCL-six programme Autotronics with touch buttons etc. Because of their decision to retain the 60-year-old winding machines and motors and of course the large quantity of electro-mechanical equipment in the controllers the lifts were worn out by 1979 owing to the excessive milage and intensive duty of having to operate 24 hours a day, 365 days a year. In consequence dealing with breakdowns became the main occupation of the service personnel and combined with vandalism becoming increasingly prevalent and lack of preventive maintenance there seemed little hope that such equipment could continue to provide an adequate lift service, let alone any prospect of improvement. Thus the central premise around which the specification for the modernized lifts was written was that operational electro-mechanical devices shall be kept to an absolute minimum in order to have some prospect of achieving adequate reliability and improvement in performance bearing in mind the onerous duty cycle. The specification also called for a group control algorithm that could guarantee to fully exploit the given reliable lift resource so as to minimize hall call response time.

Lift companies were invited to respond to the specification, a condiiton being that the refurbishment work be undertaken piecemeal on each lift the contractor being responsible for maintaining some lift service that would enable the building to operate during reconstruction. Within a short time it was clear that no company was offering equipment that would substantially comply with the specification and thus the whole task of reconstruction of the 14 lifts was handed back to LDP with the brief that some alternative scheme be proposed so that work could start as soon as possible.

2. THE EXPERIENCE

Undertaking modernization/refurbishment must always to some extent contain a measure of the unknown and commercial success would seem to depend largely upon the preparation work of getting exact dimensions and taking many photographs as well as allowing for having to custom-build a proportion of each lift so that each fits the existing structure and the style of the building.

In a 60-year-old building which originally contained Otis Micro drive with manually operated landing gates, the relative alignment of the entrance of that era would think nothing of individually fitting each landing entrance slam-post and false door panel to suit structural tolerances of those days. When refurbishing took place the task of renewing and fitting the false landing door panels etc. proved to be very labour intensive as each panel had to be fitted to each frame within a tolerance of 1 or 2 mm and there were 147 entrances!

As it was necessary to reuse the existing Waygood Otis car slings and platforms it was again discovered, when the old cars had been removed, that although nominally the same size throughout the job, each car had to be individually tailored to fit the varying heights of crown bars and the relative position of the round guides with respect to the landing entrance slam-post. As work proceeded it was quite clear that it should not have been taken for granted that automatic door gear would work satisfactorily in the refurbished scheme simply beacuse Otis in the previous modernization, had fitted power door gear. The fact that they had done so yet retained the round guides would seem to break all the rules. Measuring the entrance slam-posts showed a tolerance of 30 mm over 11 levels and the variation between lifts of the slam-post relative to the guides and thus the car sling, being up to 50 mm different from lift to lift. As there was no question of changing to modern Tee guide section and repositioning so that substantially all dimensions would be the same, the two problems of varying slam-posts from floor to floor and the varying car positions from lift to lift, could only be solved by:

(i) Custom building the entrance of each lift car to accommodate the varying guide position lift to lift.
(ii) Making universal rubber buffers that could be adjusted in length such that the leading edge of all landing doors were substantially plumb over the 11 levels even though the metal slam-posts varied.

Being a steel-framed building clad with travertine marble around every entrance, all fixtures had to be manufactured to precisely fit existing cut-outs in the marble. Whilst most of these fixtures were assembled on site without too many problems, there were a proportion which, where the depth of pocket was reduced by a $\frac{1}{4}$ in because the positioning of steel joists varied with the building height. As the design of push-box could only be reduced by 3 mm in in depth because of the electronics inside the time consuming and expensive process of cutting away the back of the interfering steel joist had to be undertaken.

In 60-year-old buildings it is inevitable that over the years additions to the building services will show up as water pipes, ducts etc. running in the lift shafts, making life extremely difficult for positioning and therefore due allowance should always be made for a proportion of custom built steelwork to accommodate these anomalies.

On the engineering design side, bearing in mind the year was 1979 and the state of the art for microprocessors, it was necessary to guess to some extent the amount of memory needed. Having successfully completed the job it can be concluded that 8K of EPROM is only just sufficient for controlling a maximum performance lift serving 11 levels. Today memory limitations are far less severe and processors are far more powerful compared with the Intel 8085 that used throughout this job.

Furthermore undertaking a similar job today would allow advantage to be taken of the cheaper and more powerful microprocessor which are in no way memory limited and thus instead of the user having to suffer the degraded lift system and the contractor having the problems of operating the initally modernized lift with the existing old system, overlay technique would be used. This provides an instant enhancement in terms of the performance reliability with minimal inconvenience to the user, the changeover taking a matter of days. All these techniques help to establish and maintain the goodwill of the user upon whom the contractor relies ultimately to treat the finished lifts with due regard.

In getting solid state drives commissioned and operational many difficulties arise which are not evident on the more conventional VV/UMV motor generator drive. SSDs quite properly only operate within close limits and respond quickly to any slight out-of-tolerance caused by maladjustment fault condition. Getting to know SSDs can therefore be expensive both in terms of fuses etc. and in man-hours of highly trained personnel. The benefit once they are working of course is the aboslute consistency of flight times and exact profiles of acceleration and speed, irrespective of variable loading or supply potential. In so far as the components were concerned commissioning of the total system confirmed very early on that the choice of English for all the messages transferred between computers was sensible as code would have made life very difficult. In the same way the decision to hard wire the circuits associated with inspection speed so that slow speed directional lift movements independent of all computers helped considerably.

3. PROBLEMS ENCOUNTERED AND PROBLEMS SOLVED

The following are a few examples.

(a) Tuning the SSD such that it is tolerant of some variations in supplies etc. and thus mitigate the otherwise intolerable nuisance produced by any slight hiccup that would otherwise cause a shutdown and consequent trapping of passengers.

(b) Taking into account the high inductance of the gearless armatures such that the drive could provide enough volts during acceleration compared with voltage absorbed by the smaller motors of the geared machines.

(c) The necessity to change plain bearings on winding units, reeving pulleys, divertors, selector drives, governors, etc. to anti-friction bearings requiring little of no maintenance.

(d) The need to use solid state controls as far as possible and minimize mechanical and electro-mechanical components. For this reason all mechanical buttons were avoided except the door open and door close pressels which were specially over engineered. Nevertheless they required further reinforcement because some users would attempt to kick them or operate them with their feet. Intensive usage of the car call and hall call buttons was anticipated and it was decided that button inserts would be bronze self-metal throughout and any engraving of legends would be filled with vitreous enamel. What was not anticipated was that the consequent pressure from people's fingers could cause the button insert to loosen and turn as its profile was circular. On future jobs the shape will be altered so that the insert will not turn.

(e) The use of a conventional centrifugal type overspeed governor to sense a 10% overspeed and shut the system down. In practise because the action of fly-weights are affected by both gravity at different angular positions and stiction forces and that the whole governor device of necessity has to be rather crude and simple, it turned out that, at the 'jerk' rate at which the lifts have been set up in order to achieve the requisite flight times, every so often an unfortunate combination of fly-weight position and stiction force combined with the high 'jerk' rate caused the fly-weights to over shoot and trip the overspeed switch. They have now been recalibrated at 25% overspeed and a 10% overspeed sensing circuit has been designed within the solid state drive which can be accurately set to police the drive and lockout accordingly.

(f) Two of the 14 lifts refurbished are basement drive. Because of the constructional difficulties of removing these old waygood Otis winding machines and the fact that Otis had retained all original machines and hoist motors 20 years before, these two lifts were retained. The machines were stripped down, the bearings renewed and backlash reduced to a minimum by so-called 'topping' the worm. This work was undertaken by a specialist contractor who guaranteed that minimum backlash would remain consistent over the coming years. Subsequently the solid state drive was unstable and would oscillate violently owing to a combination of some backlash after the worm and wheel were bedding on over a period of months, and the elasticity of the roping, the lift being a bottom drive arrangement. Of course, being an early design of machine by Waygood Otis, there are no adjustments as such for taking up backlash as there are on the other nine new winding machines and therefore the specialist team had to return to site and machine-up new bearings to effect minimum backlash. Subsequent to this work the SSD has remained stable but did require special adjustment of the derivative damping networks and also the fitting of a 10 mH air-cooled inductance in the armature circuit. The lesson learned is that solid state drives, because of their inherent fast response time and tight control should preferably be used with new hoist motors and winding units that have easy on-site adjustment for taking up any backlash.

(g) It seems to be common practice nowadays for design engineers to use standard industrial components where possible such as nylon couplings, plummer blocks, etc. and the design of these often incorporate a grub-screw type fixing on to the shafts. In practice it has been found that a grub-screw fixing is quite inadequate over the long term and therefore it is necessary to remove these fixings and substitute roll-pins by drilling the shafts. For example, nylon couplings connecting the jack-shaft to tachogenerator would come loose causing the SSD to intermittently shut down due to excess error signal. Spindles of governors would slide through the governor wheel-hub bringing it off centre and causing it to foul switches and other mechanical parts.

(h) Because small reed relays are widely used in the safety circuits of the SSD to kill any drift in op-amps, etc. these relays in some cases will be used for operating a signal relay on the main controller. Over a period of time it was found that a reed can stick intermittently in the closed position due to the contacts operating at 100 V d.c. even though the current is only 20 mA. Fly-wheel diodes and resistors have therefore been fitted across the ends of landing signal relay coils.

(i) The occasional blowing of the fly-wheel diodes across the shunt fields of a hoist motor would result in the shut down of the SSD causing possible trappings and the reason for this is still not fully understood. However, the problem has been mitigated by changing from a 600 V device (the field supply is only 200 V) to a 1000 V device combined with a snubber network which should limit the spike potential that is generated by the e.m.f. induced into the field from the armature inductance.

4. LISTENING TO LIFTS

Over the listening period of a year, it has been wise to never ignore or dismiss as of no consequence statements by passengers, users, building managers, etc. who have experienced some happening which they do not consider to be normal on their lifts. The views and the cooperation of all persons whose use the lift resource should always be welcomed and every attempt made to establish a fund of goodwill so that a degee of tolerance is shown when the inevitable happens and a lift goes out of service.

To enable the building manager to 'listen' to the lifts continuously a computer objectively displays on request the status of all the lifts over any one 24-hour period, and a printer provides a permanent record of the exact time a lift goes 'out of service' and back 'in service' as well as other status conditions such as lift on 'maintenance', lift on 'independent serive' and lifts 'returned to ground floor' under fire alarm testing. Initially it was not possible to estimate what the call rate might be and with something like

25 000 journeys undertaken per day by the 14 lifts, it was essential to respond to each break-down in a manner that would ensure that the problem was understood, cured and therefore would not be repeated. The total 'out of service' time logged on the building manager's computer over a twelve-month period was 210 hours for 11 lifts. This figure includes of course the response time of the lift fitter to get to the site during and outside normal working hours. Routine scheduled preventative maintenance of course has figured significantly during the year and will continue to do so but it has been found from experience that a high proportion of the routine maintenance does consist of cleaning and of course the regular oiling/greasing of the round guides, which may be significantly less in other buildings.

5. SOME UNEXPECTED PROBLEMS

(a) All 14 lifts are fitted with two panel side opening doors with clear openings of 1090 mm. As the opening times are less than 1 s this places considerable strain on the landing door pick-up roller assembly through which power is transferred from the car door. With the first choice of landing door equipment it was discovered that occasionally the thin die cast aluminium plate that secure the pick-up rollers to the leading landing door panel would break off. This assembly was therefore replaced with a cast iron version as the alternative, and this seems successful.

(b) All car stations for the 14 lifts were designed to incorporate a telephone compartment so that in the event of trapping communication was possible. Initially these were flush fitting panels embedded in the lower part of each car station, however it soon became very clear that such compartments were vulnerable to vandalism and consequently telephones would be found deliberately broken with the inserts removed, and even compartment covers pulled off and all manner of refuse deposited inside. As telephones and alarms are checked every 12 hours on each lift, the problem of repair became both time consuming and expensive so that an alternative had to be found. The solution was to reinforce the bronze cover of the telephone compartment and then fit a unique form of magnetic lock. Essentially this consisted of a electromagnetic assembly with a double wound solenoid fitted with remanent magnetic pole pieces. The compartment once locked against spring pressure was held locked by the remanent magnetism and thus could not be opened by hand as the cover was flush with the surrounding car station. If, however, the alarm button was pressed for more than 2 seconds current from the alarm supply, flowing through one winding of the solenoid would remove the remanent magnetism and allow the cover to flick open. Security personnel would then re-lock the cabinent after checking the alarm and telephone by simultaneously pressing the door open and door close buttons which would cause sufficient current through the other winding to saturate the pole pieces and restore the remanent magnetism in the pole pieces.

(c) The development of the software for the individual car controllers was also a learning process once commissioning of the lifts began. A lift whilst appearing to perform in a cyclical fashion say from floor to floor, by virtue of its numerous externally connected devices, sensors, etc. activated randomly by passengers, is in fact asynchronous by nature. The continual breaking up of the normal cycle, for example doors closing and lift about to proceed to next stopping floor when a late registration of a hall call must cause the doors to reopen, means that it is important to analyse each event and the necessary lift response in the light of the current state as opposed to reacting on the premise of what the lift should be doing based on the last event. At the highest level this leads to design software taking the form of many small independent modules, rather than long sequence programme statements. At the lowest level a module dedicated to lift position detection so that, for instance, the next stopping floor can be evaluated. One interesting observation that has become clearer as this approach to the software design becomes extended is how it begins to resemble in principle the form of the ladder diagram that was always used to design relay based controllers, i.e. many parallel paths.

6. RELIABILITY AND PERFORMANCE STANDARDS

In achieving lifts with the high performance combined with consistently reliable operation at Bush House, clients are asking the understandable question 'Why can we not have lifts in our buildings that perform to the same high standard?' The initial reaction to that question is to say that because of low maintenance prices charged due to intense competition in the UK lift industry adequate maintenance that would ensure high performance combined with reliability could not be undertaken. In the current climate of the lift industry radical changes to lift specification and contract conditions are needed if what has been achieved at Bush House is to become commonplace. In practical terms, the first important change in the specification is to ensure that all high performance lifts employing microprocessor control, incorporate the means by which a lift can be remotely 'listened' to via an internationally accepted link, e.g. RS 232. With objective data to hand, the building manager, user, owner and lift maker can study the print-outs so that design faults, and problems responsible for loss of lift service can be ruthlessly pursued and eliminated by the contractor.

The second which concerns the lift makers is the gradual accumulation of statistics covering lift performance and reliability that can be written into the main contract and which the lift maker can guarantee. These figures would be particularly important during the one year defects liability period after the installation of the lift system. At the end of the defects liability period, the finally agreed performance and reliability figures could be written into a form of comprehensive maintenance contract with penalties that would apply if the lifts failed to meet the agreed criteria of performance and reliability. Thus over the coming years, lift makers, having 'listened' to their lifts and gathered large amounts of data, relating to reliability and performance will be able to classify their products and offer appropriately priced lifts and maintenance contracts.

7. CONCLUSION

The experience of undertaking the design and build of high performance lifts such as those at Bush House has given a clearer insight into the problems facing lift makers. The evident success of the Bush House job is due to taking each problem that might compromise the reliability of the total system, however small, and doggedly pursuing it until a long-term solution is found. Of course there will always remain some mysteries and hiccups responsible for the occasional breakdown, and even trapping. However, this discipline of 'listening' to lifts is wholly commended and possibly should be extended by listening to all the interested parties, especially the person responsible for the regular maintenance of the lifts.

APPENDIX: SPECIFICATION DETAIL OF CENTRE BLOCK LIFTS AT BUSH HOUSE COMPLEX

A.1 Original Installation
Maker: Waygood Otis 1922

No. of lifts 4, serving 11 levels B, LG, G-8. Capacity 14 persons, 1050 kg. 400 f.p.m. Micro-drive. Attendant control. Control after 1959 Otis 62 UCL. 6 programme Autotronic w/touch buttons, 6970A door operator with saturable reactors. Original sling, safety gear, and 2.5 in round guides retained.

A.2 Modernized Lifts
Designed by Lift Design Partnership.

Winding machine: Armor No. 4B, ratio, 101:4 48 in diameter sheave. Armor brake.

Hoist motor: G. E. 40 h.p. 240 V d.c. 1200 rpm. w/G.E. tacho-drive via jack-shaft and polyvee belt.

Governor: Armor centrifugal type with nylon chain drive to 'Rotaswitch' incremental encoder.

Drive: Solid state drive 'Dynaglide'. Input 415 V 50 Hz 3 ph Supplied from G.E. Double wound transformer 53 kVA. 210 V 50 Hz 3 ph.

Selector: ILMC System 990 16-bit geared, absolute position encoder 34 turn using toothed tape drive.

Individual controller: ILMC System 990 incorporating SSD landing computer card, interface relays, safety contactors, single car computer, I/O, and parallel communication card.

Group controller: ILMC System 990 incorporating Group Computer C.G.C. control algorithm, DRC computer, I/O, and parallel communications card.

Landing doors: 2 panel side-opening 1090 mm, incorporating digital P.I., illuminated floor number on arrival, UP/DN hall lanterns, with soft tone gongs. Door facing material Rigitex ICS coloured bronze using INCO process and lacquered. GAL hangers and top tracks, GAL MO interlocks and Adams' pick-up rollers, two-speed linkages, and spring closers.

Car doors: 2 panel side-opening, incorporating modified 'Sonarray' with solid state ranging and ruggedized photo-edge, GAL clutch, Adams, Otis type hanger rollers, aircraft cord two-speed arrangement.

Door operator: Electronically controlled OTIS 6970A 100° movement with modified oil seals and incorporating analogue angular pick-off, strain guages fitted to operating arm.

Control features: Modulated door opening to produce 90% door open in 0.9 s. Modulated door closing 3.5 s with 'Sonarray' protection, energy limited within code. Modulated door reversal at any point mid-travel. Limited force on closing to comply with code. Timed out door closing and opening. Limited door opening if only one person in car for car call stop. Slow speed nudging without 'Sonarray' protection but with photo-edge operational such that the force is negligible if obstruction contacts leading edge. Infra-red detectors hidden in lift car determine empty car while load weighing with strain guages determines approximate numbers of persons. Anti-nuisance features include car call cancelling if registered car calls exceed passenger load, high security magnetic keys for changing over to 'independent' service and 'door hold', telephone compartment secured by magnetic lock so that it cannot be opened manually yet opens if sustained pressure on the alarm button. car light dimming to save energy and bulb replacement. Message display. Digital P.I. Synthesized voice.

Hall stations: Cast bronze fixtures incorporating ETA and Waiting Time display with UP/DN touch buttons. Card reader input for priority service. Triple touch cancelling of hall call.

Other fixtures: Message displays at main floors.

Dynamic performance of 4 car group. Cycle time: 8 s single floor run. Door opening time: 0.9 s. Door closing time: 3.5 s. Jump time: 4.5 s. Maximum door dwell time 6 s. For landing call, car call 2.5 s. Sustained hall call rate: 220 hall calls/30 min. 85% of hall calls responded to in 15 s or less. 440

journeys/30 min. 50 000/week. Total logged over one year 2 500 000 journeys.

Reliability of 4 car group. Down-time accumulated over one year: 0.22%. Preventative maintenace: total time for one year 380 hours. Serviceability ratio: 98.7%. MTBF (hours): 3504 hours.

17

Theoretical performance versus actual measurement

Mr V. Quentin Bates, Lerch, Bates and Associates, Littleton, USA

ABSTRACT

It is perfectly simple to calculate the theoretical performance of an elevator system. But how does this theoretical performance measure up to reality? This chapter reports a review of the actual performance of a 10-year-old building in comparison with the engineer's original design.

1. INTRODUCTION

Over 60 years ago, Bassett Jones presented a methodology for calculating elevator probable stops and round-trip time in the *General Electric Review*. This approach to analysing probable elevator performance in proposed buildings has been utilized by elevator engineers ever since. The approach is well known. Usually, heavy incoming traffic is studied first. Elevators are assumed to load to some average number of persons, then based on that loading, make a calculated number of probable stops, reverse at some upper level and return, presumably non-stop to the lower lobby to load again. The time required to make this round trip is calculated by assigning time values to various functions which occur during the trip: passenger loading and unloading, door closing and opening, car accelerating then decelerating and stopping and car running at contract speed. Additionally, the prudent analyst includes a safety factor for human vagaries.

Once a round trip time (RTT) is established, performance factors which define the quality and quantity of elevator service are calculated:

Average Interval (AI), the average time between elevator departures from the lower terminal over a designated period time (say 5 minutes), is calculated by dividing round trip time by the number of cars assumed in the group.

Handling Capacity (HC), the number of persons carried, is calculated by dividing the designated time period by average interval which yields the number of elevators departing the lobby. Departures multiplied by the average loading used to determine probable stops — yields, the number of passengers transported from the lobby during that period.

Relatively quickly, and simply, measures of performance potential for a single or group of elevators have been determined.

Finally, the values calculated for average interval and handling capacity are compared with empirical, yet subjective, standards which represent the degrees of excellence defined by those values, and judgement rendered as to the probable adequacy of the proposed elevator or group. Thus the engineers' drive to apply scientific, objective principles to problem solving has been satisfied and the resultant service is confidently predicted to be somewhere between poor and excellent depending upon the standards used for evaluating results.

The outline presented is somewhat simplistic and does not do justice to the judgement which experienced traffic engineers apply in making such calculations and evaluating the results. Nor does it imply the countless studies and enormous volume of empirical data which support this predictive approach. For 60 years, almost every structure served by elevators, has been subjected to analysis by this method.

The question to ask, however, is whether the calculation approach is valid today? If the methodology is based on 60-year-old data, does it apply to today's conditions? It might be felt that questioning the probability approach to calculating round-trip time would be like questioning Newton's law of gravity . . . yet, with the sophistication of today's analytical tools, scientists *are* questioning the absolute accuracy of Newton's law. Today, using computers, group performance is improved by substantial percentages. Also, using computers it is possible to simulate rather than calculate performance values and obtain values of waiting time and response time as opposed to average interval.

2. A 'HINDSIGHT' REVIEW

A year or so ago, several consultants in Denver, Colorado decided to do a 'hindsight' review of a representative Denver building. Conditions in that building were measured and compared with the theoretical values used in projecting requirements. Purposely a building about ten years old was selected on the reasoning that the occupancy compaction that occurs as tenants add employees without leasing more space, would have reached a balance.

Figure 1 and Table 1 indicate the information used in the final evaluation of the building's elevator system and its ability to handle traffic. Thirty

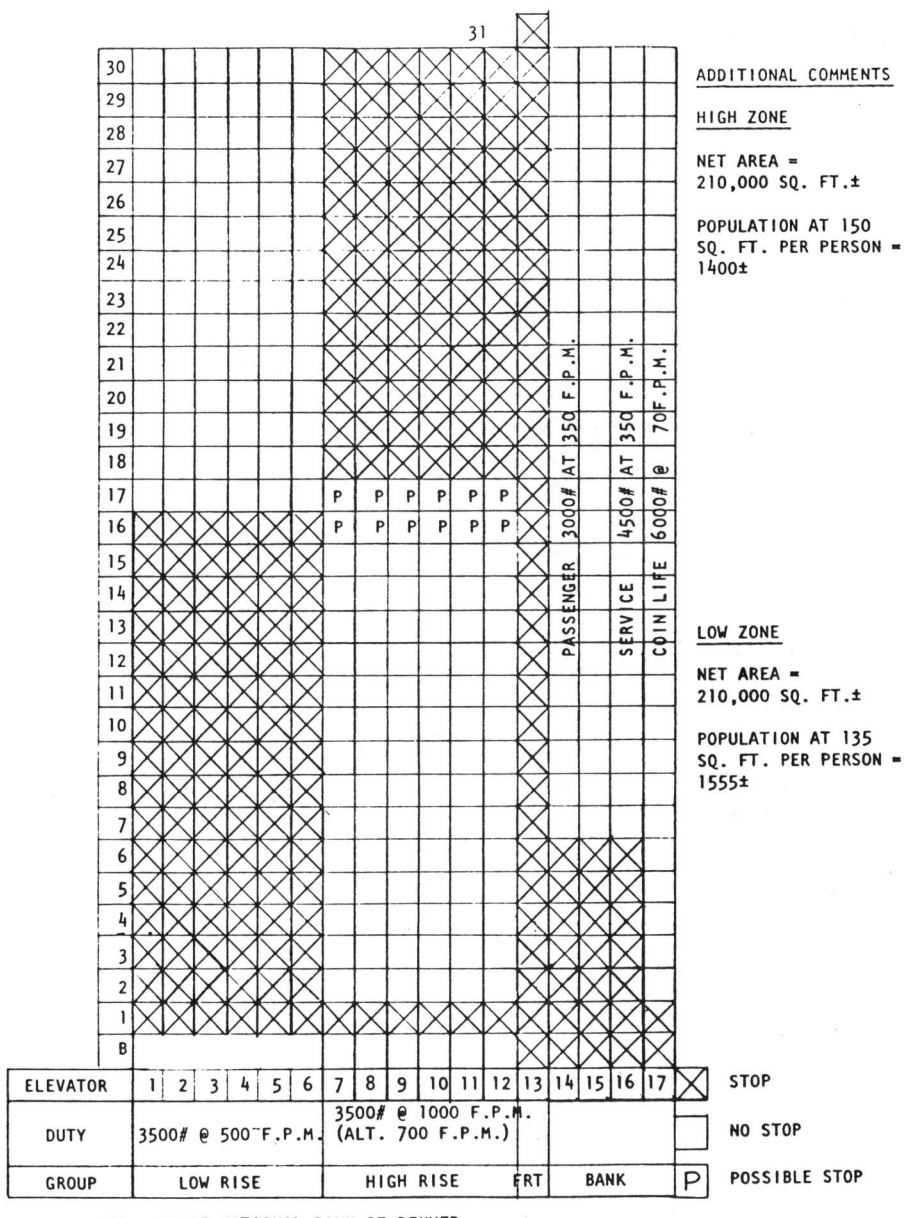

Ch. 17]   **Theoretical performance versus actual measurement**   167

Fig. 1 — Elavatoring scheme for final development

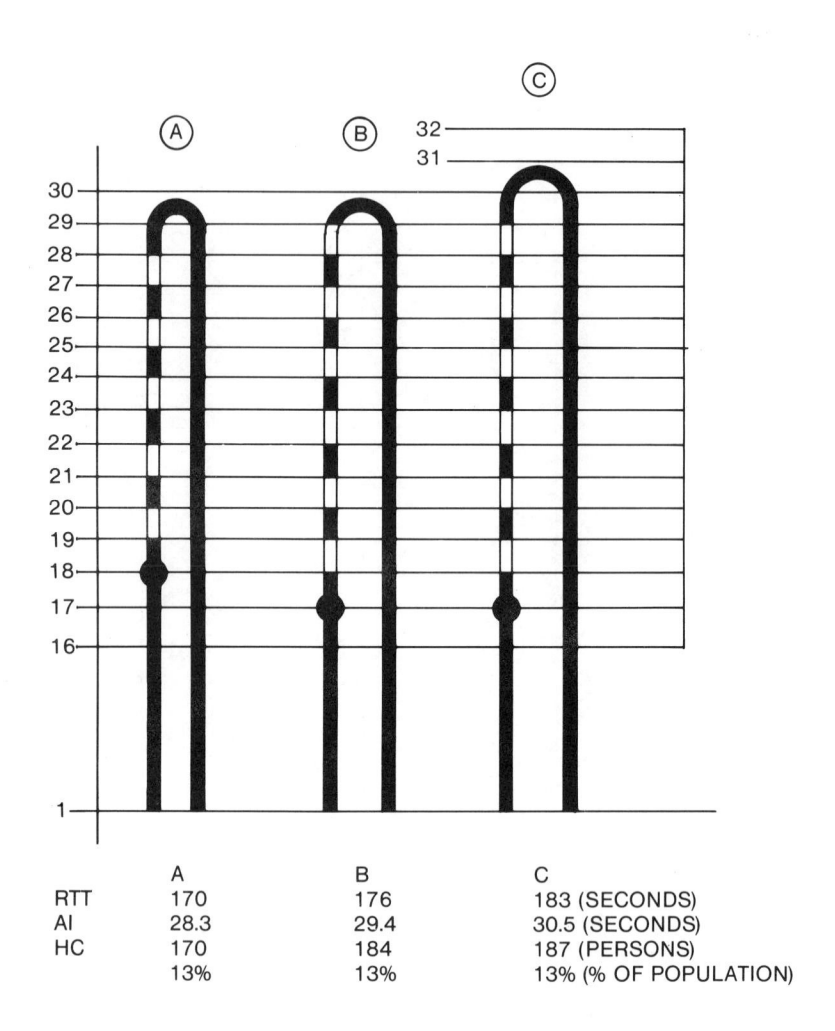

	A	B	C
RTT	170	176	183 (SECONDS)
AI	28.3	29.4	30.5 (SECONDS)
HC	170	184	187 (PERSONS)
	13%	13%	13% (% OF POPULATION)

Notes
1. The area used for usable population projection is, the gross building floor less total core and corridors (or reasonable circulation space if a full floor is developed as occupied space).
2. Provision was made at both the 16th and 17th floors for stops by high-rise elevators. At the 17th, in the event mechanical equipment did not occupy the entire floor and leasable space could be developed; at the 16th, in the event bank expansion continued eventually into the high zone and a transfer floor was desired.
3. Developing some tenant space above the 30th floor was also a possibility (Calculation C).

Fig. 2 — Review — theoretical performance.

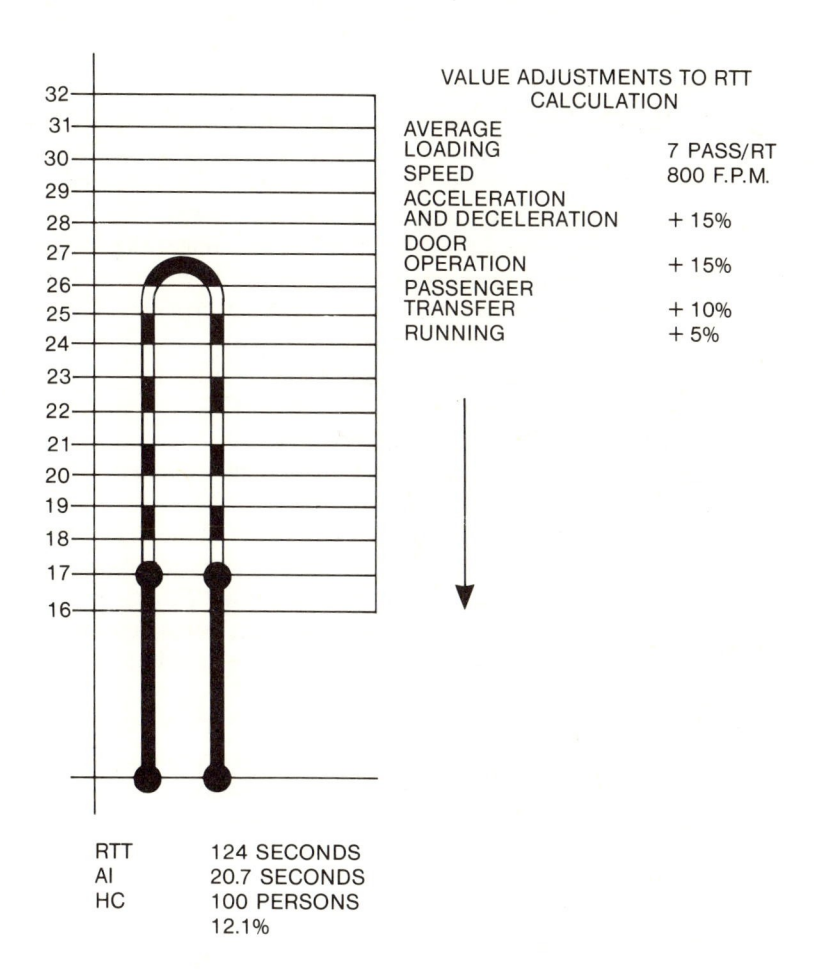

Ch. 17]   **Theoretical performance versus actual measurement**   169

```
VALUE ADJUSTMENTS TO RTT
          CALCULATION
AVERAGE
LOADING                    7 PASS/RT
SPEED                      800 F.P.M.
ACCELERATION
AND DECELERATION           + 15%
DOOR
OPERATION                  + 15%
PASSENGER
TRANSFER                   + 10%
RUNNING                    + 5%
```

RTT   124 SECONDS
AI   20.7 SECONDS
HC   100 PERSONS
       12.1%

Fig.3 — Postview — theoretical performance,

primary floors divided into two zones for elevator service, six passenger elevators serving each zone:

Low rise: 3500 lb at 500 f.p.m. serving floors 1 to 16
High rise: 3500 lb at 1000 f.p.m. serving floors 1, 18 to 30

The 17th and 31st floors are mechanical levels.

The building has a separate freight elevator serving all levels, 6000 lb at 500 f.p.m. Levels 1 to 6 interconnect with an adjacent building through a low-rise bridge structure. All the space on these floors is occupied by a single tenant. The bridge space is served by three elevators. Area and population estimates are also noted: 210 000 ft^2 in each zone with population estimated at 150 and 135 ft^2 per person in the high and low sections.

With this given and assumed data, the performance potential for the high zone was calculated as shown in Fig. 2.

In calculation A, the 18th floor is taken as the lowest level served in the high zone; in B the 17th floor was used. Calculation C reflects the same condition but without high call reversal.

Based on these calculated values, complete tenant satisfaction, that is excellent service was predicted.

How do values measured ten years later in the high zone compare with the values used for planning? Table 2 shows basic values and assumptions, Table 3 shows individual car performance and Table 4 group performance. As a post view, the performance was again calculated based on modifying values for the functions occurring during the round trip. A summary is given in Fig. 3.

Rather than expressing to lobby, local travel is assumed; values are effected as noted. Round-trip time and average interval is close; handling capacity as calculated exceeds measured capacity.

Table 1 — Area and population

	Theoretical (1973)	Measured (1985)
Gross area (ft^2)	274 000	271 455
Usable area (ft^2)	210 000	230 146
Population (persons)	1 313	828
Population density (ft^2 per person)	160	273

Notes
1. Gross area reduction is due to conversion of some space on each floor to mechanical which is not included in gross.
2. Usable area increased since space on the 17th, 31st and 32nd levels was available because of revisions in the mechanical plan. A shuttle elevator serving levels 30, 31 and 32 was added to handle tenants on these extra office levels.
3. A factor influencing the population number is that 22 530 ft^2 of usable space was vacant when surveyed. That is 9% of the zone area.
4. Population and density of occupancy were far lighter than projected.

Theoretical performance versus actual measurement

Table 2 — Basic values and assumptions

Underlying factors	Theoretical (1973)	Measured (1985)
Elevator speed (contract)	1000 f.p.m.	800 f.p.m.
Elevator capacity	3500 lb	3500 lb
Number of elevators	6	6
Average load-up trip	17/18	Average=4.2
Average load-down trip	0	Average=1.8
High-call reversal floor (average)	29	26

Notes
1. Elevator speed was reduced in the final negotiation for purchase.
2. Capacity and number of elevators were not changed. Interestingly, the first time surveyed, one car was out of service for repair and remained out for three days.
3. The difference in average loading is startling. It is based on 52 departures made during the heaviest 5-minute periods of incoming traffic observed on three consecutive days. Loads ranged from 0 to 11 persons; 11 was the heaviest loading observed leaving the lobby during 60-minute periods of incoming traffic on the three survey days. The light loading was not necessarily a lack of prospective passengers; there were other people in the lobby on many occasions who chose not to enter cars with 9 to 11 persons aboard.
4. Of 38 trips recorded during heaviest incoming traffic, 18 made stops for passengers exiting at the lobby level; an additional 10 served interfloor passengers as well. About 75% of round trips included down stops and interfloor passengers during heaviest up traffic.
5. In the 38 trips noted above, the average high call reversal floor was 26.

Table 3. — Performance — individual car

	Theoretical (1973)	Measured (average of 6 cars) (1985)
Speed	800 f.p.m.	780 f.p.m.
Door open time	1.5 seconds	1.7 seconds
Door close time	2.4 seconds	2.7 seconds
Brake-to-brake time	4.5 seconds	5.2 seconds
Floor-to-floor performance time	8.0 seconds	8.6 seconds

Notes
1. Actual speed varied from a low of about 760 to 805 f.p.m. Most cars were slightly underspeed.
2. Door open times were slightly longer than assumed optimum, ranging from 1.5 to 2.1 seconds. Doors were effectively open for passenger transfer as the cars stopped at the floor.
3. Door close average is somewhat generous — one car had a very slow final check and required 3.2 seconds to completely close.
4. Floor-to-floor performance was reasonably good, and quite consistent except on the car with the long door close time.

Table 4 — Group performance

	Recommended criteria (1973)	Theoretical (1973)	Measured (1985) (average of 3 days)
Average interval	Under 30	29.0 seconds	21.7 seconds
Handling capacity (up from lobby)	at least 13% of population in 5 minutes (170 persons)	13.4% 176 persons	6.4% 53 persons
Handling capacity (round trip):	same	same	8.7% 72 persons

Notes
1. Average interval was less than calculated. It might have been even lower if there had been more passengers. Occasionally, cars left the lobby without passengers.
2. Handling capacity was substantially below the criteria assumed for performance evaluation during design; this is true even when interfloor and down passengers are included.

3. CONCLUSIONS

What does all this mean? Probably nothing significant, except on a very general level. Looking at the results of the three individual days, there was no correlation of performance data, traffic volume or pattern, except the results observed were well below the assumptions made ten years earlier. The calculations made were theoretical, based on theoretical criteria and compared with theoretical standards.

The fact that one of the better buildings in Denver, with a very influential major tenant and a prime location, had a 9% vacancy factor meant that some vacancy in a large building might be inevitable. The substantially lower population is significant because this is not unusual in many newer buildings in major United States' cities. But, caution, do not count on it. Recently the elevator service in a good quality New York City building was evaluated. After construction, the major tenant gradually increased its space to complete the occupancy of the two lower zones. Density in those two zones was below 150 ft^2 of usable space per person.

The low percentage of population handled during 'heavy' periods of incoming traffic is also the usual situation in newer buildings with diversified tenancy. But do not count on that either. In cities with rapid transit and train service to suburbs, down-peak can be far worse than up-peak as everyone dashes for the ride home.

Would simulation have improved accuracy? No, similar initial assumptions would result in similar recommendations whether calculated or simulated performance was studied. This is not to diminish the value of simulation . . . because it is a most valuable tool that will be increasingly used for theoretical evaluations.

What can be concluded? Actual measurement varies significantly from values used for original calculations. Performance values originally calculated vary significantly from actuality. Even hindsight values do not correlate exactly with measured performance. Hindsight is not perfect because the system used is a series of approximations which yields relative values not precise prediction. The most important element, not reviewed in this chapter, is that the anaconda building in the next block, which competes for tenants, has elevator groupings which have theoretical performance values comparable to those used in the original planning for this building. Conclusion: calculation is not reality — never was, never will be.

The second conclusion is, though it may not be reality, the probable stop calculation method is an approach that can be consistent and serve as a yardstick for comparison with theoretical standards. It is simple; it has been used successfully for more than fifty years. It works. As the project was reviewed there was a final thought: 'Why make calculation more complicated by introducing facts?'

18

Elevators for existing residential buildings

Mr Hans Westling, Swedish Council for Building Research, Stockholm, Sweden

ABSTRACT

The lift group of the Swedish Council for Building Research has several projects in progress, and is engaged in the development of better and cheaper lift technology which at the same time fully complies with safety requirements.

The background is a programme for better living for disabled and elderly people. During the years 1984–1986, the group has extra aid from the Swedish Government for its experimental work. In this chapter the background to the work of the group, is described together with its objectives and results for alternative lift solutions in old buildings in different projects in Sweden. The total costs including building works have been reduced by 30% which opens a new market for lifts.

1. INTRODUCTION

The Lift Group of the Swedish Council for Building research, which has several projects in progress, is engaged in the development of better and less expensive lift technology whilst at the same time fully complying with the safety requirements. The type of lifts in question are passenger lifts suitable for installation in existing multi-family residential buildings. The background to this work is the Swedish programme for better living for elderly and disabled people. During the years 1984–1986 the group receives financial assistance for its work from the Swedish Government.

This Chapter contains a presentation of the work of the group, and the results which have been achieved in terms of alternative passenger lift solutions which have been installed in existing multi-family housing buildings in different pilot projects in various parts of Sweden. The total lift installation costs, including building work, have been reduced by 30%, which opens up a large new market for passenger lifts.

2. BACKGROUND

The Swedish Government has declared that everyone regardless of age or disability, has the right to a dwelling which fully satisfies the requirements of good accesibility. One of the major problems today, both in Sweden and in other developed countries, is that the existing environment is badly adapted to the changes taking place in the population structure, whereby an ever-increasing number of people live to an old or very old age. If it is not possible for this group to be able to continue to live on in their own homes then the only remaining solution is most frequently some form of long-term institutional care. Obstacles to accesibility in the form of stairs or steps are often the factors which create this situation. In Sweden it has been shown that if elderly and old people can continue to live on at home for a few more years, then it pays, from a national economic viewpoint, to install lifts in three to four storey multi-family housing buildings.

In 1983, Sweden commenced a ten-year Government programme for the modernization, alteration, and repair of nearly 300 000 flats. Sweden has a total of about 3.5 million dwellings and a population of slightly more than 8 million. Since 1977 the Swedish Building Code requires that all multi-family housing buildings on which alteration work is carried out shall be provided with lifts if the buildings have more than two storeys. These regulations are based on Nordic recommendations. The Swedish Government and Parliament has allotted special funds for the development of lifts. A Lift Group has been given the task of initiating both experimental and development work and forms for bulk purchase of better and less expensive lift technology.

The Lift Group works in co-operation and collaboration with representatives of tenants, disabled people, various authorities, lift manufacturers, building construction firms, housing administrators, and research workers. By means of various pilot projects the group has stimulated the following activities:

(i) A development of today's technology and know-how in order to be able to install lifts without the tenants having to move or be faced with very high increases in rent.
(ii) The development of lifts which are not only safe and attractive, but which at the same time are inexpensive to manufacture, install, service, and maintain.

The results of this development work opens up a large market for passenger lifts for existing multi-family housing.

In order for a lift to be suitable for use in existing multi-family housing buildings it must be;

 (i) inexpensive
 (ii) functional,
 (iii) take up as little space as possible,
 (iv) be safe,
 (v) be able to be speedily installed.

3. INEXPENSIVE

The reason why many existing three-storey multi-family housing buildings are without lifts is because it has been considered both too difficult and too expensive to install lifts when the buildings were being altered and modernized. Very considerable building operations were necessary earlier in connection with the installation of the actual lift.

Until now the normal total cost of installing a lift in an existing three-storey multi-family housing building has been about 70 000 US dollars (SEK 550 000). The main reasons for this high figure are:

 (i) That the lift and building construction solutions employed were designed to be used in new and considerably taller buildings. As a result of this, for example, the capacity and performance of the lifts has been unnecessarily high for existing 3–4-storey buildings.
 (ii) That various authorities' regulations and requirements, particularly with regard to safety, have also been primarily formulated with new buildings in mind. The safety clearance for the shaft top and the shaft pit have proved to be especially expensive when applied in connection with the installation of lifts in existing 3–4-storey housing buildings.
 (iii) That the whole lift installation process has been carried out using what could roughly be described as 'traditional' or 'manual' methods;
 (iv) That there has been a division between the responsibilities of the lift contractor and those of the building contractor.
 (v) That the work has taken a long time, with the result that it has been necessary to evacuate the tenants for long periods — a difficult and expensive procedure.

It was found that the cost of the actual lift amounted to only one-third of the total costs involved, the remainder consisted of building costs and tenant evacuation and building owner's costs. This lead to the objectives that the costs must be cut by half and that a new technology must be developed.

The basic objectives of the Lift Group are (1) to cut the cost of the installation of a lift by half – from 70 000 US Dollars down to 35 000-40 000 US Dollars (SEK 300 000) by 1988, (2) to achieve low running costs, and (3) to develop lift installation solutions which only entails the evacuation of the tenants for a few days. Figure 1 illustrates this plan.

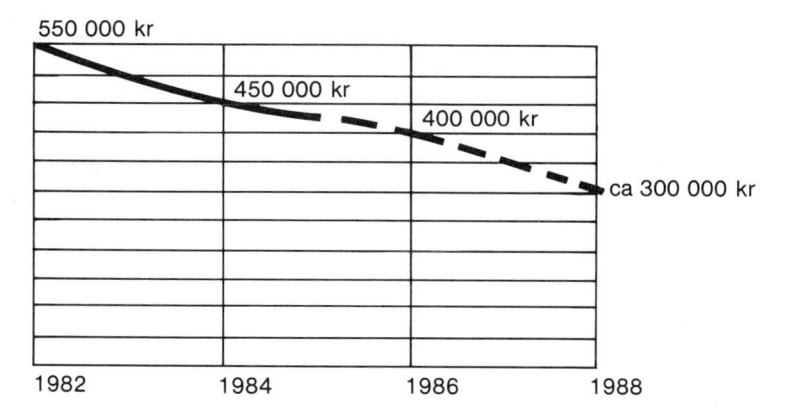

Ch. 18]   **Elevators for existing residential buildings**   177

Fig. 1 — Reduction in total lift costs including building work, in 1984 value of money
(1 US Dollar = approx. SEK 8, exchange rate used).

The technical goals are that the lifts:

(i) Shall be functionally adapted to meet the requirments of tenants and visitors, the transport of sick people, fire safety etc; see Fig. 7(b).
(ii) Shall require as little space as possible when being installed in a staircase or in a space which was previously part of a flat, see Fig. 2.

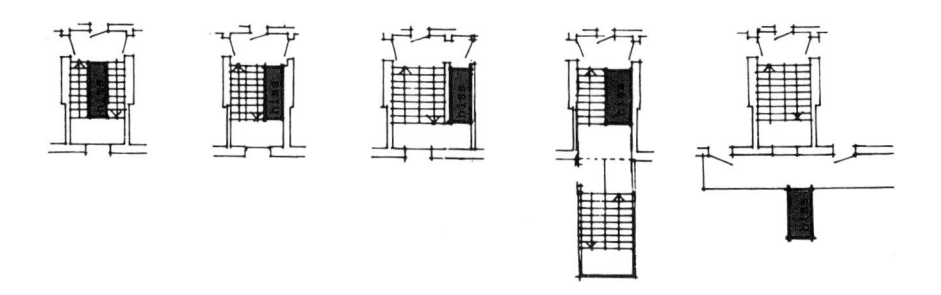

Fig. 2 — Various installation arrangements.

(iii) Shall have a simple construction.
(iv) Shall require as little building work as possible.
(v) Shall be able to be quickly installed.
(vi) Shall be able to be installed as an independent measure, i.e. not necessarily in conjunction with any modernization or alteration works to the building;
(vii) Shall be safe for users and service personnel.

Up to now the total lift installation costs, during the development work has, been reduced by 30%.

4. FUNCTIONAL

The dimensions of the lift car are the most important pre-conditions in order that a lift shall be able to function properly for all different categories of people, e.g. a lift shall be able to be used by someone sitting in a wheelchair. Apart from the dimensions of the lift car this also places certain demands on the placing and design of control buttons and lift doors. This applies not only with regard to people with impaired mobility but also with regard to people with impaired vision or hearing. A growing group of people suffer from various forms of allergies. Someone suffering from asthma, and who is therefore unable to climb stairs, may not be able to use a lift if somebody has been smoking in it.

A small but important point which must not be forgotten is that a lift is one of the links in a transport system. The existence of a single step can destroy this system, and for this reason the existence of a half flight of stairs before the lift can be reached at entrance level, or between a lift stop and a proper floor level, cannot be considered as being functionally acceptable solutions.

Stretcher transport shall be able to take place either in the lift or by means of the remaining staircase (Figs 7(a) and (b)).

It can be necessary in many cases that the shaft doors (side-hung for reasons of space) need to be mechanically operated in order that someone in a wheelchair can use the lift. If the lift is not fitted with a door-opening mechanism then it should be prepared for retrofitting when the necesity arises.

5. ECONOMICAL USE OF SPACE

A lift which is installed in a 2.4 m wide stairwell (common in Sweden) shall not have an external shaft wider than 0.95–1.0 m, including metal plate shaft walls. The remaining width of the staircase is then 0.7 m.

In existing buildings it is frequently very difficult to be able to build a deep shaft pit and a high shaft top in a conventional manner Fig. 3(a) and to build a lift machine-room above or alongside the lift shaft; Fig. 3(b). This has led to the development of lifts with shallow shaft pits (0.1–0.2 m), low shaft tops (2.5 m), and with no, or only a small lift machine-room under a staircase or similar, Fig. 3(c).

The level of efficiency for an ISO lift is less than 40%. Standardization is a useful aid but it should not be allowed to stand in the way of development and competition. The level of efficiency of the lifts which are a result of the work of the Lift Group is approximately 80%.

The level of efficiency gives the usable transportation volume in relation to the total shaft and lift machine-room volume. This applies to the required dimensions in connection with the actual lift car, where the present standard is not adapted to the situation in existing buildings.

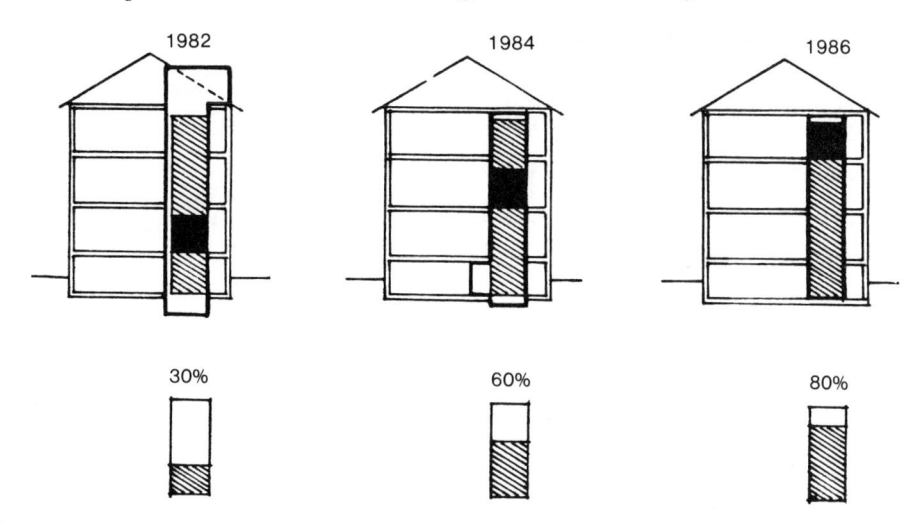

Ch. 18]   **Elevators for existing residential buildings**   179

Fig. 3 — Usable transport volume in percent of total volume (shaft and machine-room, including walls).

6. SAFETY

6.1 Safety requirements

The basic pre-condition in this connection is that the safety level shall be at least as high as in the case of conventional lifts. The National Board of Occupational Safety and Health (the responsible authority in Sweden), has the following safety system demands in connection with shallow shaft pits and low shaft tops in order to achieve the required rescue spaces:

(i) Lift speed should not be greater than 0.3 m/s (for three to four buildings).
(ii) When someone enters the shaft pit a mechanical barrier to the downward movement of the lift shall be automatically activated and the lift control current circuit shall be broken. This can either be done by means of a separate contact when the bottom shaft door is opened, or by means of a maximum 15 kg load being put on a tramp mat or tramp plate located in the shaft pit.
(iii) When someone goes onto the roof of the lift car then the control current circuit for the control buttons both in the lift car and at the different floor levels shall be broken. Inspectional movement of the lift from the roof of the lift car shall be blocked until a mechanical barrier to the upward movement of the lift has been brought into operation, either manually or automatically. The safety system is set in operation either when a lift shaft door is opened, by means of a special door contact, or when a tramp mat or tramp plate is loaded.
(vi) The return to normal functioning for the safety system for both the shaft pit and for the shaft top takes place in the lift's electrical panel.

The Swedish National Board of Occupational Safety and Health allows lifts without a car door or a folding gate at the opening to the lift car provided that the lift speed does not exceed 0.4 m/s and provided that there are photo-cells or a safety threshold.

Lifts with the lift machinery on the roof of the lift car, may, after special permission has been received, be installed in order to gain experience.

6.2 Main data for the lifts
Rated load 4 persons/325 kg.
Speed 0.2–0.3 m/s.
Electric connection 16 A at 380/220 V.
Depth of shaft pit 0.1–0.2 m.
Height of shaft top 2.5 m.
External breadth of lift shaft 0.95–1.0 m.
Free internal width of lift cab 0.8 m.
Free internal depth of mini-lift lift car which has room for a person in a wheelchair min. 1.2 m.
Free internal depth for a lift car which allows stretcher transport 1.9 m.
Photo-cells or safety threshold at opening to lift car (no door nor folding gate).
Slide-hung hinged doors in the lift shaft.
Door-opening mechanism, or pre-prepared for door-opening mechanism.

The above data and safety requirements etc. cannot be regarded as final. The development work continues, and the experience gained from functioning lifts and the results of investigations and studies etc. can lead to alterations.

6.3 Drive system
The demands of low costs, space economy, and low speeds has furthered the development towards more unusual drive systems. These systems are, screw drive, rack and pinion drive, rack and worm-wheel drive, and fixed roller chain and sprocket drive. In all of these sytems the drive machinery is placed on the roof of the lift car. In other words, no lift machine room is necessary. All that is necessary is a space for the electrical cabinet.

Other drive systems used are drum drive, indirect acting hydraulic drive with ropes, and indirect acting hydraulic drive with chains. In the case of a lift installed in a stairwell the machine and the electrical panel were placed in a space under the stair at the entrance level to the building, see Fig. 7(c).

7. SPEEDY INSTALLATION

The installation of a lift should not necessarily mean, as is so often the case, that the tenants have to be evacuated for a long time. The Lift Group's pilot projects have shown that two days are sufficient in order to be able to guarantee safety in the stairwell. Figure 7(d) depicts an installation being carried out.

8. THE MARKET

If the functional and technical goals and requirements can be achieved at reasonable costs then a large market opens up for passenger lifts specially adapted for installation in existing multi-family housing building. In Sweden there are some 200 000 stairwells in buildings of this type, with three storeys or more, which do not have lifts. Many of the buildings without lifts are in housing areas where all the stairwells are rather similar. There are, for example, 50 000 stairwells in Sweden like the one illustrated in Fig. 4.

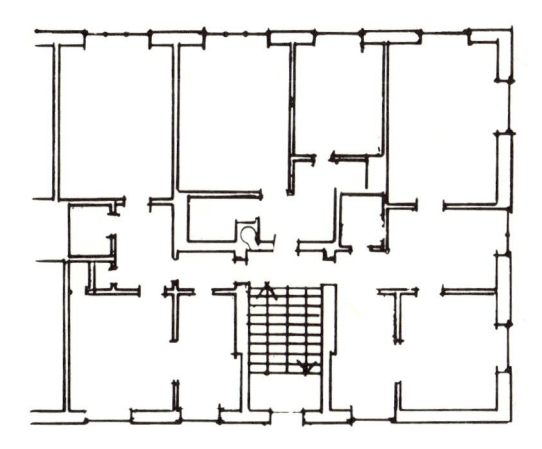

Fig. 4 — Typical stair plan at flat level.

Today there are a total of 73 000 lifts in Sweden of which approximately 50 000 are in multi-family housing buildings. A modest first step is to provide all housing areas with a few more buildings with lifts, no housing area should have no houses with lifts. Today a few hundred lifts per year are being installed in existing housing buildings. The plan is to increase this number to around 1000 lifts per year — a 70% increase on the present demand for lifts in Sweden.

9. PILOT PROJECTS

Since 1983, a number of lifts with varying technical solutions have been installed in different places in Sweden.

The Lift Group has initiated the pilot projects, and the experience gained in these pilot projects can be applied to new lifts with a high level of volume efficiency, placed either in stairwells, in areas which were previously part of flats or outside of the building. As a result of solutions with an economic use of space there is less need for alteration work to existing flats, and this means that the possibilities increase for the tenants to be able to continue to live in their flats whilst a lift is being installed. A chart of current and planned projects is given in Fig. 5 and the type of lift and its manufacture is shown in Table 1. A diagrammatic representation is given in Fig. 6.

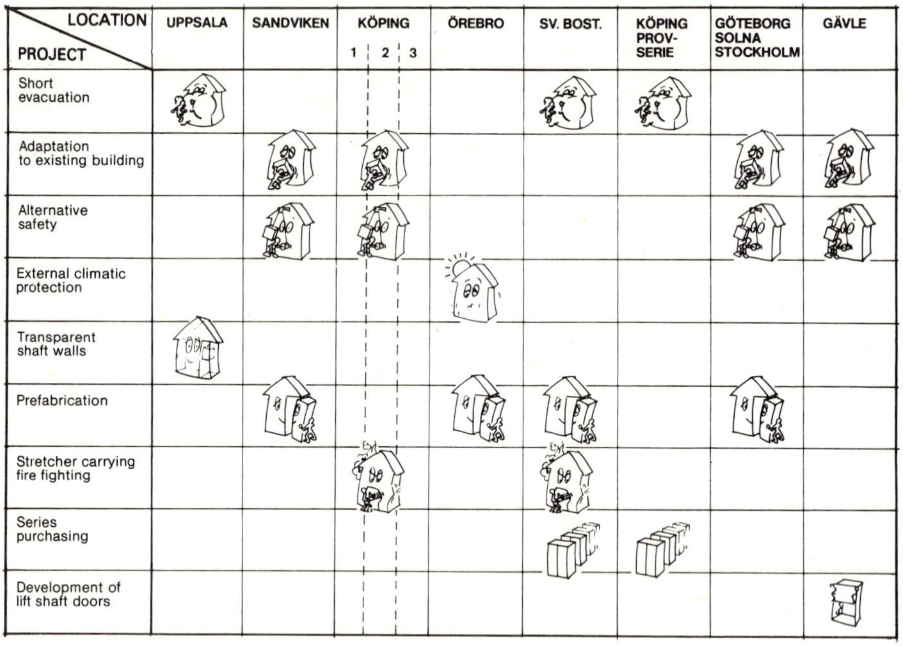

Fig. 5 — Current and planned pilot projects for less expensive lifts, with the most important tests set out.

10. EXPERIENCE GAINED AND REMAINING PROBLEMS

The experience of the Lift Group suggests that the minimum free car size, which makes it possible for a person in a wheelchair to use a lift is 80 × 120 cm. In order that a stretcher can be transported in a lift then the minimum free car size must be 80 × 190 cm. A stair-landing depth of 130 cm is satisfactory provided that the lift is correctly placed and that the doors are suitably designed. These dimensions are at present being tested and evaluated. A lift speed of 0.2–0.3 m/s is sufficient in a three to four storey house and facilitates the use of alterntive safety systems. The different lift solutions are at present under intensive mechanical and other tests.

Ch. 18] **Elevators for existing residential buildings** 183

Table 1 — Types of lift and their suppliers

Lift suppliers and installers 1983–85	Type of lift
KONE Hissar AB	Indirect acting hydraulic lift with ropes (Fig. 6(c))
Devehissar AB	Direct acting hydraulic lift
C.E. Söderlunds Hiss AB	Indirect acting hydraulic lift with roller chains (Fig. 6(a))
ÅKAB Maskinkonstruktion HB	Roller chain drive lift
Hissmontering KR Foresberg AB	Drum drive lift (Fig. 6(b))
Alimak AB–Fastec AB	Rack and pinion drive lift, outside of building (Fig. 7(d))
NID Hiss AB	Screw drive lift
Kalea Hissar AB	Rack and worm-wheel drive lift

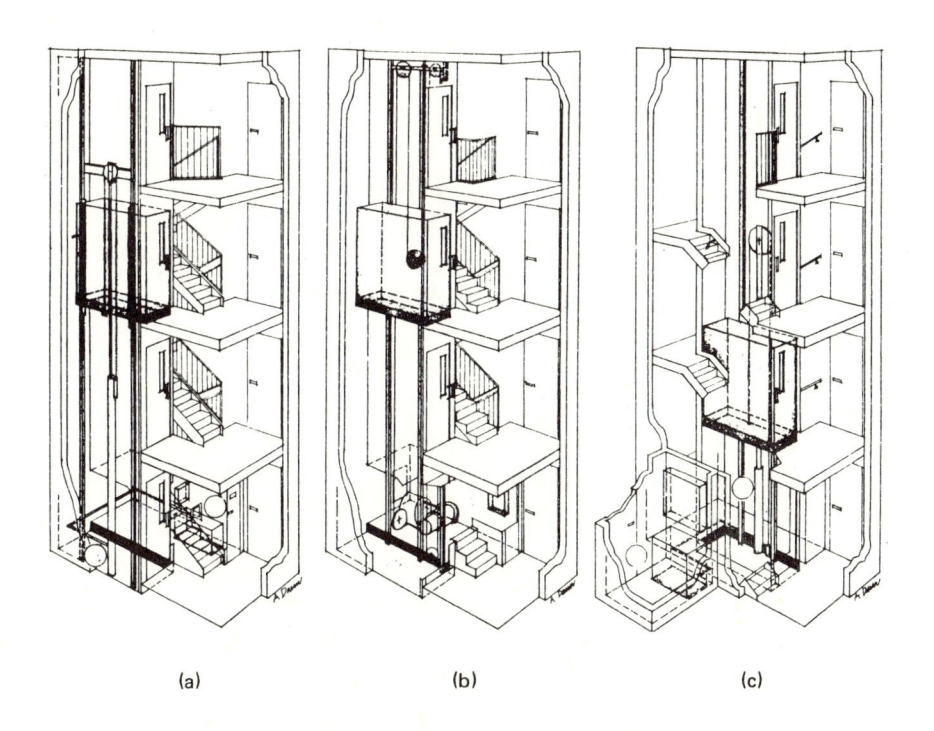

(a) (b) (c)

Fig. 6 — Types of lift installations (see Table 4).

Fig. 7 — (a) Stretcher capacity; (b) Stair capacity.

Ch. 18]	**Elevators for existing residential buildings**	185

(c)

(d)

Fig. 7 (cont.) — (c) compact location of equipment; (d) Installation on the outside of a building. (Alimak–Fastec).

The remaining problems are primarily of an administrative character namely property costs and national economic factors. the principles for the division of costs between the property owner, the local or municipal authority, and the government with regard to the national economic advantages involved, have not yet been defined. there are, however, some technical lift problems and building problems remain, but they are being investigated. Bulk purchasing during 1986 is expected to further speed up the development.

11. ACKNOWLEDGEMENTS

The work presented in the chapter is the result of efforts of many persons and organizations including Sten Söderström, Assistant Professor, Royal Institute of Technology, Stockholm, with long experience of planning for disability, Bertil Ulfward, retired from the Swedish Board of Occupational Health and Safety and earlier responsible for safety regulations for lifts in Sweden and Hans Ornhall, the Swedish Board of Physical Planning and Building, responsible for the Swedish Building Code.

Part 6
Lift management

19

The energy consumption of elevators

Dr.-Ing. Joris Schroeder, Schindler Management AG, Lucerne, Switzerland

ABSTRACT

Energy consumption has received an increasing amount of attention in recent years. This is true in the elevator industry as much as in other industries.

This chapter describes a method of determining the daily energy consumption of elevators in the early planning stages subject to type of drive, motor rating and starts per day.

Comparing the energy cost of elevator operation with the rental cost of office space leads to the conclusion that elevator energy cost represents less than 1% of the total cost of office space.

1. INTRODUCTION

During the last ten years, the world has become very energy conscious: the cost of energy has been steadily increasing, and energy consumption has been given more and more attention. This is also true of the elevator industry, whose product is very efficient, with low energy consumption. Studies have indicated that the average elevator system, even in a tall building of 20 to 40 floors, consumes less than 2% of the total energy required to operate the building, including heating, cooling, ventilation, lighting and all other services. One of the problems in defining elevator energy consumption has been finding a general formula, applicable to any building. Such a formula has been developed by H. Bosshard, Schindler Management AG, Switzerland, and will be covered in this chapter.

2. THE ENERGY CONSUMPTION PLOT

To understand fully the effect of car loading and trip length, a number of tests were conducted and energy consumption was recorded for a range of load and travel. The results were plotted, as shown in Fig. 1, floors travelled, with loading as a secondary parameter. The plots were very similar in shape (though differing in scale) for slow speed geared equipment.

The energy flow becomes very transparent when a 'lifting work' line (dash–dotted) is added; the following categories can be distinguished:

(i) Friction losses incurred while travelling.
(ii) Dynamic losses while starting and stopping.
(iii) Lifting (or lowering) work, potential energy transfer.
(iv) Regeneration into the supply line.

The plot is not symmetrical about the kW-axis, if an overbalance of less than 50% is used.

The energy consumption plot (Fig. 2) is slightly different for high-speed equipment, such as gearless, which only reach full rated speed after two or more floors of travel. For this type of elevator, the consumption line is not a straight line over the whole travel, the slope changes at the distance at which full speed can be attained.

As will be noted, there are now two intersecting points for the consumption lines. If the car does not reach full speed, the dynamic losses are decreased, and the intersection of the steeper lines occurs at a lower energy consumption level.

The actual consumption plots were similar to the typical plots shown here, with one major difference; the consumption line intersections were not distinct. In some cases, the intersections covered a small range. It is believed that this is due to errors in overbalance, variations in friction and other conditions.

3. THE TYPICAL TRIP

The energy consumption plots were used to analyse the daily consumption of an elevator. The result of this study was rather surprising, as it indicated very clearly that there is a simple, single, representative or *typical trip*, which can be used to define the energy consumption of elevator systems.

The following sample case was examined (see Fig. 1):

Data
1250 kg at 1 m/s, high mass AC servo, serving six floors above ground; Probable stops served during up-peak: 5.5

Up-peak consumption:
Travelling up, trip distances average 6/5.5 = 1.1 floors
Energy consumption
    Full load up   = 175 kWs
    Empty car down   = 33 kWs
    Average up trip   = 104 kWs (1.1 floors)
    Empty car down   = 520 kWs (6 floors)
Up peak average   = (5.5×104+520)/6.5=168 kWs

Ch. 19] The energy consumption of elevators

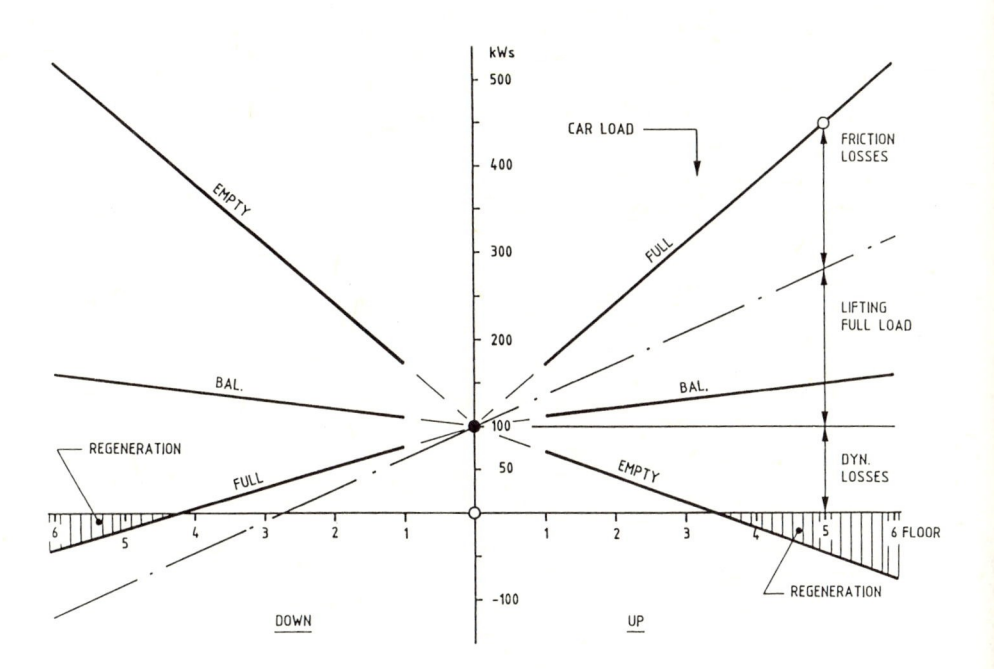

Fig. 1 — Typical geared energy consumption plot. (Assume 1250 kg × 1 m/s, AC servo, high-mass).

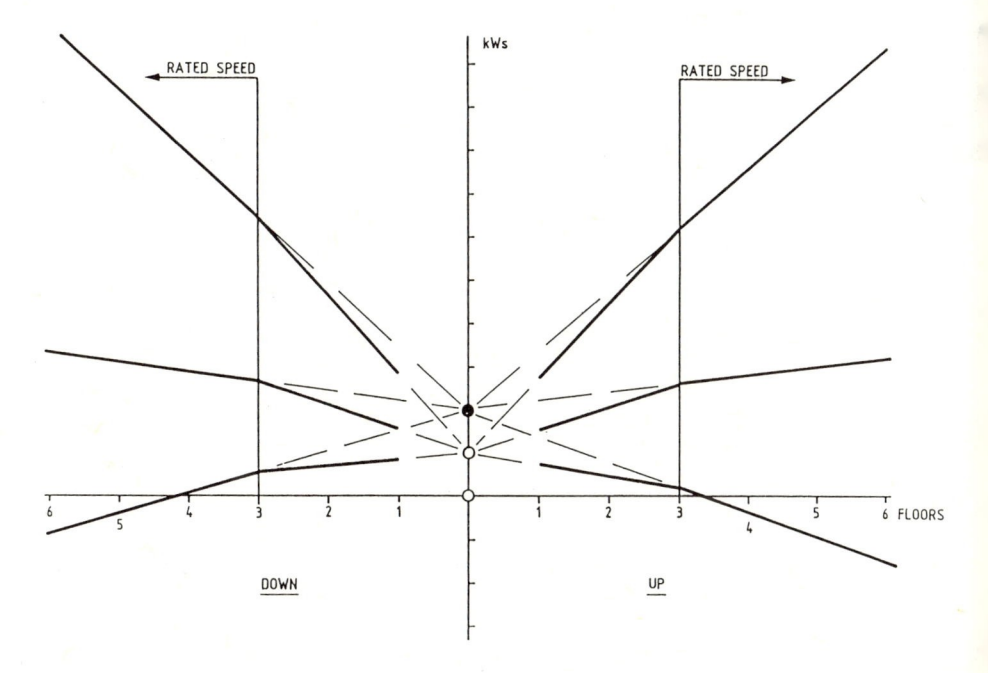

Fig. 2 — Typical gearless energy consumption plot.

Down-peak consumption:
Travelling down, trip distances average 6/5.5 = 1.1 floors
Energy consumption
    Full load down   =   70 kWs
    Empty car down   = 175 kWs
    Average down trip   = 122 kWs (1.1 floors)
    Empty car up   = −75 kWs (6 floors)
Down-peak average (5.5×122−75)/6.5   =   92 kWs

The mean value of up-peak + down-peak (each up-peak must be complemented by a down-peak!) is:

Average peak consumption = (168 + 92)/2 = 130 kWs

What, then, is the average day-time trip and its consumption? To answer that question Fig. 1 should be examined again. As a first step, assume that cars travel an average of three floors at balanced load. The energy consumption would then be 130 kWs, up or down. As a second step, assume that cars travel three floors at 25% load. The energy consumption would then be 21.5 kWs down, 65 kWs up, i.e. an average of 140 kWs, which is not far from the previous average. This could be continued for many other loads, but the outcome would always be the same. It is, therefore, reasonable to conclude that the average peak consumption of 130 kWs also is the typical consumption per trip:

$$130 \text{ kWs} = \text{Typical trip consumption (for sample case)}$$

The typical trip consumption applies to a specific drive system only. It increases, if a larger motor or a different drive system is used. Consequently, it is more practical to define a new typical value:

$$1TC = \text{Typical trip consumption kWs/motor kWs}$$
$$= 130/15 = 8.7 \text{ s in the sample case.}$$

This *typical consumption factor* (TC) varies, subject to the efficiency of the drive system used, but it is not subject to motor size. The typical trip time *(TP)* can be converted to a typical, relative *motor-on-time (m)* by multiplying with the *starts/hour (ST)*:

$$m = (ST \times TP/3600) \times 100\% \tag{1}$$

If a gearless installation ($TP = 4$) and heavy traffic ($ST = 150$) are assumed m becomes:

$$m = (150 \times 4/3600) \times 100 = 16.6\% \tag{2}$$

This means that energy consumption of this gearless machine operating under heavy traffic is equal to the consumption of the motor running at rated load for 16.6% of the elapsed time.

    Generally values for the relative motor-on-times for the various types of drives at intensive traffic (150 starts/hr) are given in Table 1.

Table 1 — Values of motor-on-times (m) for various drives.

Drive	Floors above ground	m(%) Range‡	Mean
Hydraulic w/o cwt.	3–4	22–28†	25†
Geared			
AC 2-speed	4–8	37–50	44
ACVV, high mass	6–12	29–33	31
ACVV, low mass	6–12	21–33	27
Gearless			
MG	12–18	17–25(33)*	21(25)*
thyristor	12–18	12–21	17

* For low traffic intensity (<1000 starts/day)
† Motor kW = 3 × AC 2-speed, at equal load + speed!
‡ Lower end of range applies for 1:1 roping, large motor and few stops (upper end vice versa).

4. THE ENERGY EQUATION

The tests mentioned before were evaluated as described in the previous paragraph to produce a table (Table 2) of typical trip factors.

Table 2 — Typical trip factors

Type of drive	Floors above ground	TP(s) Range	Mean
Hydraulic w/o counterweight†		5–7†	6†
Geared			
AC 2-speed	6	9–12	10.5
AC VV, high mass	12	7–10	8.5
AC VV, low mass	12	5–8	6.5
Gearless			
MG	18	4–6(8)*	5
VV thyristor	18	3–5	4

* For low traffic intensity (<1000 starts/day)
† Motor kWs = 3 × geared AC, at equal load!

The lower end of the range applies if 1:1 roping and a relatively large motor is used; the upper end of the range applies when 2:1 roping and a relatively small motor is used. If the number of stops is higher than shown,

TP increases slightly, and vice versa. The *TP* values apply to local zone elevators only, if express zone arrangements are examined, the following express elevator by-passing multipliers should be used:

    1 zone : 1.5
    2 zones: 2.0
    3 zones: 2.5 etc.

The *TP* factors from Table 2 above can be used to calculate the daily energy consumption (*E*) of an elevator:

$$E = \frac{R \times ST \times TP}{3600} \quad (3)$$

where *E* is daily energy consumption (kWh/day), *R* is motor rating (kWs), *ST* is number of starts per day and *TP* is typical trip factor (s).

The number of car starts/day which should be used in eqn (3) is dependent on traffic intensity, i.e. light traffic 750, medium traffic 1000 and heavy traffic 1500.

5. ELEVATOR ENERGY CONSUMPTION IN A BUILDING

Though the energy consumption of elevators is very low, it may be of interest to examine some real numbers for an assumed building of 14 floors, 1400 persons day-time population, and an elevator handling capacity of 15% of population in 5 minutes. There are three possible elevatoring solutions:

    (1) 6 Cars, 2000 kg × 2.0 m/s, gearless 25 kW
    (2) 6 Cars, 1600 kg × 3.15 m/s, gearless 31 kW
    (3) 7 Cars, 1600 kg × 1.6 m/s, geared 22 kW

If it is assumed that very intense traffic exists the cars will perform 1500 starts/day, except for solution 3), which will only require 1500 × 6/7 = 1286 starts/day. The energy consumption could then be, for a 10-hour day:

(1) $E = 25 \times 1500 \times 6 \times 3.25/3600 = 203$ kWh/day
(2) $E = 31 \times 1500 \times 6 \times 3.25/3600 = 252$ kWh/day
(3) $E = 22 \times 1286 \times 7 \times 6.5/3600 = 358$ kWh/day

Obviously, the gearless solutions (1) and (2) are better from an energy consumption point of view owing to the greater efficiency of the gearless drive.

Energy consumption in a building is usually related to floor area in m^2, and expressed in annual amounts (*e*). To convert to this standard, space utilization must be defined. Assume an average floor space of 20 m/passenger and a 0.85 ratio of net (office)/gross floor area, and 250 days per year, then the elevator consumption of the sample building (Solution 2) would be:

$$e = \frac{\text{kWh/day} \times \text{days/year} \times 0.85}{\text{population} \times \text{space/person}} \quad (4)$$

$$e = \frac{252 \times 250}{1400 \times 20} \times 0.85 = 1.9 \text{ kWh/m}^2 \text{ per year}$$

The basic calculation presented, permits the evaluation of taller buildings, too. For this purpose, elevatoring calculations were made and the machine rating kWs was determined. Applying the energy equation has then resulted in the general pattern shown in Fig. 3.

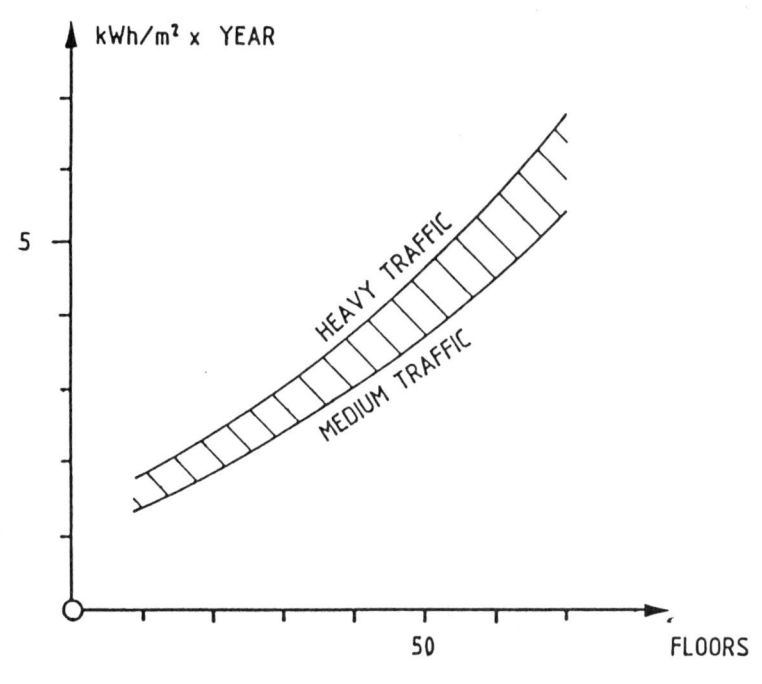

Fig. 3 — Typical energy consumption of elevators. (Assume gearless, VV thyristor).

To bring these numbers into perspective they can be compared with the rental cost of space:

Assumed energy cost	0.10$/kWh
Building height	50 floors
Elevator energy consumption (from Fig. 3)	4.2 kWh/m^2 per year
Elevator energy cost	0.42$/m^2 per year
Rental cost	100$m^2 per year
Percentage elevator energy cost	0.42% of rent.

Based on the same assumptions, the daily cost of elevator service per office can be calculated. With office space of 20 m^2 and 250 working days per year the daily elevator service cost per office becomes 0.03$/day. This pays for an average of 6 rides/day.

6. CONCLUSION

Elevators provide efficient, low-cost, transportation!

20

Towards improvements in lift maintenance

Mr A. M. Godwin, Lift Design Partnership, London, UK

ABSTRACT

Increased competition within the lift industry to maintain market share of lift maintenance has caused prices to fall. In consequence standards of lift maintenance have also fallen whilst Building Managers and users at the same time are becoming more vocal and demanding improved standards of performance and reliability from their lifts. A conflict thus arises. One solution the author proposes is for the lift makers and independent suppliers to install information systems which allow a Building Manager and his tenants to objectively assess the reliability and performance of his lift installation. Lift companies could then offer 'guaranteed performance' maintenance contracts at appropriate price levels.

New ideas are presented for planned maintenance and 'performance guaranteed' maintenance contracts with reference, in part, to a fourteen car intensive duty lift installation in London. The author will try to persuade the multi-national lift makers, in particular, that in remotely interrogating lift installations for their own benefit they should also include equipment needed for lift management by Building Managers and tenants of buildings. Standards of reliability and performance can then be objectively assessed and therefore maintenance contracts differentiated and appropriately priced.

1. BACKGROUND

Clients, the owners and future tenants of high-tech buildings which are to be constructed in the next two years are indicating that they are concerned about the levels of reliability and performance currently being obtained from recently installed lifts in their new developments. With rents the equivalent of £30 per square foot and upwards, the tenants of such buildings rightly expect high levels of reliability and efficiency from all their building services. Essentially what they are demanding are guarantees for the ongoing performance and reliability of these services and specifically the lifts.

Too often consultants are drawn into tangled situations where owners and tenants are at loggerheads with lift companies disputing the reliability and performance of their lifts. The same problem often occurs during a twelve-month defects liability period where unfortunately there is no hard evidence one way or the other to prove how reliable a given installation has been during that period. A wind of change is blowing through the lift industry and lift makers should start accommodating this new demand for user friendly lift systems that report their performance and reliability.

The list companies should bear in mind that this is an era where complex electromechanical machinery such as office copiers, printers and computer disk drives, etc. do work and provide very high levels of realiability. Since the control systems for lifts are now solid state and what remains of the mechanical parts such as the door-gear and mechanical locking systems should need only a minimum amount of maintenance then lift systems should similarly be able to provide the same levels of reliability.

One solution to both the problems of unreliable equipment and the present downward spiral in maintenance prices and quality in general, is to specify, contractually, levels of reliability to be achieved from the outset. These levels of reliability will be specified and during the twelve-month defects liability period will have to be compiled with before substantial sums of money, as opposed to small retentions, are released to the contractor by the engineer. From that point on the maintenance contractor will be tied down to a performance guaranteed maintenance, or PGM, contract. This will inevitably mean that monitoring euipment is either built-in or added-on after construction in order that the relevant information is available on site. This chapter will discuss a number of parameters that might need to be achieved to meet the stipulations of such a contract.

2. QUALITY OF PRODUCTS AND IMPERFECT RELIABILITY

Lifts are like any other products, that is there are wide variations in the quality of equipment available and correspondingly different levels of reliability are attributable to individual components and complete systems. When the lifts themselves cease to function satisfacrtorily they will need to be repaired, adjusted or replaced. To some extent, however, the need for repairs may be reduced through prior effective maintenance. Thus arises:

(a) *Preventive* and *planned maintenance* which is provided in order to sustain satisfactory operation of items and equipment.
(b) *Breakdown maintenance* (i.e. repair) which is undertaken in order that items and equipment might be returned to satisfactory operation.

(c) *Replacement* which occurs when items and equipment wear out and are no longer capable of safe and satisfactory operation.

Because of this imperfect reliability, a system has to exist to put lifts back into safe working order, and this is the reason for lift maintenance contracts. These at present offer the customer a somewhat limited range of options for the way in which he can choose to look after his capital investment.

The sort of questions that are going to arise are: What level of reliability should be considered reasonable for a lift installation, 95% or 99.5% serviceability or somewhere in between? How many manufacturers have this kind of information to hand? Would they be prepared to publish it and could they, if asked, guarantee it? These are the questions being posed to our industry. These are also clearly items requiring internationally accepted standards to be developed.

3. OBJECTIVES OF MAINTENANCE

3.1 The Theory

The purpose of maintenance is to attempt to maximize the performance and reliability of equipment by ensuring it performs regularly, efficiently and safely by preventing breakdowns or failures and thereby minimizing the 'loss of service'.

Each of the following will contribute towards this goal:

(i) Improvement of the quality of components through improved design.
(ii) Improvement of the quality of components through 'tighter' manufacturing standards.
(iii) Improvements in the level of skill used in the construction and installation of the equipment.
(iv) Improvements in the design of equipment so as to facilitate the replacement of broken items and to facilitate inspection and routine maintenance work.
(v) Improvements in the layout of equipment maintenance work, i.e. providing space around or underneath equipment.
(vi) The provision of 'graceful degradation' or built-in 'back-up' in equipment.
(vii) The provision of a stock of spare parts close to the installation to ensure that the failure of equipment is not reflected in excessive equipment down-time because of a shortage of materials or parts for immediate replacement.
(viii) The establishment of a repair facility, so that, through speedy replacement of broken parts, equipment down-time is reduced.
(ix) The carrying out of preventive maintenance, which, through regular inspection and/or replacement or adjustment of critical parts, reduces the occurrence of breakdowns.

As far as long-term reliability is concerned, increased reliability of existing equipment will come largely from:

(a) Preventive maintenance.
(b) Proper repairs.
(c) Substitution of troublesome components by more reliable counterparts.
(d) Improvement of the feedback channel by which any equipment reported as unreliable is quickly modified by the designer and these changes are adopted by the department responsible for manufacturing the equipment.

Increased effort in preventive maintenance should reduce cost of repair maintenance.

3.3 Preventive Maintenance: What Does it Mean?
Preventive maintenance is used to delay or prevent the breakdown of equipment and also to reduce the severity of any breakdowns that occur, two aspects can be identified.

(i) Inspection. Inspection especially of critical parts is the most important direct means of increasing equipment reliability. Inspection will often indicate the need for replacement or repair well in advance of probable breakdowns and is normally called for by all maintenance schedules.
(ii) Servicing, routine cleaning, lubrication and adjustment may significantly reduce wear and hence prevent breakdowns. Schedules should be constructed from both operating experience and manufacturers' recommendations.

No matter how much preventive maintenance is conducted, failures will still occur. It will therefore always be necessary to carry out breakdown maintenance. These situations may require a number of attentive strategies including:

(a) Use of specialist subcontractors.
(b) Immediate or subsequent repair of defective equipment.
(c) Replacement of part or sub-assemblies.
(d) Replacement of whole pieces of equipment.
(e) Temporary repairs or use of standby euipment.
(f) *In situ* repairs or removal of equipment.

Inspection is usually carried out at regular intervals to gauge what repairs or maintenance are required in the near future. Such inspections can, however, be disruptive, i.e. loss of lift service. On the other hand, it might reasonably be expected to reduce the amount of time lost (down-time) through breakdowns. One problem is to decide how much time to devote to inspection such that total down-time is minimized.

In so far as preventive maintenance as a whole is concerned the contractor and the customer should seek to do the minimum amounts of preventive maintenance becuase it is clostly in terms of labour and materials. Getting the right balance is a subtle art gained mostly from detailed recording of breakdowns, the ascertaining of MTBF figures, etc. The whole area of obtaining optimum reliability is one which we will now dicuss.

4. PERFORMANCE AND RELIABILITY MEASUREMENT

Actual system performance will always need to be compared with planned or intended performance, any resulting variance being used to determine appropriate action aimed at securing performance more closely corresponding to that intended. Performance is measured in order that incentives, objectives and targets may be set of modified.

From an operations management point of view two objectives are identifiable:

(i) Provision of customer service, i.e. planned maintenance, minimizing of down-time and response to call-backs.
(ii) Resource productivity, i.e. minimization of hall call response times.

These two objectives may be measured and hopefully achieved by the use of a performance guaranteed maintenance or PGM contract.

Before a contractor sets out to achieve guaranteed levels of performance from equipment should be a 'fact file' of information on it available. Amongst that information should be data on reliability. Unfortunately the sort of figures for reliability that are so common in other industries for equipment are difficult to obtain for lift systems. This information is MTBF and serviceability ratios which are defined below.

$$\text{MTBF (hours)} = \frac{\text{MP (hours)}}{\text{Nf}} \qquad (1)$$

$$\text{SR (\%)} = \frac{\text{MP} - \text{MT} - \text{CUT} - \text{PMT} - \text{DT}}{\text{MP} - \text{MT} - \text{CUT}} \times 100 \qquad (2)$$

where MTBF is mean time between failures, MP is an appropriate monitoring period, Nf is the number of failures in the MP, SR is the percentage serviceability ratio, MT is any modification time, CUT is any customer usage time, e.g. independent service, PMT is preventive maintenance time and DT is down-time.

If this sort of data were to hand then deciding on a base point for reliability of lift systems would be a simple task. Instead it is somewhat 'hit or miss' as to whether a specified level of reliability can be met with any given system, as lift companies appear to be very secretive.

5. SERVICE CONTRACTS PRESENTLY OFFERED

Lift maintenance contracts, are offered to customers with a large variety of inclusions and exclusions applying depending on clients' special requests or contractors' preferences and the price the client wants to pay. Broadly speaking, however they fall into one of four categories.

(i) Inspection only.
(ii) Service contracts — basic type A, oil and grease with replacement of some parts included. These are also known as P.O.G. contracts.
(iii) Similar to (ii) but includes for breakdowns that require adjustment only.
(iv) Comprehensive contracts, known as full maintenance or FM contracts in the US.

All of the above normally exclude call-backs outside normal working hours, repairs due to vandalism and work required as a result of changes in legislation, codes of practice, etc.

There are usually no stipulations on response times to call-backs, number of visits or ongoing reliability in any form. Typically lift companies argue they will normally be able to attend a call-back within two hours but there is no guarantee of this. From a brief survey it is estimated that comprehensive maintenance is now being offered on a typical 10 floor gearless commercial development for anything in the region of £1500 to £2500 per annum per lift of from under 1% of the capital cost per annum to perhaps 2%.

6. A NEW CONCEPT: PERFORMANCE GUARANTEED MAINTENANCE (PGM) CONTRACTS

It could be argued that if excellent reliability and performance were achieved the contractor should receive possibly up to 10% of the capital cost of the installation per annum. This charge would be comparable with, for example, the computer industry where charges of this level for maintenance contracts are the norm. Since lifts are complex electromechanical equipment with many moving parts, not just a box that sits on a desk, this price might not seem unreasonable. However, to imagine that owners will be prepared to pay up to five times their current lift maintenance cost may be going too far too quickly even if such excellent lift reliability and performance was available.

It might be helpful, however, to look at some key factors in defining a PGM contract.

6.1 factor 1 — Call-backs

Call-backs, whether in or out of normal working hours should be included for in the price of the contract. Thus the onus is now back with the contractor to ensure reliability or extra costs will be incurred paying staff money for call-backs without receiving any additional income. At any rate for a reliable installation the call-backs should never exceed more than five per annum per lift, excluding vandalism and therefore the contractor should be able to budget for these occurrences. Response times to call-backs should be minimal because of the interest of the contractor to restore service. A minimum reponse time should be set with a financial penalty for not responding within the stated time. There should, however, also be a contractual liability on the owner, unless the lift company is operating some form of remote monitoring, to inform the contractor within fifteen minutes, say, of any lift becoming out of service. Again a parallel can be drawn with the computer industry where response times are of the order of one hour if the customer so wishes.

6.2 Factor 2 — Down-time

First of all comes the absolute necessity to have a genuine uniform means of reporting lifts out of service. That is why with today's computerized lifts a form of reporting system is easy to implement with a simple remote printer or alarm as a minimum in the building manager's office.

The person nominated to be responsible for reporting lifts 'out of service' needs to have a built-in time-lag during which he carefully investigates each situation reported by the printer before instructing the maintenance company to respond. Assuming a system is devised acceptable to both parties, each quarter figres would be accumulated for any down-time incurred at the amount payable for maintenance each quarter bearing in mind the other four factors we are proposing.

For one lift available for 365 days per annum, 24 hours per day, this is 8760 operational hours. Suppose, there are six call-back/repairs per annum some of these will be minor and require only a few minutes attention but there will also be some, especially involving electronic component failures, that will take much longer to identify and may even result in delays whilst spare parts are located and fitted. Making due allowance for such events assume 87.6 hours down-time in total. This 87.6 hours is exactly 1% down-time per annum, and in the absence of more informed data might be chosen as the base point for the PGM contract.

The sliding scale could be a linear one with say a cut-off point above 2% down-time or it could be some other relationship that, in effect, makes a higher and higher payment available for each 0.1% less down-time over the range.

6.3 Factor 3 — Planned Maintenance Activities

The customer should be informed of the planned maintenance activities, as far as possible, well in advance. The contractor therefore must give the customer details of which lifts will be put out of service, and roughly for how long, for these activities to be carried out. If the contractor bills quarterly in advance the client should be handed the dated planned maintenance activities with the bill so that the customer has ample warning of the loss of lift service for maintenance. The customer should also be informed of the total number of hours that planned maintenance activities will take up and hence the loss of lift service that will result in a given year. The visits of maintenance personnel to the site should be recorded carefully to correspond with each planned maintenance activity. Again a financial penalty would be incurred if more than a certain number of visits were missed.

6.4 Factor 4 — Intensity of use

To cater for the wide variation in the intensity of use of lifts in different installations a table of duties could be used (Table 1.).

Table 1 — Intensity of use multiplier.

Lifts with journeys per qtr	Multiplying factor
$\geqslant 125\,000$	1.3
$\geqslant 100\,000$	1.2
$\geqslant 75\,000$	1.1
$< 75\,000$	1.0

With journey counters fitted to each lift the customer can check every quarter the accumulated total and number of journeys performed by each lift during the previous quarter. An appropriate multiplying factor could then be applied to the bill. Lifts used more, i.e. doing more starts, will therefore cost more to maintain. The above table is perhaps an oversimplification and once again some form of non-linear relationship should be applied. Sufficient to say, however, that this is an important factor that must be included in the type of contract being discussed.

6.5 Factor 5 — Satisfactory Group Control

Finally there remains the importance of ensuring that not only are lifts in service but that they are responding to hall calls in the manner in which they were intended to. That is to say that the group control policy or algorithm is fully operational and that waiting times are thus not excessive bearing in mind the equipment installed and the demands being placed in the system. Adequate performance from this aspect is extremely difficult to ascertain manually. However the manufacturer or the maintenance company must demonstrate at regular intervals the ongoing performance of the lifts from this aspect. Thus if the equipment and data is not available from the system directly, an alternative analyser type of equipment will either have to be fitted permanently or brought to the site and connected regularly to satisfy this requirement. Of course if the customer so wishes this aspect could be left out of the contract but he takes a risk in doing so. By making regular analyses of hall call response times the customer can be reassured that the system continues to give satisfactory service and that the lift resource is being fully utilized by the control system employed. If the system does not continue to give satisfactory performance from this aspect then, once again, a financial penalty should be imposed upon the contractor until such time as the equipment operates satisfactorily.

7. AN EXAMPLE OF A PGM CONTRACT

The Bush House, London, complex contains 14 lifts in four groups and it has been possible, in conjunction with the building management, to monitor the reliability and performance of these lifts over the past twelve months. From the outset of thier period, the tenant made a number of stipulations to the owners concerning what they deemed acceptable performance these included:

(i) That two lifts in any one group should never be 'out of service' simultaneously, and that means even for five minutes!
(ii) That for three of the groups or eleven lifts, they stipulated that the downtime during the said twelve months shall not exceed 1752 hours, i.e. out of $11 \times 24 \times 365$ of 96 360 operational hours the downtime should not exceed 1.8%.
(iii) Normal working hours for the lifts and for the conditions stipulated is 24 hours of every day of every week.

The on-site building managers' lift management computer has the following facilities available.

(a) Lift status display.
(b) Waiting time display.
(c) Return lifts to ground (fire precaution).
(d) Message commands.

A remote printer is located in an office that is manned 24 hours a day by security personnel where action can be taken to call in lift maintenance staff subsequent to checking that the 'out of service' message received is confirmed by their routine investigation. This is always necessary because the monitoring system will for example mark a lift 'out of service' even if the only problem is someone holding the doors for an excessive period and under such circumstances the lift will return to 'in service' status once it is

Ch. 20]   **Towards improvements in lift maintenance**   205

allowed to close its doors, make a journey and open its doors at another floor.

With this system in operation the building manager can log accumulated 'down-time' from the print-out. The results after one year of monitoring are given below:

For the eleven lifts, total down-time, as stated by the building manager, and agreed by the maintenance contractor, was 214 hours.
Thus:

$$\frac{214}{96\,360} \times 100 = 0.22\% \text{ downtime}$$

which includes the response time of the contractor.

During the same period the call-backs day and night are given in Table 2.

Table 2 — Types of call-back.

Category	Quantity
Vandalism/Misuse*	15
Requiring mechanical adjustments	4
Fuses requiring replacement	7
Nylon chain drives split	2
Unexplained incidents*	16
Overspeed governor switch	1
Objects in car or landing tracks*	2
Faults or repairs to electronic equipment	6
Double journey timer required resetting	4
Other, sundry	4
Total	61

Ignoring vandalism, misuse and unexplained incidents marked (*) there were just 28 call-back or on average 2.5. per lift/year.

The preventative or planned maintenance time (PMT) recorded was 95 hours per lift. Therefore the serviceability ratio and MTBF are:

$$\text{SR}(\%) = \frac{(8760 - 95 - 19.45)}{8760} \times 100\%$$
$$= 98.7\%$$

$$\text{MTBF (hours)} = \frac{8760}{2.5}$$
$$= 3504.0 \text{ hours}$$

From the above figures it can be justifiably argued that the customer could be offered a PGM contract on an ongoing basis having determined that the sort of reliability and downtime figures are well within the guidelines originally proposed earlier in this chapter. It is worth bearing in mind that it is figures such as these that can be the only genuine measure of both how well maintenance is being carried out and the equipment reliability.

8. CONCLUSION

With the advent of more and more information from lift systems and the possibilities for presenting this information to the customer in a concise manner, the author believes that lift companies should take the lead and use this information to benefit themselves and the owner. Owners and tenants will doubtless in time be demanding this information anyway and along with this will come stipulations on reliability. By taking up the challenge lift companies that know their products are dependable and reliable, can offer to undertake PGM contracts and be paid more money.

This enhanced income will be partly recycled towards the organizations' needs to have skilled mechanics and engineers available to maintain very high standards of performance. This in turn will be another stimulus to the proper training of personnel which, of course should lead to more systems being able to come under PGM contracts. In this way the downward spiral that has been present for so many years could be broken. The maintenance company that takes on servicing just for the income, at almost any price, without the skilled knowledge, will become exposed to his customer and to achieve the reliability objectives they will doubtless have to be disposed of in favour of perhaps, the OEM companies or maintenance companies with their own in-house skills that are capable of providing the necessary back-up to meet the targets.

In any event the owner should win and have a more reliable lift system in operation and the contractor will win by receiving greater income and will be able to devote, in turn, more skill and time to maintain and be more profitable in return. In due course then it will become accepted practice to pay for performance and reliability.

21

Remote monitoring of lifts

Dr J. R. Beebe, Lift Innovations Ltd., Bolton, UK

ABSTRACT

Large numbers of unattended lifts distributed over a wide geographical area presents a difficult management problem. The travelling public expects that a lift should be available and operational at all times, and that prompt remedial action should be taken if and when a failure occurs. Even a daily inspection of lifts becomes impossibly time consuming and therefore expensive if more than a few units are to be maintained. Thus remote monitoring of lifts is highly desirable. This chapter describes the development of a lift monitoring system, the Lift in Service Indicator (LISI) and its integration into a distributed Lift in Service Management (LISM) network which reports the 'in service' status of up to 1000 lifts on a central computer. Experience is described where such a system has been installed for two Public Housing Authorities.

1. INTRODUCTION

The task of lift management can be considered as comprising four levels (Beebe, 1980), in order of increasing sophistication:

(i) Lift-in-service indication — i.e. indication of serviceability of individual lifts for normal passenger transportation.
(ii) Equipment diagnosis — i.e. to determine whether the installation is functioning in the same manner as when it was initially installed.
(iii) Monitoring and analysis of passenger traffic i.e. passenger demands versus lift-system response.
(iv) long term optimization of lift group supervisory system — i.e. to

achieve the best utilization of the lift resource against varying and unpredictable passenger demands.

The availability of reasonably priced modern computing and communication equipment makes facilities at all these levels available to management staff at their office rather than in the motor room. Remote access to information at all levels greatly improves the effectiveness of lift management strategies. Additionally, as the number of lifts to be managed increases, the requirement for remote monitoring becomes important. This chapter will be confined to the basic level of lift-in-service indication (Level (ii)).

2. LIFT-IN-SERVICE INDICATION

2.1 Scale of the problem

The fundamental problem of lift management is to determine if lifts are in service and capable of carrying passengers normally.

If this is to be achieved by visiting the lift installation the task becomes more difficult as: (i) number of lifts increases, (ii) locations of the lifts become more dispersed.

Example 1. Medium scale problem. Consider 14 lifts in four groups in an office complex.

The lift could be monitored by constantly visiting each group in turn and travelling in each car of the group at least once per visit. this requires a great deal of staff time which could be more efficiently used. (Estimated staff time 1 hour per day.)

Example 2. Worst case problem. Consider 1000 lifts in simplex or duplex configurations installed in single residential buildings or estates, spread over an area of 80 square kilometres.

It is quite impracticable to have these lifts monitored by staff visiting them and remote monitoring is the only viable solution. (Estimated staff time 100 hours per day.)

2.2 Nature of the problem

Apart from the technical and financial problems of transmitting and receiving information over long distances another question which must be answered before lift-in-service monitoring can be attempted is: 'When is a lift in service?' This problem was considered by Clarke (1978), who reasoned as follows: 'A lift can only perform its most basic function — transporting passengers from one floor to another — *if* it is able to open and close its doors (to let passengers in and out) *and* travel from one floor to another.'

This can be expressed as a Boolean statement (the sort of logic used to design digital electronic circuits and computer programs):

    IF   doors can open and close
    AND lift moves from floor to floor

THEN lift is in service
ELSE lift is not in service

This says nothing about how well the lift works, whether it levels properly to the floor, whether the ride is rough or whether the destination floor is that requested by the passenger.

2.3 A practical solution

To designate a lift as being in service therefore requires only two signals to be monitored: Lift Moving (LM) and Door Opening (DO). Door Opening is the most positive of any that might be derived from the door gear equipment since it implies that the passenger is able to leave the lift at his destination floor.

The identifying LM–DO sequence which must be defined so that the many possible valid variations will all cause the Lift-In-Service (LIS) signal to be generated, whilst illegal sequences will not (See Fig. 1).

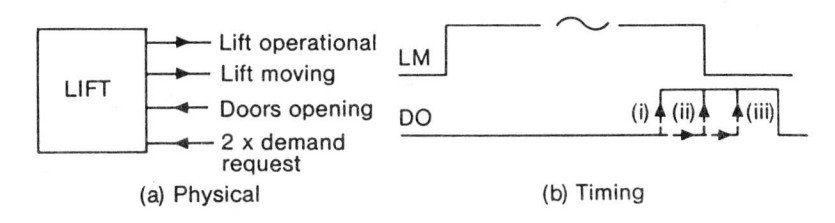

Fig. 1 — LM–DO sequence. Duration of the LM signal varies with the distance travelled as door may open: (i) before; (ii) as; (iii) after lift stops.

This identification process can be achieved quite easily using common digital electronic circuits or a simple computer program.

2.4 Lift-in-service indicator

If the LM–DO sequence is not detected for some period of time, then the reason for this could be because either the lift is out of service or because no one is using the lift. To determine the true status of the lift it must be commanded to make a trip. This can be done by a Demand Request (DR) signal which generates a car call after a predefined period has elapsed without an LM–DO sequence being detected. If the lift is in service then a correct LM–DO sequence will be detected and the LIS signal confirmed. Otherwise the lift will be declared out of service. In fact, two car calls must be generated to ensure that the lift moves since a single car call may coincide with the current position of the life and cause only a door cycle to occur with no car movement. If the terminal floors are chosen for these car calls then only one trip will be made as the second call will be cancelled when the car reverses.

Around these principles, a lift-in-service indicator (LISI) can be designed using simple electronic circuitry (LIL, 1985). A simple interface to the lift controller is effected using a set of contacts on the lift motor and door motor contactors for the LM and DO signals respectively and a set of contacts across the terminal of a car call button for each of the Demand Request car calls.

Some additional features of the LISI:

(i) Lift Operational (LO) input signal: This is valid only when a lift is available for passenger service, i.e. *not* independent service, maintenance, repairs etc.
(ii) Demand requests disabled during *known* periods of low demand, i.e. during hours of darkness, or after four consecutive demand requests producing correct LM-DO sequences.
(iii) A battery powered clock to measure out-of-service time.
(iv) A counter to record number of demand requests.
(v) Local indication of lift status to intending passengers by a flashing Jewel at each landing entrance.

3. REMOTE SIGNALLING

With equipment to report the status of one lift the problem of signalling that information possibly from hundreds of different lifts to a central site and then displaying it, must be considered. Wareing (1983), studied possible solutions to this problem as presented below:

3.1 Considerations
(i) The physical nature of interconnection medium must be: widely available, easily installed, inexpensive to run and able to cross public roads.
(ii) Transmission unit. Many are required so these must be simple and inexpensive
(iii) Receiver unit. With a large number of LISIs transmitting information to the central site some processing is required to sort and record reports to simplify interpretation by management staff.

3.2 Solutions (See Fig. 2)
(i) *Nature of interconnection medium.* A line to the public switched telephone network (PSTN) is available in most countries. It can be run directly into the motor room by the telephone authority at minimal expense, is reasonably priced to maintain and charges are only made for connection time.
(ii) *Transmission unit.* Proprietary systems are available from a number of manufacturers for signalling infrequent events to remote monitoring station (typically used for intruder/fire alarms) using PSTN or private communication links. The transmitter units can: (a) Autodial a preprogrammed PSTN line number. (b) Transmit a three digit site identification number. (c) Transmit a single-digit code to identify a particular event type.

Ch. 21] **Remote monitoring of lifts** 211

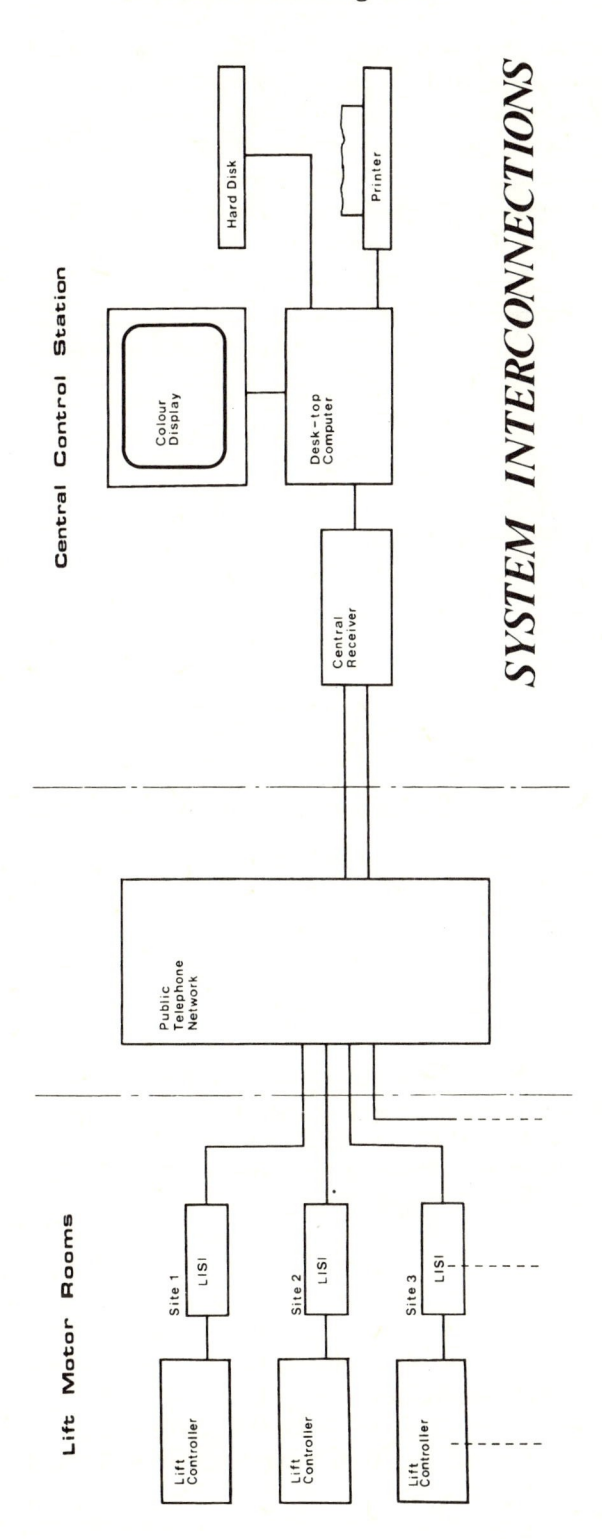

Fig. 2 — Schematic of lift remote monitoring.

The transmission is triggered by the closure of a set of contacts, in this case a relay operated by the LISI.

(iii) *Receiver unit.* A compatible receiver unit accepts incoming calls on a PSTN line at the central management station, validates the incoming message and then prints or displays the site code of the calling station together with the event code and the time and date. A single digit event code allows 10 possible event types to be signalled. An unofficial international standard for these codes is given in Table 1.

Table 1 — Signalling Codes

Code	Interpretation
0	Supply/battery voltage low
1	Fire
2	Intruder
3	Personal attack
4	⎫
5	⎬ Not used
6	⎪
7	⎭
8	Communication test (e.g. every 24 hours if no other events).
9	Restore code

The unused codes 4, 5 and 6 can thus be used to signal information from the LISI:

4	Lift out-of-service
5	Indicator Jewel circuit failure
6	Passenger trapped in lift (not sent unless 4 is true)

Suppose 200 lifts with LISIs are connected to a management system. Even if all the lifts remain in service then every 24 hours there will be 200 calls to the central station. If each lift suffers from, say seven faults per year then on average there will be a call received every 7 minutes approximately.

This has two implications:

(i) The number of PSTN lines to the central station may need to be increased so that an out-station has a reasonable chance of making a successful call to a free receiver (up to nine attempts to make an autodialled call are allowed before the out-station shuts down).

(ii) The quantity of incoming data is too large to be handled manually. Merely checking to ensure that each out-station has reported within the last 24 hours presents a major time-comsuming task.

The solution to the latter problem is the use of a computer system to log and analyse all calls which are received. This process requires little sophistication and is highly repetitive and this is an ideal job for a computer.

Once the decision to use a computer has been made, analysis of data can be improved to provide a greater depth and faster access to historical records. In particular, features such as colour graphics can be used to present data in an informative and attractive way, and the computational power of the computer can be used to provide long-term trend analysis. The example system (Fig. 3) uses an Apple IIe microcomputer, 10 megabyte hard disk, colour monitor and 80 column printer and can perform the following functions:

(i) Display all sites with current malfunction.
(ii) Display status of all sites.
(iii) Display all site details of a specific site (site name, address, reference number, maker's name, type of equipment etc.).
(iv) Display current last failure synopsis (time, date and duration) for all fault types.
(v) Display for one site, for a specified period — number of failures, duration, failure type etc.
(vi) privileged operator facilities to allow configuration and adjustment of parameters and site details of the system.

3.3 Experience

Two such systems that have been installed for local authority housing departments provide some interesting results:

(i) Lifts have been found that go into and out of service for short periods of between 20 minutes and several hours owing to such faults as malfunctioning safety edges on doors. These faults would generally go unreported until finally the lift was brought to a complete standstill because the problem had worsened or another more serious fault had occurred. However, the disturbance and nuisance to the tenants of such faults is significant and can in itself lead to vandalism of the lift and its fixtures.
(ii) Mischievous use of the lifts and doors being jammed to hold the lift can also be seen as very short out-of-service periods.
(iii) Routine maintenance can be observed and recorded to ensure that at least the site is attended at regular intervals and that sufficient time is spent on site.

4. TRENDS FOR THE FUTURE

The LISI monitoring system is applicable to all automatic lifts regardless of make and design of lift controller. However, with the advent of microprocessor-based lift controllers, it has become possible to incorporate not only the LISI function into the controller itself but also to identify and record much more detailed event and error information. This information can be interrogated on site or can be signalled over PSTN lines using modems installed within the controller either as faults occur or on demand from the central station.

Some discussion has taken place between manufacturers and other interested parties on the formulation of a standard format for the information to be transmitted. However, in the absence of agreement many manufacturers have chosen to tread independent paths and implement mutually incompatible communcation protocols.

Fig. 3 — Lift-In-Service Manager.

5. CONCLUSION

Remote lift management can provide sophisticated and powerful analysis of the performance of many lift systems, although the most basic and commonly required function is that of lift-in-service indication. The philosophy and implementation of a lift-in-service indicator, which can be attached to any make and design of automatic lift, has been described. Many lifts can be remotely monitored by connecting LISIs which can automatically dial-up a central station using the public telephone network. This central station uses a small dedicated computer to log received calls and provide graphical-tabular trend analysis of lift availability.

6. REFERENCES

Beebe, J. R. (1980). Lift management, PhD Thesis, University of Manchester Institute of Science and Technology.

Clarke, M. C. (1978). Lift status monitoring, MSc dissertation, University of Manchester Institute of Science and Technology.

LIL (1985). LISM: Lift Management System, Lift Innovations Ltd, Bolton, England.

Wareing, M. (1983). A network for lift status monitoring, MSc Thesis, University of Manchester Institute of Science and Technology.

*Part 7*
Standards and safety

22

International standardization in the lift industry

Eng. F. de Crouy-Chanel, Otis Elevator International, Paris, France

1. INTRODUCTION

Until about 1950, lifts were 'tailor-made' in most countries, and the car dimensions of lifts installed in different buildings were rarely identical. In the early 1960s, some companies began to have their own standards. A first attempt for setting some common dimensions and characteristics between manufacturers was made in Europe with the Federation Europeenne de la Manutention (FEM), Section VII, which is an association of European lift manufacturers. At International level, standardization work started in 1971 within the International Standards Organisation (ISO), beginning as a working group, later as a sub-committee and now — since 1980 — as a specific technical committee for lifts and escalators, the TC 1978. In the same way as dimensions, the safety rules were also found to be very different from one country to another, and work has been carried out for about 15 years on common safety codes within the Comite Europeen de Normalisation (CEN), which is an association of the national standards organizations of those European countries which are in the Common Market or in the Free Trade Association.

A starting point, was to take the existing Commission Internationale pour la Reglementation des Ascenseurs & Monte-charge (CIRA) safety directives published in 1971 by the International Labour Office in Geneva. CIRA is a group of lift experts in safety, inspection and design from eight European countries.

2. STANDARDIZATION WORK WITHIN ISO/TC 178

2.1 The ISO participants
The following countries are *P* (participating) members of the Technical committee:

Austria	Finland	South Africa
Belgium	France	Spain
Bulgaria	Germany (FR)	Sweden
Canada	Hungary	Switzerland
China	Italy	United Kingdom
Czechoslovakia	Netherlands	USA
Denmark	Poland	USSR

2.2 The Organization
The organization comprises Technical Committees (TC) and Working Groups (WG) as shown in Fig. 1.

ISO / TC 178
Lifts, service lifts, escalators, passenger conveyors and similar apparatus

- W.G.1. Lifts on ships
- W.G.2. Guide rails
- W.G.3. Lifting platforms for handicapped persons
- W.G.4. Safety standards comparison
- W.G.5. Escalators and passenger conveyors

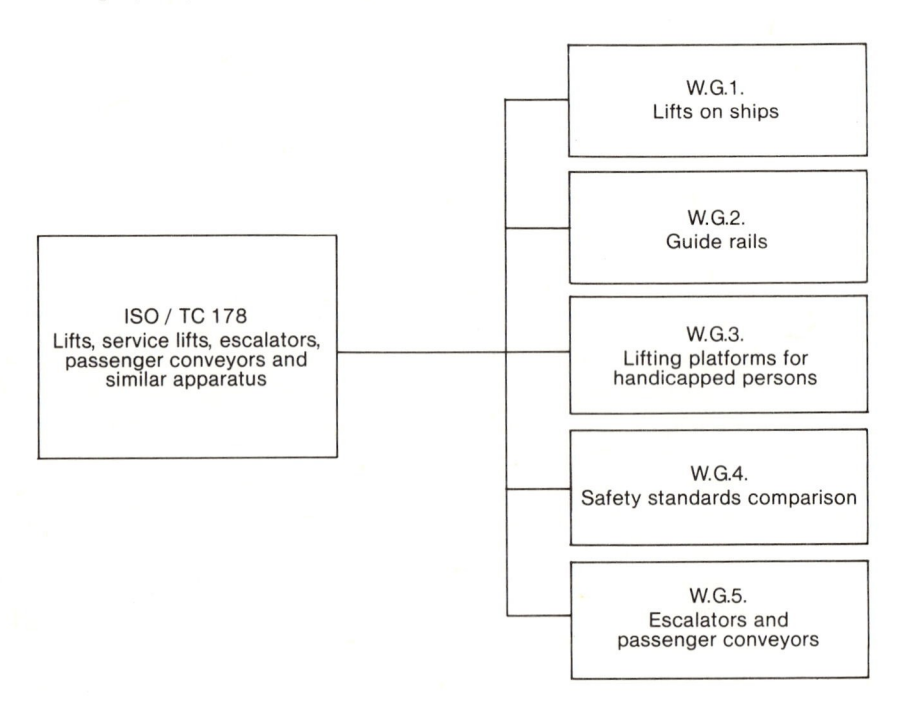

Fig. 1 — ISO committee structure.

2.3 Progress of the Work
The ISO standards which have been developed and published or are under development one given in Table 1.

Table 1 — ISO Standards progress.

Standard	Standard Number
Dimensions of passenger lifts	4190/1
Dimensions of bed lifts	4190/1
Dimensions of goods lifts	4190/2
Dimensions of service lifts	4190/3
Control devices and signals	4190/5
Planning and selection of lifts to be installed in residential buildings	4190/6
Lifts on ships — specific requirements	8383
Guide-rails for lifts	7465
Under development: Lifting platform for handicapped persons, Main safety standards' comparison (technical report), Dimensions for hydraulic lifts, Building dimensions for escalators.	

3. HIGHLIGHTS OF ISO ACTIVITY

Rather than review in detail the contents of all the standards, the following is a summary of some particular features, and the background of some of the conclusions within ISO.

In the standard 4190/1 for dimensions of lifts, a short sentence indicates that loads have been selected from the R5 and R10 series of ISO preferred numbers and the speeds from the R5 series. The letter 'R' stands for Renard, who first recognized the advantage of such a series. The R5 is given by eq (1):

$$\text{R5: Ratio } \sqrt[5]{10} \text{ (approximates 1.6)} \tag{1}$$

Rounded-off terms of the R5 series are shown below:

$$1.00 \quad 1.60 \quad 2.50 \quad 4.00 \quad 6.30 \quad 10.00$$

The R10 is given by eq (2):

$$\text{R10: Ratio } \sqrt[10]{10} \text{ (approximates 1.25)} \tag{2}$$

Rounded-off terms of the R10 series are shown below:

$$1.00 \quad 1.60 \quad 2.50 \quad 4.00 \quad 6.30 \quad 10.00$$
$$1.25 \quad 2.00 \quad 3.15 \quad 5.00 \quad 8.00$$

These series of geometrical progressions are widely used in many industries.

What led to the conclusion that loads and speeds of lifts could be selected from these series? Consider first the car dimensions of lifts for residential buildings, as shown in Fig. 2.

The three-cars correspond to the three main categories of traffic in

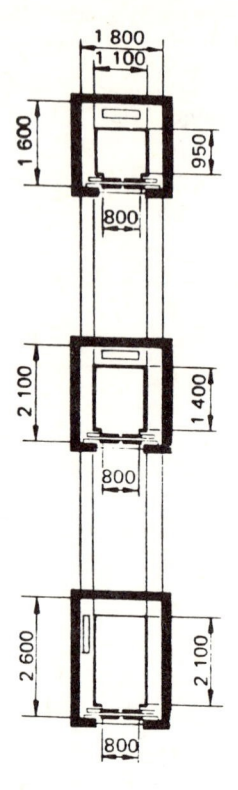

Fig. 2 — Car dimensions for residential lifts.

residential buildings for (a) passengers (standing), (b) handicapped persons on wheelchairs, and (c) stretchers (and also furniture or other bulky objects).

A few comments can be made on these dimensions. As traffic peaks are not usually heavy in residential buildings, a unique door opening of 800 mm was selected, common to the three lifts. This permits the transfer of stretchers and wheelchairs and favours large scale production. For the car depth, it was found that 2100 mm was necessary to accommodate stretchers and 1400 mm for wheelchairs.

For low-rise and small residential buildings in which some countries do not require wheelchairs or stretchers to be transported in lifts, a small car with a depth of 950 mm was defined.

With a common car width of 1100 mm the corresponding rated loads were 400 kg–630 kg–1000 kg, in accordance with the main safety standards and with the R5 series of ISO preferred numbers as shown in Fig. 3.

The 75 kg per person to be included in the ISO standard was disregarded, although it is commonly used in Europe, because different values are used in many countries, such as Japan, USA and USSR.

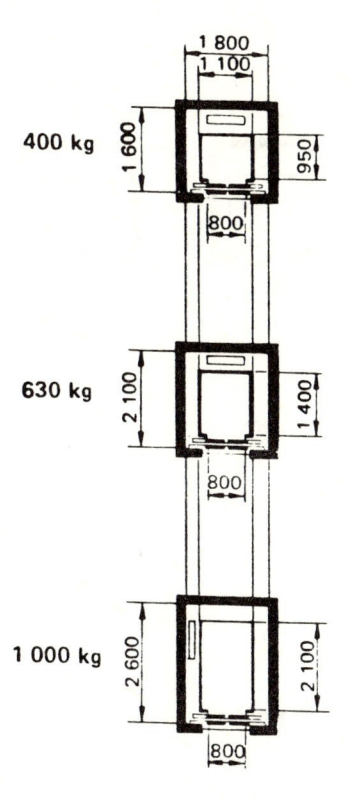

Fig. 3 — Rated loads: residential buildings.

For non-residential buildings, the load gap between two numbers of the R5 series was found to be too great and consequently, the load range for these lifts was selected from the R10 series as shown in Fig. 4.

Notice that for lifts designed for non-residential buildings, where heavy traffic peaks often occur, a second door width, namely 1100 mm has been standardized, which is necessary to permit simultaneous transfer of two persons using the larger cars.

In these types of buildings, such as offices, hotels and stores, only centre opening landing and car doors have been selected, in order to reduce door times, whereas in residential buildings, lateral opening doors can also be used as an alternative.

For bed lifts, lateral opening doors have been standardized with a free opening of 1300 mm. This is necessary in order to transfer beds and hospital equipment (see Fig. 5). The rated loads are also in the R10 series of preferred numbers for bed lifts.

As far as rated speeds of the whole line of lifts are concerned, they have been selected from the R5 series, which were found convenient to most of

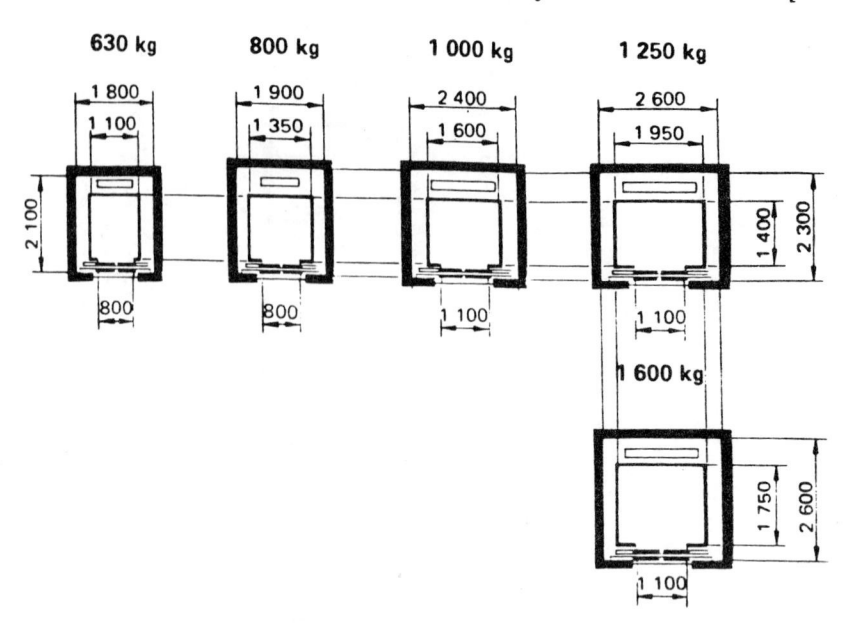

Fig. 4 — Rated loads: non-residential buildings.

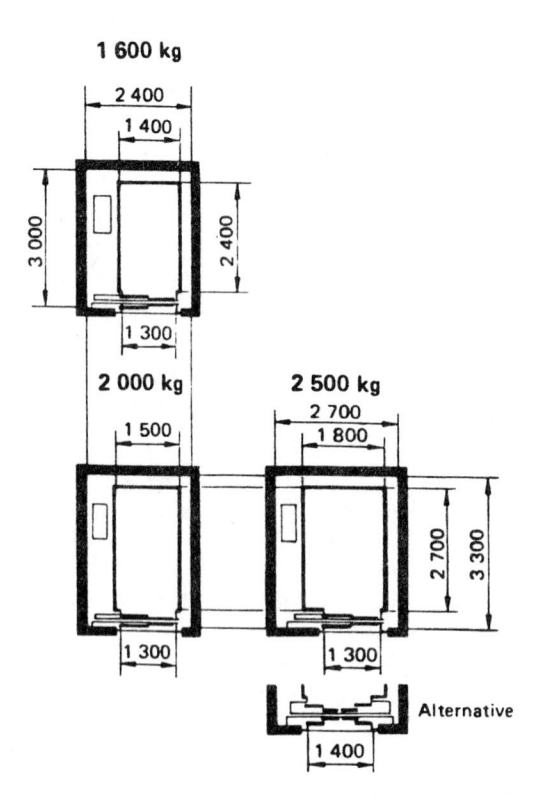

Fig. 5 — Rated loads: bed lifts.

the countries. These are given below in m/s:

$$0.25 - 0.40 - 0.63 - 1.00 - 1.60 - 2.50$$

The speeds of 0.25 m/s and 0.40 m/s are generally limited to goods passenger lifts.

4. STANDARDIZATION WORK WITHIN CEN/TC 10

ISO has been dealing mainly with *dimensions* and *characterstics*, whilst CEN, has been working exclusively on the harmonization of the *safety* codes.

4.1 The CEN members
The following countries' Standards Associations are members of the CEN:

Austria	Greece*	Portugal*
Belgium*	Ireland*	Spain*
Denmark*	Italy*	Sweden
Finland	Luxembourg*	Switzerland
France*	Netherlands*	United Kingdom*
Germany (FR)*	Norway	

*EEC Member States

4.2 The Organization
The organization is illustrated in Fig. 6.

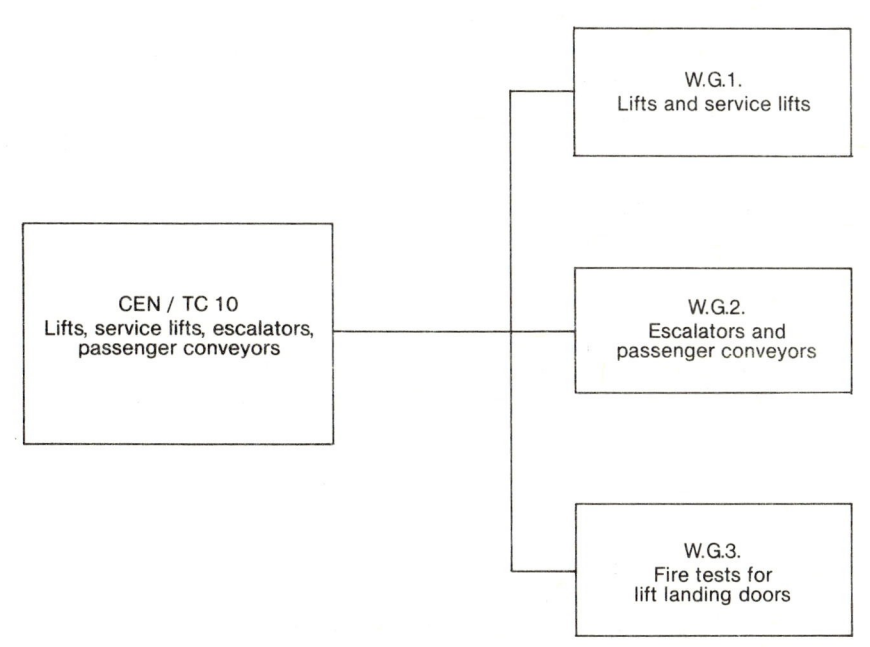

Fig. 6 — CEN committee structure.

4.3 Progress of the work

The CEN standards which have been developed and published are given in Table 2, together with those under development.

Table 2 — CEN Standards Progress.

Standard	Standard Number
Safety rules for the construction and installation of electric lifts	EN 81–1
Safety rules for the construction and installation of escalators and passenger conveyors	EN 115
Under development:	
Safety rules for the construction and installation of hydraulic lifts	EN81–2
Fire tests for lift landing doors	

When a CEN standard is developed and published, it does not necessarily mean that this standard is also implemented in each country. Indeed, in many CEN member countries, safety regulations have been made mandatory through laws and governmental decrees, which cannot be changed quickly and easily. However, the safety code for electric lifts is already implemented or will be implemented within a short time in 11 or 12 of the 16 CEN member countries.

As far as the safety rules for escalators and passenger conveyors are concerned, an almost full implementation or acceptance is expected, because there are very few countries having existing governmental decrees or laws in this field.

4.4 Coordination with ISO

Maximum rated speeds whenever compatible with safety were selected in line with ISO standard 4190, and are highlighted by some examples:

0.63 m/s: maximum rated speed permitted for:

> inspection for lifts,
> instantaneous safety on the car,
> goods lifts without car door in some countries.

1.00 m/s: maximum rated speed permitted for:
> instantaneous safety gear on the counterweight,
> instantaneous safety gear with buffered effect on the car,
> spring buffers (or energy accumulation type buffers).

1.60 m/s: maximum rated speed permitted for:
> energy accumulation type buffers, with buffered return movement.

5. CONCLUSION

During the work within ISO and CEN, when standardizing dimensions, characteristics and safety rules, the wishes and needs of architects, builders, consultants, users, and lift manufacturers were kept constantly in mind. The common priorities are for:

(i) Safe design construction, installation and operation.
(ii) Reasonably large choice of lifts, without however jeopardizing manufacturers' objectives for pre-engineering and series production.
(iii) High level of performance and quality.

23
Safety gear and European standards

Eng. Carlo Distaso, IGV/ELEVATORI, Milan, Italy

ABSTRACT

The European Community has approved the EN81/1 Standard in order to bring into harmony the technical regulations among all Member States. The most interesting novelty in such regulations, at least from Italy's point of view, is the institution of the homologation system. This is no doubt considered the most suitable to shed light on both technical and functional qualifications made compulsory at least for those components on which depends the safe use of elevators. The most discussed among these components certainly is the safety-gear and strong differences of opinion exist on whether it is useful or necessary, even though everybody agrees that this device excercises a positive effect, from a psychological point of view, to the end user's tranquillity. The wide range of safety-gear devices that are being marketed has been divided into three different categories and each of them is subject to specific qualifications and sphere of use.

This chapter intends to analyse the possibility that accidental and casual external parameters may affect the performance of some categories of equipment; also it intends to develop a comparative analysis between equipment with a shock-absorbing effect and equipment with a progressive effect, in order to investigate the degree to which the former could be usefully employed in the latters' application field.

1. INTRODUCTION

The subject of this chapter takes its starting point from the EN81 standard, created to bring into harmony the different European regulations concerning the manufacture, installation and maintenance of elevator systems.

The EN 81 regulation refers to a new procedure which was originally called homologation, but is now called type approval. It applies also to the safety gear, which has always been considered one of the most important elevator components both because it is regarded as a vital mechanism to safeguard the safety of persons and because its presence alone gives security and confidence to the end user.

Starting from the idea that the main objective of the safety gear is to safeguard the passengers' safety, and by the knowledge that the human organism, is not sensitive to speed (at least within the limits directly tested by man), but is strongly sensitive to speed changes, causes the regulation to rightly favour progressive safety gears.

Old safety gears of the instantaneous type have been equally rightly confined to the low speed range, namely 0.63 m/s. which, by allowing 1 m/s as the maximum limit of operation for the overspeed governor, when they lock the car, they case an effect equivalent to that of an object in free fall from a 50 mm height. A further allowance has been made for these safety gears which although they operate with an instantaneous setting, as they have a shock-absorbing power. It is known, that for such gears a deceleration average value equal to 1 g was fixed, with maximum peaks, which may be in excess of 2.5 g, provided their operation time is not longer than 1/25th of a second. Their field of use is limited to units having a nominal speed not exceeding 1m/s, controlled by a speed governor adjusted so as to operate at a speed not exceeding 1.5 m/s. Under these conditions the use of an instantaneous type gear would produce an effect corresponding to that of an object in free fall from an 115 mm height.

In spite of the rule allowing this possibility, the majority of elevator manufacturers have directed their approach towards the progressive safety gear, intending to use it also in the speed range where recourse to the gear with a shock-absorbing effectis allowed. This choice is justified by the necessity to contain the proliferation of products as much as possible, so that manufacturing costs are better controlled. The first discussion is therefore, given to progressive safety gears.

Simply, a progressive safety gear consists of a jaw system, which by sticking to the two sliding surfaces of the guides tightens them with an effort, which gradually increases depending on the slide, up until the car stops. and locks. Reading the definition quoted in item 3 of the European Standards, provides some very important information:

> The progressive safety gear is a safety unit which provides deceleration by means of a braking action on the guide rails, and is suitably conceived in order to limit the induced forces on the suspended system to an allowable value.

This definition enlarges the number of components belonging to the safety system up to the inclusion of those where the biting effect exercised by the jaws, due to the insertion of component between a rigid section and an elastic one, which remains steady during the whole braking period.

The evenness and steadiness of behaviour of such a system, which is more correctly called *safety gears with a friction effect,* rely on the reproduction of the rolling friction effect between two metal surfaces. And this is just what worries the designer the most.

2. PROGRESSIVE SAFETY GEARS

2.1 Physical-chemical effects

It is common knowledge that the friction coefficient does not depend only on geometrical conditions, but is strongly influenced also by both kinematic and dynamic conditions of the two solids in relative motion between themselves. In addition it is extremely sensitive to the physical-chemical state of the two surfaces in contact.

Practice shows that changes in the physical-chemical state, so small that they can hardly be measured, may provoke changes in the friction coefficient by a factor of ten times. This is the reason why all tests based upon the friction effect prove to be difficult to repeat. Difficulties that are met in a laboratory in order to reproduce the same test conditions over two consecutive tests, become partically impossible to overcome, when the safety gear leaves the laboratory to be installed on an elevator system, no matter whether lubricated or dry rails are planned. In the case of lubricated rails it is virtually impossible to guarantee the stability of the physical-chemical features of the thin fluid coat covering both the rail and the stopping component, even if it is made compulsory to always use the same type of lubricant.

The seasonal thermal changes, the extensive exposure of the lubricant layer to the oxygen contained in the air, the effect of moisture, dust and gas containing suspended solid residuals, strongly alter, in the long term, the behaviour of the friction of the fluid layer. Where the gear is properly defined to work only on dry surfaces, the difficulty remains to always keep the guides perfectly clean as well as the stopping component. Even small amounts of lubricant substances in the traction component is sufficient to cause vapours. Although these are very rarefied they consist of substances containing very lengthened molecules with one end strongly active, through which they establish a cohesion link with the guide rail surface. The rail is covered entirely like a thin velvet carpet and is highly resistant to abrasions, like a solid surfce. Whenever abrasions should occur, the surface would not remain dry for long, but would be quickly covered with a new stratum of molecules.

The wide range of chemical products having these properties, together with the physical-chemical changes due to temperature, oxygen from the air, presence of dusts or solid substances etc. makes the behaviour of the friction factor of this monomolecular coating extremely uncertain. To complete the picture of all the parameters involved in the change of the friction coefficient it is necessary to mention the pressure exerted between the surfaces in contact. Practice shows that if between two steel surfaces a 500 kg/cm^2, the friction coefficient goes up to 0.12 (a 50% change) and at a pressure of 2000 kg/cm^2 (118% increase).

In the light of all these considerations what might happen when a safety gear previously set and tested in a laboratory is operated? When working in a laboratory it is not hard to establish ideal conditions both with regard to the materials used, guides and safety gear (new, neat and even), and with regard to the environmental point of view (absence of vapours and other poisoning substances). Under these conditions the friction factor could be approximately 0.1.

2.2 Theoretical considerations

If F is the braking force which the safety gear induces on a free fall then the gear is suitable to be used on an elevator whose total load is:

$$C = P + Q = F/1.6 \tag{1}$$

where P is the weight of the car and Q is the load. To obtain a normal effect for F, with a 0.1 friction coefficient, a normal effort $N = 10\ F$ is needed. Suppose now that the gear is installed on an elevator where, for various reasons, the friction coefficient is 0.3. The braking effort becomes three times as great as the setting value. Therefore, if the car is descending in free fall, fully loaded, the safety gear locks it by means of a constant deceleration.

$$a = 3.8\ g = 37.3\ \text{m/s}^2$$

Also in the event of a car descending in non-free fall, as the deceleration is higher than g, the effect produced by the safety gear would be the same. The counterweight, though, would continue its run towards the top, until it exhausts its kinetic energy, and then would fall back giving a jerk to the ropes, which could unlock the safety gear, and the car could precipitate a second time.

Consider the opposite hypothesis and suppose that in a laboratory higher friction conditions could be simulated with the friction coefficient equal to 0.3. (This value ensures that the average value of deceleration is lower than 0.6 g.) It is also necessary to suppose that the gear is installed in humid environments the friction coefficient may decrease to values ranging from 0.7 to 0.1, and should the car descend in free fall whilst fully loaded, not only could it not be stopped by the safety gear, but it would continue its run with an acceleration of about 0.46 m/s^2. The gear would work in a normal manner only, if the car should descend in non-free fall. Then, supposing that the car's own weight is almost the same as the capacity with a 50% balancing, the deceleration induced by the safety gear would be about 0.54 g.

To conclude this first part, devoted to progressive gears, recall what Professor Aberkom (1984) from Delft University of Technology, Netherlands, said at the Amsterdam Symposium in 1984.

> In view of the results described above, it is not surprising that not one of the progressive safety gears tested fully met the demands of NEN–EN 81–1. One time the deceleration was too high, the next time too low (causing the load to slide down to the bottom of the

pit), or not all four prescribed tests could be executed, or the safety gear was deformed or cracked, or the difference between the momentary and the main brake force exceed the allowed 25%.

From a point of view of repetition of results, both the instantaneous safety gear and the instantaneous gear with buffered effect prove to be far more reliable. Of these, the former, apart from the limited braking load which depends either on larger or smaller sizes of lock and on the nature of the material, has a foreseeable behaviour, the wedge slope or the roller diameter have been fixed, and the finish of the surface of the friction components. However, since it is unthinkable to use these gears outside the low speeds field, it is better to turn our attention to gears with buffered effect and examine them in deal.

3. HYDRAULIC BUFFER

3.1 Description and basic theory

The hydraulic type buffer (probably the most tested), consists of two coaxial cylinders of which the internal one has a number of small connecting holes between the hollow space and its inside. The inside of the smaller cylinder contains an airtight piston, which is connected with a rod, that passes through a head-piece. The working of this gear is based upon the process of energy transformation for one form into another. The sum of the kinetic energy and that of the position of the car in free fall is transformed into pressure energy inside the smaller cylindre providing the braking action. The pressure energy is then transformed into the kinetic energy of the oil, flowing into the hollow space through the discharge holes. Lastly the kinetic energy of the oil is dissipated in the form of thermal energy, as it whirls inside the hollow space.

To understand the mechanism imagine an infinitely long cylinder, filled with oil, and provided with a series of small holes at the bottom, containing a piston on which a load is applied.

Let the starting speed of the piston be U_0 and U be the instantaneous speed, then the change of speed depends on time as given by the following expression:

$$\frac{V+U}{V-U} = \frac{V+U_0}{V-U_0} \cdot e^{t/\phi} \qquad (2)$$

where ϕ is an empirical parameter with the dimensions of time.

Equation (2) shows that, whatever the starting speed U_0 may be when time (t) approaches infinity the speed U tends to a value V, which depends on the following:

(i) The load on the piston. This is directly proportional to the square root of the internal pressure and it is an essential parameter in order to dimension the gear according to the load and the impact speed.
(ii) The type of oil used. This is inversely proportional to the square root of

the specific weight and the oil viscosity. A multigrade oil, with a high index of viscosity, guarantees a negligible change of speed V over a wide temperature range.

(iii) The ratio between the cylinder section and the discharge holes. The piston, as it moves down, reduces the number of active holes and the limit value of speed V gets smaller. Thus, the impact maximum speed may be brought down to low values by suitably selecting the diameters and position of the discharge holes.

it can be noted that maximum speed V does not depend on the starting speed U_0; therefore the same final speed is reached, whether with $U_0 > V$ or with $U_0 < V$. The rapidity with which U_0 approaches V, namely by deceleration or acceleration, decreases faster in the initial stage, and more and more slowly as time passes, in an exponential fashion. Acceleration based upon time is expressed by the following equation:

$$a = \frac{2\zeta e^{t/\phi}}{\phi(\zeta e^{t/\phi}+1)} \cdot V \qquad (3)$$

where ζ depends on the starting speed U_0, whilst ϕ depends on the same parameters as V.

The deceleration value decreases with time. Its maximum value is obtained in instant $t=0$ when it equals:

$$a = g\left(1 - \frac{U_0^2}{V^2}\right) \qquad (4)$$

from which it is possible to see that if $U_0 > V$ acceleration is positive, but as already explained, speed U will never be able to exceed the limit value V. Once again the important point is to choose the quantity and the diameter of the discharge holes so that the starting deceleration and its average value do not exceed the limits provided by the code. Experiments carried out in the test tower show that deceleration develops in a way sufficiently close to the project expectations.

Problems occur at the time when the discharge holes are all covered, especially at low speeds. At those instants of time deceleration may reach values ranging from 3 to 4 g with a duration which is never longer than one hundredth of one second.

3.2 Further analysis

To be in a position to proceed with a more precise and correct sizing of the buffer it would be necessary to have more exact information concerning the behaviour of the oil passing through a small diameter hole at high speed. It is important to determine as precisely as possible the losses of energy suffered by the oil, when it passes from inside the smaller cylinder to the hollow space, expressed in the percentage of the energy of the oil unit of weight descending at the same speed as the car (as suggested by the Bernoulli's equation) i.e:

$$\Delta E = Y \cdot \frac{U^2}{2g} \tag{5}$$

where the unknown parameter is Y, a dimensionless quantity which may be called impedance.

In order to fix the value of Y, a series of experiments should be carried out. Suppose that the value of Y depends only on the following parameters:

(a) The section of the internal cylinder A1 (sq.m.);
(b) The section of the discharge holes A2 (sq.m.);
(c) The sliding speed of the piston U (m/sec);
(d) The dynamic viscosity of the oil μ (barsec);
(e) The specific weight of the oil γ (N/m³).

It is reasonable to think that the relation linking Y to the above parameters is of the following type:

$$Y = A_2^m \cdot A_1^n \cdot U^p \cdot \mu^q \cdot \gamma^r \tag{6}$$

Through dimensional analysis a system of three equations for the five unknown exponents can be obtained.

By means of such a method three of these exponents can be obtained which are dependent on the remaining two. The previous equation is changed into:

$$Y = A_2^{-(q+n)} \cdot A_1^n \cdot U^q \cdot \mu^q \cdot \gamma^{-q} \tag{7}$$

The parameters may now be separated in groups with equal exponents:

$$Y = \left(\frac{A_1}{A_2}\right)^n \cdot \left(\frac{U\mu}{A_2\gamma}\right)^q \tag{8}$$

Therefore Y depends on the two dimensionless groups of which the former is a ratio between surfaces that are indicated with S. The latter the author has not been able to identify and is indicated as X.

The expression then becomes:

$$Y = S^n \cdot X^q \tag{9}$$

If a certain quantity, Q, of which the values of μ_1 and γ are known, is pumped up in a pipe of section A_1, provided with holes having a total area A_2, and a pressure p_1 inside the cylinder is measured, then values S_1 and X_1 are known from the beginning. From the knowledge of the oil speed U_1 and the pressure p_1, through the Bernoulli equation, Y_1 is calculated and the relation is:

$$Y_1 = S_1^n \cdot X_1^q$$

By repeating with different values of X, with S_1 unchanged then:

$$Y_2 = S_1^n \cdot X_2^q \tag{10}$$

From the logarithm of the ratio between the two above equations the following equation is obtained:

$$q = \frac{\log Y_1/Y_2}{\log X_1/X_2} \tag{11}$$

If experiments are repeated again with the S changed and X unchanged, then:

$$n = \frac{\log Y_1/Y_2}{\log S_1/S_2} \tag{12}$$

These parameters indicate with more precision the behaviour of the shock absorber, subject to the action of a given load, and a known starting speed.

4. CONCLUSIONS

It can be concluded that because the behaviour of the instantaneous safety gear with buffered effect can be foreseen more easily than a gear with a frction effect, the author feels it is a restriction to limit the former to elevators having speeds lower than 1 m/s. The limit that may be imposed on a gear with buffered effect depends exclusively on its loading.

But if the average value of the deceleration is not to be higher than 1 g, then an elevator with a 2.5 m/s speed will result in a rod movement of just over 300 mm. This means that for a load which includes the rod, the cylinder and all the accessories, the distance of one metre is not exceeded. And this is an acceptable condition for a lift which, when travelling at such a speed will require a pit of suitable depth. Above that speed limit there is justification in the use of progressive type gears, even though they are more expensive and more sophisticated, and have all the defects of gears with a friction effect. But, at least, they do not allow the unlimited slide of the car, and at a certain point stop the car and lock it on the guides.

5. REFERENCES

Aberkrom, P. and Gaakeer, A. C. J. (1984). Experiments with safety gears, *Proc. of Int. Lift Symposium Amsterdam, June 1984.*

24

Elevator safety

Mr D. A. Swerrie, Principal Safety Engineer, Elevator Unit, San Francisco, California, USA

ABSTRACT

More elevators move more people with more safety than any other people-movers made by man in the world today.

Some 130 years ago the words 'All Safe Gentlemen, All Safe' were spoken and the era of elevators began. For the following sixty to seventy years buildings and structures have gone higher and higher as the use of elevators made reaching the heights possible. The elevators became more and more complex but always the safety from falling has been maintained. As more elevators came into use it was found that there were other hazardous factors and exposures.

During the past sixty to seventy years the elevator industry has taken these factors one by one and eliminated them. Today we are proving that the words Elisha Otis spoke were true.

Examples of accidents are described. The importance of competently supervised maintenance is stressed. The evolution of new technologies suggests the development of new inspection techniques.

1. INTRODUCTION

The State of California has about 10% of the elevators in the United States (about 2% of the world's elevators). The Californian elevator code dates back to 1915, with enforcement since 1920. The code is applicable to existing as well as new elevators. California has elevator records, including accident reports some dating back to 1920, that are as complete as any agency in the United States or possibly the world. When investigations show a trend, that trend is used to revise the code and to eliminate that type of accident causing conditon.

There are about 25 million elevator rides taken each working day on

California's 40 000 passenger elevators, which is about 6500 million rides each year. In 1984 21 accidents were reported on those passenger elevators (one fatality and 20 injury accidents). During that year there was only once chance in over 300 million of being involved in an accident riding an elevator, and only one chance in 6500 million of being involved in a fatal accident.

World-wide there are about two million elevators providing 312 000 million rides every year, with very few accidents. No other man-made means of transportation moves the equivalent of the world's population every week. Elevators are safe! At the same time, it is recognized that there are accidents involving elevators, possibly, 1000 a year. Most important, however, is that many of the occurrences are avoidable either by better inspection or maintenance.

Pause for a moment's reflection, remember that the fear of heights, the fear of falling is the oldest and largest of all human fears. Since the beginning of time, mankind has striven for elevators and constructed baskets and boxes, hung them on ropes or vines, applied power, and lifted or lowered people or inert objects from one level to another. It started in antiquity and continues today, but until recently, always used with fear and never used by everyone.

Then at a time when the world was ready when land costs soared when density of populations were expanding when wire rope had become available when mankind was entering the industrial age and learning manufacturing techniques came the man of the hour: Elisha Graves Otis. The place was the New York Crystal Palace Exhibition and the year 1853, when Otis demonstrated his 'Safety Elevator'. Elisha dressed in long coat and top hat stood on a platform as it was hoisted aloft. The suspension rope was cut — nothing happened; the safety ensured the platform stayed in place aloft. Otis swept the top hat from his head, bowed to the audience and proclaimed, 'All Safe, Genetlemen, All Safe'. And the elevator was possible. By 1857 passenger elevators were being installed.

So from the very beginning, the word elevator was all by synonymous with the word safety. What other means of transportation was introduced by proclaiming safety. Elisha did not say 'it works' he did not say 'it goes up and down'. He said, 'All Safe'. Was this to mean that no one would ever get hurt? No!

2. HISTORY

For the first forty or fifty years as buildings went ever higher and more elevators were installed, it was a time of growth and expansion. The industry was developing and emerging. As hazards were recognized, the industry responded and overcame each hazard in turn. But then, as now, older elevators with their hazards remain in use.

Elevators are safe, but that is not to say that accidents do not happen. As this chapter was being prepared the author was notified of an accident. A college professor approached a freight elevator with bi-parting doors. He opened the doors and stepped in to turn on the elevator light as it was dark. The professor fell into the pit. The accident occurred because the elevator was not at that landing. The elevator was installed in 1968 and new interlocks were put on the doors in 1984. In 1986 that type of accident should not be happening as it is out of the past, before interlocks were developed. The reason here was a faulty interlock.

Falling cabled elevators have never been a problem. The classic fall occurred 40 years ago on a Saturday morning, when a plane crashed into what was then the world's tallest building. An elevator fell some 75 floors because the hoist ropes and the governor rope failed. This industry should not spend time trying, to prevent such occurrences.

Elevators of long ago, and before codes were written, caused problems. Some examples are given in Table 1.

Table 1 — Problems and their solutions

Problem		Solution
They went too high and went too low	ACCIDENT	Final limit switches
Elevators got stuck on the final limits	ACCIDENT	Normal limit switches
Elevators went through the limits	ACCIDENT	Bumpers
Elevators hit bumpers too hard	ACCIDENT	Buffers
Elevators bottom out the buffers	ACCIDENT	Buffer stroke criteria
Ropes fail, counterweights fall	ACCIDENT	Counterweight safeties
People get struck by descending elevator	ACCIDENT	Hoistways are enclosed
People push a hand through openings in car	ACCIDENT	Solid car enclosures
People put a hand through openings in hoistway enclosure	ACCIDENT	Close hoistway enclosures

By 1920 elevators were starting to be recognizable by today's standards. The high-rise, high-speed, smooth-running roped hydraulic, that at the turn of the century was predicted to be the elevator of the future, was on the wane. In 1925 in San Francisco a victim fell into the pit from about the 3rd floor of the building he worked in. The accident report stated 'This accident would not have occurred if the doors had been equipped with electrical interlocks' — A year or two later the code was revised to require interlocks. By the late 1920s interlocks were being required and the accident caused by elevators running with their doors open was on the way out.

By the 1930s hoistways and elevator cars were being solidly enclosed and solid car dorrs were becoming standard. The 'caught in', or 'caught between', accidents were on their way out. No longer were people to be exposed to contact with a stationary wall or hoistway door as a high-speed elevator hurtled aloft in a high-rise building at 3 to 5 m/s.

Such a victim was a young girl. The elevator was a 3 m/s car switch passenger elevator in a hotel with a trained operator, no car door, and broken arm hoistway door closers. The girl had just been named the winner of a beauty contest. She, her mother, and another contestant were going up to her room in the elevator. They were dressed formally in full billowing skirts and the young girl was quite excited and very happy. She pirouetted and her skirts billowed at the instant they were passing a floor. The handle of the broken arm closer snagged the flaring skirt and held. The celebration ended in tragedy as the young queen was drawn into that little space between the platform edge and wall and disappeared before the operator could stop the car.

By the late 1940s fright elevator hoistway gates, and gate locking and contact devices were being required, and the days of the worker opening a gate and falling into a hoistway were on the way out. Prior to 1950 there were few or no locks and people fell in regularly. The elevators of the 1950s and the 1960s were, but for some operation sophistication, the same as the elevator of today.

3. THE PENDULUM SWINGS

The pendulum is swinging into time. An example is the unenclosed elevator runway, they are called observation elevators. Now, as in the past there are accidents with people trying to climb into the hoistway or onto the car. Such a case was where a young man wanted to get into the elevator. He climbed over the rather low enclosure, and lowered himself to the pit area at the back of the car. He then went to the front and climbed up in the door return area between the car and the hoistway enclosure. At about that time someone above called the car up. As the car went up the car sill raised the young man, and the hoistway door header held him from above. The result was a fatal accident.

It is necessary to learn from what has happened. The insurance companies, the elevator companies, and the building owners must insist that the juridictional authority investigate accidents and document causes. Then they all meet and confer with the inspectors, and together work on eliminating the causes of accidents.

It is well and good to reminisce on what has happened, but more important, what is happening? The trend must be seen to practice prevention. As preventative maintenance pays good dividends, the so does safety. Thus, preventative safety programmes are aimed at avoiding the first accident. Some trends today are:

(i) Buildings are going higher and elevators are going faster.
(ii) Space is getting more expensive.
(iii) Architects resist 10–15–20 ft overhead or over-car clearance.
(iv) A buffer that measures 25 to 30 ft from its base to the tip of the plunger demands more pit than is being provided.
(v) A 10 ft plunger on a buffer needs a follower to ensure it does not buckle or bend when struck.

Thus there is more cost, more equipment, more room, more resistance.

Cabled elevators in high-rise buildings can fall up as well as down. Elevators are going into the overhead with increasing frequency, even fairly low rise elevators. Is it maintenance or is it the new controls? Nederbragt (1985) makes some valid points. Is the elevator industry ready to consider eliminating the car holding safety device and substitute an up and down overspeed braking device? Again, various authorities are considering the elimination of the car 'emergency stop switch' and emotions can become very strong. Safety, however, dictates that these topics must be faced and resolved.

The question has been asked:

'How many people are killed in elevator accidents in a year?'

The estimate for the world is

'About 25. Most fatalities happen to people who force open the doors and fall down the shaft.'

This occurrence can be prevented but how many years will elapse before all elevators in use are arranged so it cannot happen?

It is known that other accidents occur when people are entering or leaving the elevator when they trip and fall or the doors bump them. Should not these occurrences be reduced as to accept them is a defeatist attitude? If these accidents account for 70% of all accidents just a little improvement would be significant.

Consider the direct plunger elevator as some 70% of the elevators are hydraulic or oildraulic. In the United States they have no car holding safety, and thus they can and do fall at (approaching) free fall speed. California has 25 000 to 30 000 direct plunger elevators and accident records and reported incidents indicate, that about two of the hydraulic elevators, are failing and falling each year.

In one case several people got in an elevator at the street level floor. As the elevator was ascending, the bottom of the cylinder failed, and the car fell to the basement level: all the persons went to the hospital. In another case a mother and daughter stepped out of an elevator that was making strange noises and would not quite reach floor level. Almost the instant they left the car the packing gland failed and the car dropped and it struck the bottom hard enough to bend the bolster It is time to at least consider the idea of a car holding safety device for direct plunger elevators.

The records show that more and more elevators are being installed and more and more accidents are happening, but is the rate going up? Supervision of the maintenance and service mechanic appears to be diminishing, is it because of competition and the rising cost of the mechanic's labour? There is very little (maybe none) independent inspection by knowledgeable persons during the time the elevator is being installed. At least in California, there is not; it rests with the installing company, often only the elevator mechanic, to independently evaluate the installation for safety and compliance with codes. It is getting increasingly difficult to hire inspectors of the quality that is desirable. In some places in the United States the salaries paid to inspectors is falling too far short of the wages paid to mechanics. Lastly, elevator equipment is getting to be so complicated and sophisticated that company trained mechanics must be present for any inspection to be meaningful. (Never should anyone but a fully trained and competent mechanic make any changes or manipulate the controls.)

4. THE ANSWER TO SAFETY?

If elevator safety is to be improved what should be done? The answer has four parts;

   (i) The elevator code. Codes are adequate in the overall picture, but now is the time to work on code uniformity, i.e. take the best of each, come to agreement and settle on one standard.
  (ii) Elevator inspection. Will more inspectors bring more safety? In areas where there are no inspectors, yes. The number of inspectors is dependent on the number of elevators and the type of inspection programme. There will be those who recommend more inspection. In those areas where there are no inspections, the commencement of inspections is in order. The inspectors must be properly trained, properly supervised and properly motivated. The move to select persons with the desirable qualifications has started.

  Change is advocated. It is time to bring inspection procedures up to date, and to develop 1986 techniques. The reliability of equipment being manufactured today if once properly adjusted and operating should allow for much less frequent inspections. The inspections conducted must be thorough and complete (more like the original acceptance inspection).
 (iii) Elevator maintenance. Elevator safety ultimately depends on competent, professional preventative maintenance. The elevator maintenance mechanic must be properly trained, properly motivated, and properly supervised. The elevator service company must make careful selection, must insist on proper training. The mechanic must be properly supervised.

The maintenance performed between inspections must be of a type to prevent accidents, as well as prolong the life of the components and provide for trouble-free operation. The inspection forces can work with, and assist the elevator service companies, in this endeavour.

The elevator service mechanic should accomplish various items of maintenance on a regular basis and be held responsible to do such.

(iv) The elevator safety programme. The Elevator Safety Law in a jurisdiction determines the elevator safety programme that is in use. The elevator safety programme should be as up to date, current and sophisticated as are the elevators being installed in that jurisdiction. At the same time the programme must allow for ongoing supervision of the oldest and uncomplicated elevator still in use. The programme should provide for each type according to its needs. As the laws governing the use of motor vehicles and aircraft are updated to keep abreast of the changes in equipment, numbers, and innovation, so should the laws and programmes governing elevators.

A carefully planned elevator safety programme, properly administered, can work to the benefit of the elevator company, the elevator owner, the insurance company, and the jurisdictional authority. It will ensure safer elevatoring, better and more complete maintenance, and much less downtime for inspections. It will weed out the incompetent and wrongly motivated mechanic. It should be less costly than today's inspections and should not have any negative effect on the service companies' financial outlook.

This is not the time to discuss the fine points of each type of safety programme best suited for each locality, or even a given locality. This is the time and place to state that some sort of a programme is mandatory if elevator safety is to be maintained or enhanced.

5. CLOSING REMARKS

Remember that it was *this* industry, working with the insurance industry, and the building owners, and the local authorities, that gave life to the elevator safety programme that brought the elevator safety existing today. If this industry now allows the authorities to independently revise the programme the industry may find the programme less to their liking. Revisions in elevator safety programmes are going to occur. Safety Authorities, at present merely enforce the codes that the experts have indicated are in the public's best interest. It is necessary that the Safety Authorities work with the Industry in developing a better programme.

It is important to always remember that any elevator safety programme needs to be tailored to the laws in the given jurisdiction. Whereas the safety codes that are enforced may all be similar or even identical, the enabling law of a community or country is unique and any programme must conform with the enabling laws that apply.

Every accident avoided or prevented is to the industry's benefit. Prevention saves money, reduces stress, saves time and most importantly, it reduces pain, misery and saves lives.

6. REFERENCE

Nederbragt, J. A. (1985. Uncontrolled speed up or down, *Elevator World,* vol. xxxiii, No. 12.

25

Explosion protection of electronic hoist controls

Dipl.-Ing. Paul Schick, R. Stahl Switchgear, D-7118 Künzelsau, West Germany

ABSTRACT

Hoisting equipment for the chemical, petrochemical and related industries has to be partly installed in hazardous locations. Due to its nature and properties, electrical apparatus for operation and control of this type of equipment constitutes a potential source of ignition for potentially explosive gas–air or dust–air mixtures.

To avoid these risks, special regulations for construction and installation were issued within the scope of the electrochemical standards that at an international level, although containing similar basic elements, differ essentially in practice.

The chapter gives a survey of the international standards for explosion protection of electrical apparatus and the various designs and their installation. Another focal point is the special application to hoisting equipment under the aspect of various tasks and the possibilities for their solution. For this not only the protection of equipment against explosion risks, but also the safe function in accordance are instrumental in regard to safety.

1. EXPLOSION HAZARD AND PROTECTION

In industrial plants, where larger quantities of flammable materials are being produced or required for production (e.g. gasoline, solvents, paints, adhesives etc.) there is a danger of fires or explosions, if the flammable materials mix with the oxygen in the air and if this mixture is then heated up. Potential sources if ignition are, for example, open flames, pumps etc. as well as electrical apparatus and installations due to temperature effects of electrical current (sparks, temperature loss of cables, motor windings etc.).

A potential explosion hazard can be eliminated by keeping the flammable material away from th source of ignition (e.g. by circulating fresh air through or oil immersion of the ignition source). This is called 'primary explosion protection'. The 'secondary protection' affects construction of the apparatus itself.

2. CONSTRUCTIONAL REQUIREMENTS AND CLASSIFICATIONS FOR EXPLOSION PROTECTED APPARATUS IN HAZARDOUS AREAS

The construction of electrical apparatus for hazardous areas is based on the physical properties of the flammable materials and the apparatus. This has resulted in national as well as international constructional requirements for this apparatus in the last decades.

The IEC Recommendations No. 79, published by the worldwide 'International Electrochemical Commission' are only 'recommendations', serving as an example for international standard work. In the United States the 'National Electric Code' (NEC) applies, and in Western Europe the 'European Standards' (EN).

Depending on the properties of flammable materials in regard to ignition temperature, ignition spark in 'flameproof encapsulations' and ignition energy in 'intrinsically safe' circuits, the construction of the respective electrical apparatus are as given in Tables 1(a) and (b).

Table 1 — Properties of flammable materials.

(a)

Ignition temperature (°C)	Temperature class
>450	T1
>300	T2
>200	T3
>135	T4
>100	T5
> 85	T6

(b)

Max. flame gap (MESG)	Explosion group
>0.9 mm	IIA
0.5 mm–0.9 mm	IIB
<0.5 mm	IIC

246   **Standards and safety**   [Pt. 7

According to the IEC Publications 79-10, hazardous areas are divided into Zones 0–2. In Zone 0, only specially designed electrical apparatus may be installed, whereas for Zone 2 the requirements are somewhat eased. Zone 1 is the main classification for the application of explosion protected electrical apparatus (Table 2).

Table 2 — Hazard definitions.

Zone	Definition of hazard
0	Hardous explosive atmosphere is constant or long term
1	Potentially hazardous explosive atmosphere sometimes
2	Potentially hazardous explosive atmosphere only seldom and short term

3. TYPES OF PROTECTION AND INSTALLATION

Table 3 shows the types of protection for electrical apparatus in respect to the most important international standards. For modern apparatus type of protection d, i and e are mainly applied.

Table 3 — Types of protection for electrical apparatus to IEC and NEC.

Type of protection	Applicable International	Applicable Europe/Germany	Applicable US
Flameproof enclosure, d	IEC 79-1	EN 50018/ VDE 0170/0171 Part 5*	UL 698 (ANSI C33-30) UL 886 (ANSI C33-27)
Pressurized apparatus, p	IEC 79-2 IEC 79-13	EN 50016/ VDE 0170/0171 Part 3*	NFPA 496 (ANSI C106.1) ISA S 12.4
Intrinsic safety, i	IEC 79-11	EN 50020/50039 VDE 0170/1071 Part 7* and Part 10*	NFPA 493 UL 913 ANSI 4913† FM 3610
Oil immersion, o	IEC 79-6	EN 50015/ VDE 0170/0171 Part 2*	UL 698 Part II
Increased safety, e	IEC 79-7	EN 50019/ VDE 0170/0171 Part 6*	Not recognized
Powder filling, q	IEC 79-5	EN 50017/ VDE 0170/0171 Part 4*	Not recognized

* Furthermore EN 50014/VDE 0170/0171 Part 1 is to be observed.
† Requirements for the 'installation of Intrinsically Safe Instrument System in Class I Hazardous Locations' are given in ISA RP 12.6.

Internationally, the types of installation systems have developed differently. In the United States and many other countries the conduit system in 'flameproof encapsulation' is preferred. In Europe, however, cable systems are mainly used, whereby the cables are either directly entered into the flameproof chamber of the apparatus or indirectly via an 'increased safety' terminal chamber. Figure 1 illustrates.

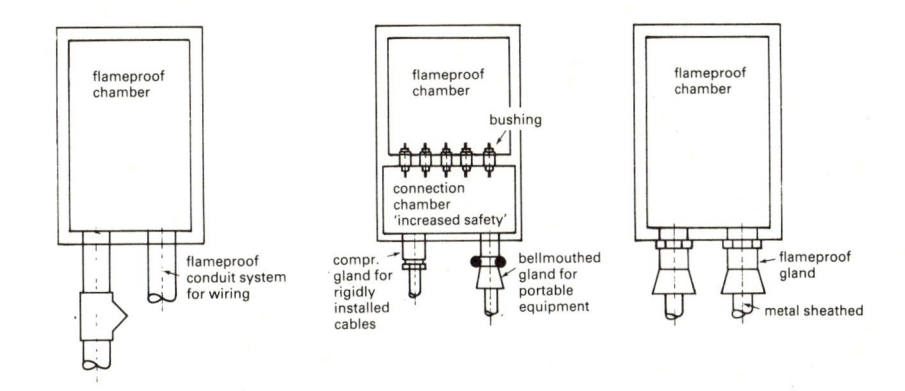

Fig. 1 — Types of flameproof chamber.

4. EXPLOSION PROTECTION OF ELECTRICAL APPARATUS FOR LIFTS

In order to keep the extra costs for an explosion protected lift installation at a low level, the lift control is usally installed in the safe area ('Primary explosion protection'), as shown thus not requiring any special protection measures, as shown in Fig. 2. The shaft and machine-room of a lift are usually considered Zone 1 and therefore require explosion protected apparatus to be installed there. This applies to the motor, control and signal equipment as well as junction and terminal boxes for the connection cables.

Up to now most lift installations in explosion hazards areas were designed in the 'classic' type of protection 'flameproof encapsulation'. Recently, however, 'intrinsic safety' type of protection as illustrated in Fig. 3 offers considerably more elegant and rational solutions. The heat to be converted to electrical energy is limited to safe values, thus enabling the use of standard industrial control and indicating equipment. This is an advantage with special parts, e.g. door locking mechanism. The electronic components in the safety circuits of the lifts are, however, subject to a special test and approval (in Germany carried out by the technical factory inspectorate (TÜV) in order to guarantee safe operation). Besides the

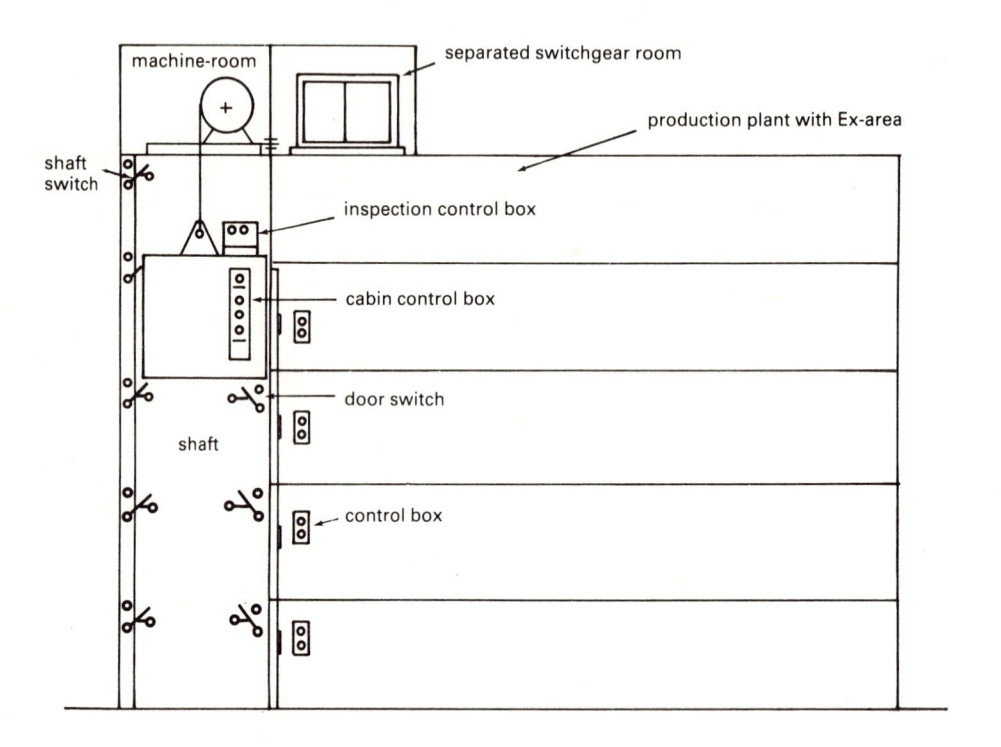

Fig. 2 — Illustration of a lift control in an explosion hazardous area.

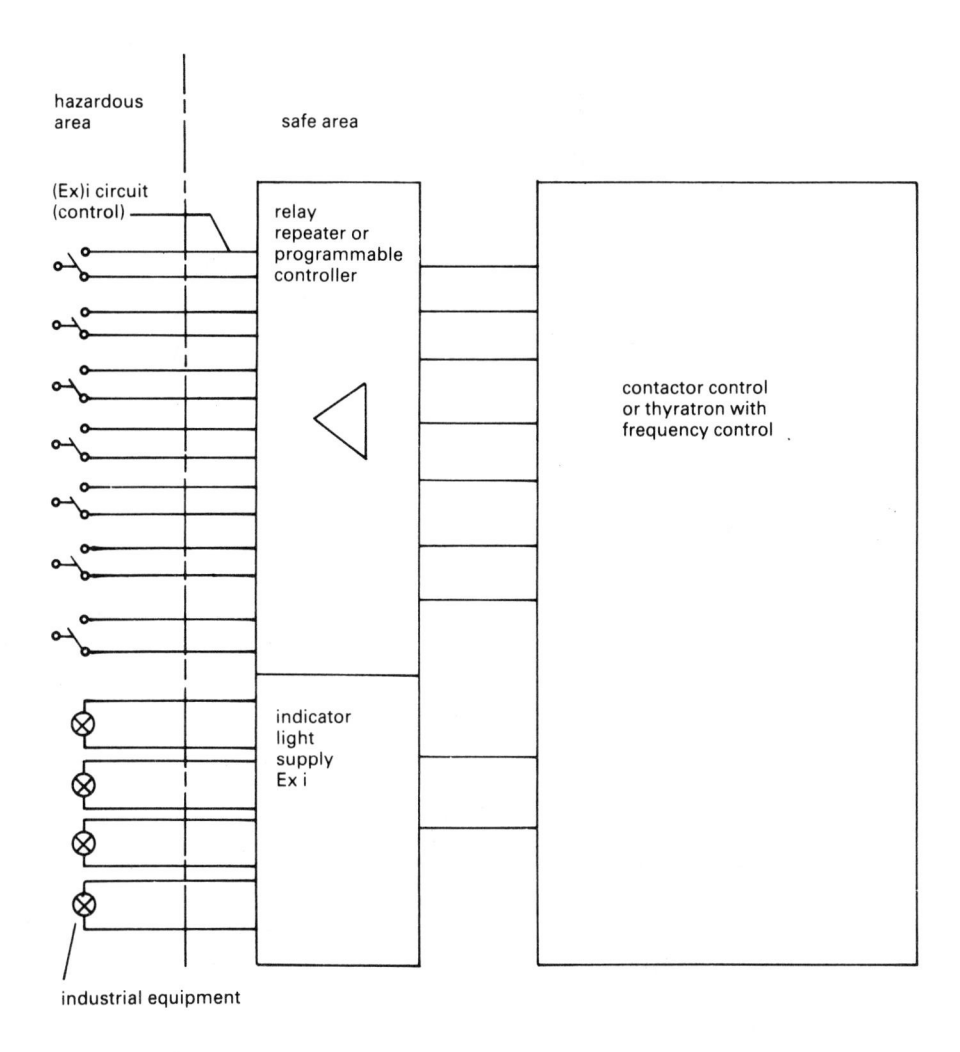

Fig. 3 — Illustration of a lift control with intrinsically safe circuits.

advantage of 'intrinsic safety' the application of electronic assemblies also offers the possibility of using modern microprocessing technology with programmable controllers. Figure 4 presents a number of examples of explosion protected lift controls.

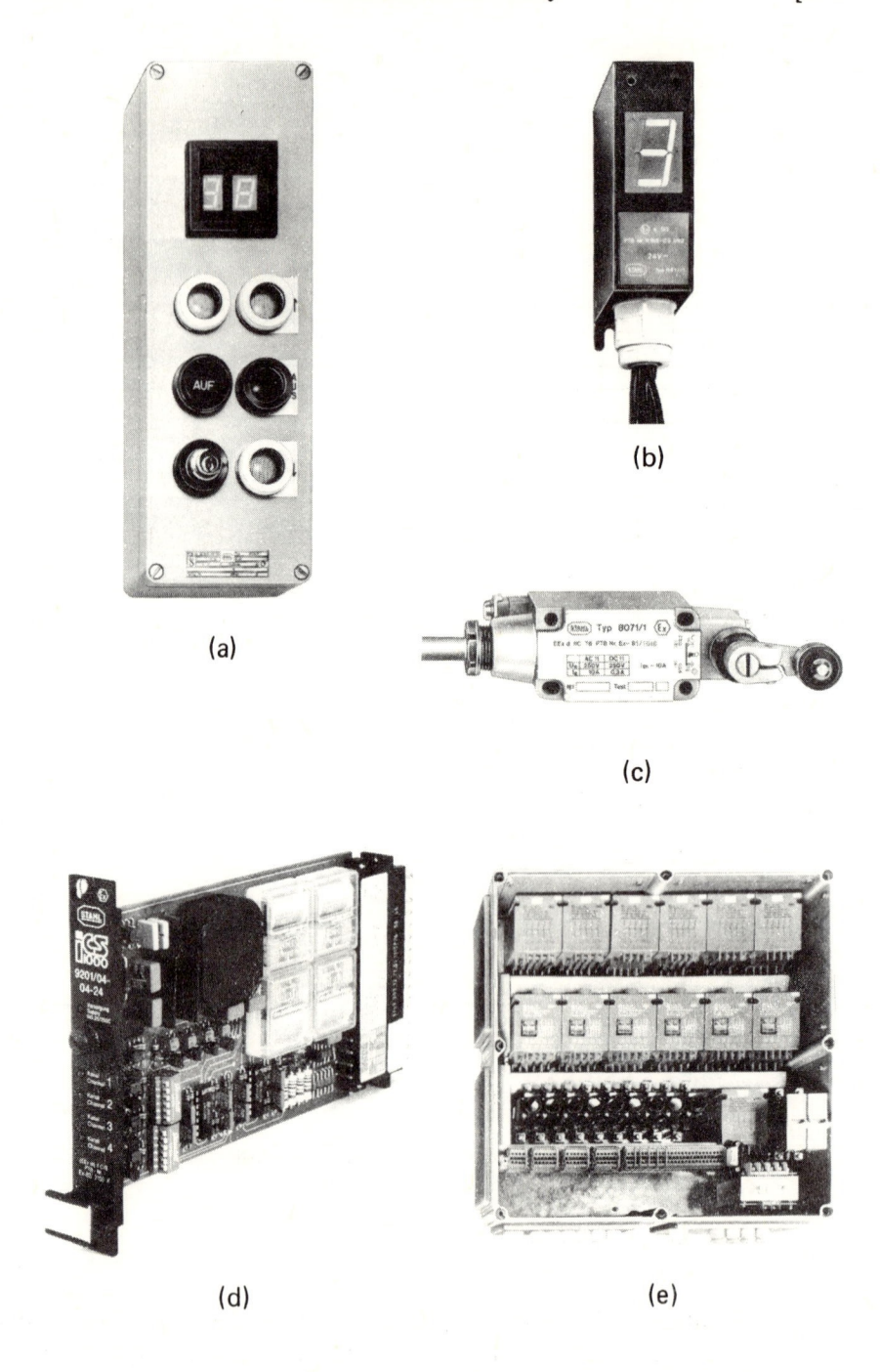

Fig. 4 — Examples of explosion protected lift contrals. (a) Control box with 7-segment floor indication. (b) Seven-segment indictor module for cage position. (c) Shaft switch. (d) Four-fold relay repeater unit with IS control circuits on Europa-card. (e) Explosion protected control panel with individually encapsulated flame-proof units installed in one 'increased safety' enclosure.

5. SUMMARY

In conjunction with the large plant extensions in the chemical and petrochemical industry the demand for explosion protected lifts has increased. The lift technology for explosion hazardous areas moves slowly in regard to the development of modern electrical equipment, since additional requirements for special construction exist, with only small quantities being required.

The explosion protected protected electrical apparatus currently on the market and built to the increasingly coordinated international standards offer new solutions for technically high class and economical designs for lift installations in hazardous areas.

6. REFERENCES

IEC: Electrical apparatus for explosive gas atmospheres — Classification of hazardous areas — Electrical installations in explosive gas atmospheres, IEC Publications No 70-0 to 79-14.

EN: Electrical apparatus for hazardous areas, *EN 50014 to EN 50020*.

Schick, P. and Leffrang (1984). International installation techniques for explosion protected switchgear and plant, *Middle East Electricity*, October 1984. London.

Schick, P. (1983). The new generation of control and indicating elements for explosion protected lifts, *List-Report No. 1/83*.

Schick, P. (1984). 'Explosion protection for electrical apparatus for use in the chemical industry, *Lift-Report 5/84*.

26

Availability, reliability and safety of lifts and their components

Dipl.-Ing. H. Streng, Consultant, Stuttgart, West Germany.

ABSTRACT

The chapter will review reliability, the probability of failure and the influence of elevator failures on safety. Particular attention is given to the analysis of accident reports. The requirements of safety codes is discussed and the measures taken to avoid damage to persons, material and property.

1. INTRODUCTION

The general understanding of the expressions *availability, reliability* and *safety* is not very precise. In daily life these expressions are used in very different ways and with different meanings. It should help to evaluate the moral, formal and legal consequences of the special considerations for electric and hydraulic lifts if definitions are proveded here.

1.1 Safety

In general, safety means that there is no danger. Thus, safety and danger are by definition complementary expressions. The word safety has a very wide use in expressions like safety paint, safety equipment, safety glass, safety belt, safety pin, safety lock, safety match, etc.

What is considered as safe or dangerous is never absolute. To have 100% safety is impossible as is 100% danger (100% safety would mean to be safe against all possible risks). Risks may come from a specific technical device, the environment, personal behaviour or a combination. The smaller the risk the higher the safety and the higher the risk the higher the danger. What is considered as a limiting risk depends on our personal evaluation or on a convention such as safety codes.

In mathematical terms a risk may be quantified, but practically a risk may only be qualified or evaluated. It depends on the extent of probable danger and the frequency of such a possible risk. If the risk is smaller than the limiting risk then the situation is considered safe, and if the risk is higher than the limiting risk then the situation is considered dangerous.

In daily life the personal evaluation of this limiting risk may differ. It depends on personal experience, knowledge and constitution (e.g. children — adults — disabled persons). It also depends on what is feared, this is a very strong psychological component. It is possible to feel safe, if a danger is unknown or, if a person is familiar with a risk and has not had a negative experience.

1.2 Reliability and availability

Reliability indicates within a defined time the probability of full functionality. The criteria or the conditions for the failures considered must be agreed and an evaluation made of the probability.,

Availability indicates the probability that the system or the component provides full functionality. Repair and redundant design increases the availability.

2. RELIABILITY AND PROBABILITY OF FAILURES

Each characteristic of a system or component may have a defect (or fault), which leads to a deviation, with a consequential malfunction (temporary) or a change (permanent). Malfunction and change both lead to a failure.

Reliability depends on the quality of a component and is given by the probability of failure over a given time. Failures occur for different reasons. Catastrophic or sudden failures do not follow any probability distribution. Very often it is an exceptional cumulation of conditions, where each condition alone would not have caused a failure. A failure can happen occasionally, e.g. a bursting car tyre. Degradation failures follow a statistical law and are predictable, e.g. the normal wear of the tread of a tyre.

3. PROBABILITY OF FAILURES

Catastrophic and sudden failures are not predictable. For any component or system its lifetime is the time from start to failure (repair and maintenance may prolong the lifetime). The probability of a failure is the probability for the component to fail and depends on the assumed distribution of failures which can be expressed by the basic equation for the probability of survival (reliability)

$$R(t) = 1 - F(t). \tag{1}$$

The so-called 'bath-tub curve' (Fig. 1) shows three typical phases. (1) The initial or early failure. (2) The random failure which is dependent on time. (3) The wear-out failure where the failure rate grows with the time and may be reduced by preventive maintenance.

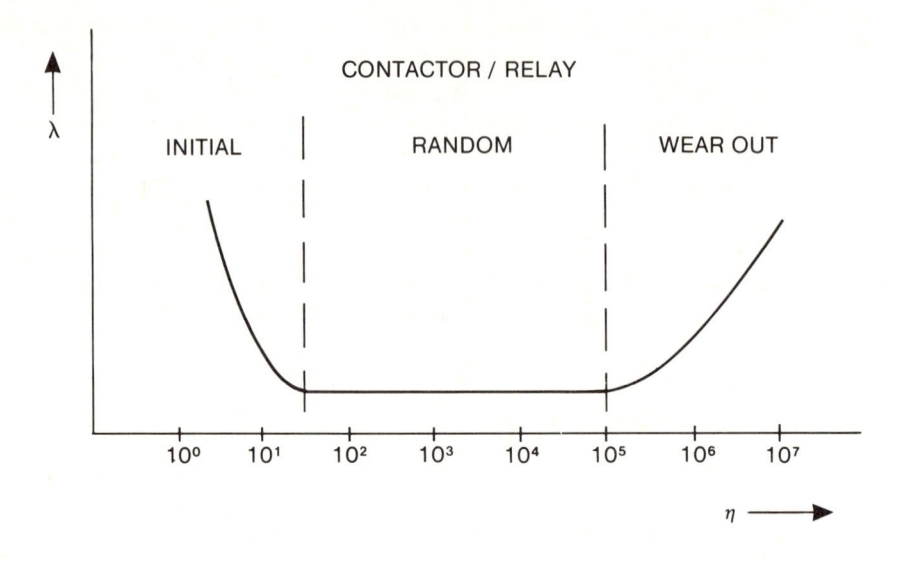

Fig. 1 — Curve of failure rate (λ) as a function of number of operating cycles (n).

3. THE INFLUENCE OF FAILURES ON SAFETY

Every defect and failure does not have an influence on safety. A defect in the paintwork or a scratch in the paintwork are for instance only an aesthetic failures with no influence on the safety. If a failure prevents the lift from operating there is also no influence on safety. But for an aircraft such a condition would caus a dangerous situation.

So it is necessary to distinguish between failures, which have an influence on safety and those which do not. With those which influence safety it is necessary to evaluate the form and estent of the unsafe influence. This weighing of the failures is very important.

A dangerous situation may cause injury and damage to persons, material or property. A system is considered safe if the risk of danger, which may cause damage, is lower than the limiting risk. In technical systems very often the consequences of a failure are beyond the influence of the user. The user may even not realize a potentially dangerous situation exists. Many technical installations and especially lifts have therefore special safety devices, in order to prevent dangers operating conditions in the event of a defect.

4. SAFETY AND AVAILABILITY

The availability of an installation or a system may be reduced if a safety device operates. This is the case if a blocked operating condition is also the safe condition. In such a situation special safety devices detect failures, which could lead to dangerous operating conditions, and keep the whole system in a safe operating condition. If a defect does not cause a total failure (black-out) of the system it is only a partial failure. Very often partial failures are acceptable in order to increase the availability even in the case of lowered safety.

If availability is directly linked to safety (e.g. aircraft) the operating system must have redundancy so that in the case of a failure in one system another system may maintain the safe operating condition. (Redundancy means a deliberate duplication or partial duplication of a device, circuit or information, in order to decrease the probability of a system or communication failure. Sometimes multiple redundancy is used to increase safety.)

5. PRACTICAL BENEFITS FROM THE ANALYSIS OF ACCIDENT REPORTS

Practically an installation like a lift cannot be considered entirely in a theoretical way. It is not possible to consider all potential influences which may cause a failure with the possibility of damage. Thus the limiting risk cannot be determined without solid experience. Any discussion about safety has no base if the accidents and damages that have occurred are not properly reported and studied. Only the quantity and extent of accidents and damage can give guidelines for improving the design, erection, use and maintenance of lift, Reports should show in detail which kind of lift on which the accident or damage occurred. It should also be related to the total number of similar installations.

West Germany damage and accident reports are made by the VdTÜV, and the detailed causes are listed according to special codes. Figure 2 shows as an example the types of accident and Fig. 3 shows the distribution of the main causes.

The reports separately consider the different types of lift and the reasons for the accidents are also detailed with their consequences. Study of these reports has for many years been used by the German code committee as guidelines to the updating of the German lift code and testing procedures.

6. MEASURES TO AVOID INJURY TO PERSONS, MATERIAL AND PROPERTY

With the knowledge from the accident reports special measures may be taken to avoid further accidents and damage. These measures cover the design, erection, use and maintenance of a lift.

Besides considering the injury of the damage itself, it is necessary to apportion the blame in the criminal and civil law.

Accidents are very low considered with the number of lifts in service and the number of journeys made. Nevertheless, any injury to a person should be avoided. Here it is necessary to face the manufacturers product liability even in cases of careless use. This situation is already known in the United States and is now also of concern in Europe.

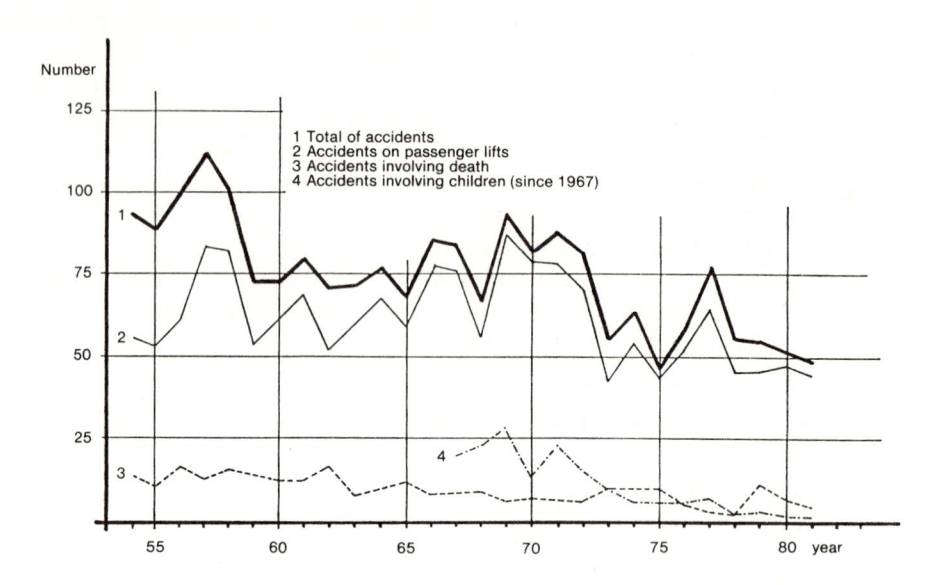

Fig. 2 — Development of accidents.

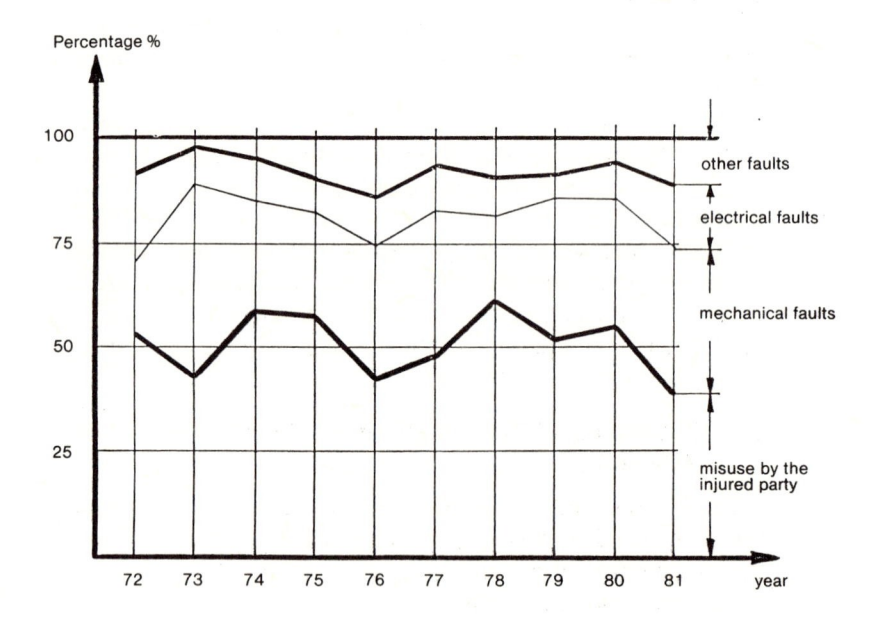

Fig. 3 — Distribution of accident causes.

Ch. 26]   **Reliability and safety of lifts and their components**   257

Each manufacturer has to consider very conscientiously accident reports concerning his products and to follow up all relevant information. The corresponding conclusions are important for the operation and maintenance of his lifts.

In each case the following questions have to be answered:

(i) How did this happen?
(ii) How could this happen?
(iii) How did this happen?
(iv) Is it exceptional?
(v) How will it be avoided in future?
(measures for the future or for immediate action)

The lift safety codes (e.g. CEN standard EN81) give guidelines for the safety philosophy.

7. REQUIREMENTS IN SAFETY CODES

There is a failure convention in each lift safety code in order to prevent accidents and injury, i.e.:

(i) List of dangerous operating conditions.
(ii) List of considered faults.
(iii) List of excluded faults.

This convention defines the limiting risk and expresses the expert view of specialists who elaborated or updated the code.

CEN standard EN 81 deals with safety rules for the construction and installation of lifts. Tabulated below is the possible accident types considered in this standard and the persons and objects to be safeguarded.

Possible accident types (clause 0.5.1):

Shearing
Crushing
Falling
Impact
Trapping
Fire
Electric shock
Damage to material
Due to wear
Due to corrosion

Persons to be safeguarded (clause 0.5.2):

Users.
Maintenance and inspection personnel.
Persons outside the lift well, the machine room or pulley room (if any).

Objects to be safeguarded (clause 0.5.3):

>   Loads in car
>   Components of the lift installation
>   The building in which the lift or service lift is installed.

It is also state (clause 0.7) that this standard has been drawn up, to take into account any careless act of a user.

The standard EN 81 describes also type-tests for:

   (i) Door locking devices
   (ii) Landing doors
   (iii) Safety gear
   (iv) Overspeed governors
   (v) Energy accumulation buffers

The directive of the European Community also refers to these type-tests except for landing doors.

The standard EN 81 represents the opinion of an international group (working group) of lift experts. All their experience and knowledge has been incorporated into this standard although sometimes the different opinions have resulted in compromise. Nevertheless, the standard cannot cover all possible solutions. Therefore, each manufacturer or service specialist has to be conscious of the fact that he is responsible for his products and work, even in the case of an official supervising organization.

8. INSTALLATION DESIGN AND COMPONENT SELECTION

The design of an installation should take account of the user requirements, i.e. an installation in an office building will be different from an installation in a petroleum plant. Not only may the installation be different but also the components may be different. The selection of the components is always a compromise between the technical requirements and the cost.

Few lift companies manufacture all the components used.

So today when components may be brought together on site it is important that the selection and composition of the components are properly considered from both theoretical and practical viewpoints.

It is very important that all the components are suitable for the installation. The characteristics and the range of application must be considered, especially if the components come from different sources. Special care should be given to type-tested material where the use must be in line with type-test certification.

9. LIFT COMPONENTS AND SAFETY DEVICES

The list of the defined faults in the electric control of lifts in the CEN standard EN 81 is shown below. The objective is that one single fault shall not cause a dangerous malfunction of the lift.

Defined electrical faults (clause 14.1.1.1)
    Absence of voltage
    Voltage drop
    Loss of continuity of a conductor
    Insulation fault in relation to the metalwork or the earth
    Short circuit or open circuit in an electrical component such as resistor, capacitor, transistor, lamp
    Non-attraction or incomplete attraction of the moving armature of a contactor or relay
    Non-separation of the moving armature of a contactor or relay
    Non-opening of a contact
    Non-closing of a contact

For electrical faults (including the safety circuits) it was assumed that if one fault combined with a second fault could lead to a dangerous situation, then the lift shall be stopped, at the latest, at the next operating sequence in which the first faulty element would be used. Further operation of the lift should then be impossible until the faults are cleared.

The possibility is not considered of the second fault occurring after the first, and before the lift has been stopped. The electrical safety circuits and safety devices like safety gear, interlocks etc., are considered in separate chapters in nearly all standards and safety codes.

Safety devices prevent dangerous operating conditions occurring, if there is a failure in a lift component. There is, however, a possibility that a defect may occur in a safety device. Component and safety device failures have different effects as a failure in a safety device only creates a dangerous operating condition if there is also a failure in the component itself.

Alternatively a failure in a component can only lead to a dangerous operating condition if the corresponding safety device does not work.

10. OTHER FACTORS

Even if the design and the manufacture of all components is carried out in the best possible way they are only part of the quality of a lift, although a most important part. Quality, safety and availability are very much influenced by the quality of erection. At acceptance the performance of the equipment and the standrad of erection are checked as well as the proper functioning of the safety devices, in order to make sure that the handed-over installation is according to the requirements.

In order to preserve the original quality it is very important for the lift maintenance specialists, to know the equipment and the problems of an installation in detail, in order to safeguard the equipment and limit the risk of accidents and damages. Periodic checks of the function of some components and the safety devices are good maintenance and prolong the lifetime of the installation.

In some countries periodic checks are required by the authorities. Such checks look only for certain specific items in respect of safety and are not concerned with availability. However, they can never replace a good maintenance.

11. CONCLUSION

The risk of an accident and injury is very low, if all work is properly executed. In the codes of today some failures are not considered as they do not justify additional precautions. By correct material selection, functional checks and by correct maintenance such failures are reduced to low levels so that no special safety devices are required.

The code safety requirements are basic but considered by experts as necessary. In a country where a code or standard is compulsory, it has to be followed in detail. The standard EN 81, part 1 has been taken as the technical definition in a directive of the European Community. After the transition time, which ends in September 1986, no country of the European Community can refuse material which conforms with EN 81, part 1 and its clauses.

In the future lifts will have more electronic components. A combination between the electronic control and electronic safety devices will emerge for those installations, where integrated microprocessor control gives economic advantages, or where specific environments may justify the higher cost.

The major part of a lift will remain mechanical as the car has to carry the load, together with the suspension, the drive and whole hoistway structure. The hoistway encosures including the doors have to prevent unauthorized access. Therefore, the main safety aspects will not be altered. The means of achieving the intended safety levels may change in line with technical possibilities.

12. REFERENCES

DIN 31004 T. 1. Grundbegriffe der Sicherheit. Beuth-Verlag, Berlin 30.
EN 81. Safety Rules for the Construction and Installation of Lifts and Service Lifts — Part 1: Electric Lifts. National Standards Organizations (UK BSI, France AFNOR, Germany DIN).
Hoseman, G. (1981). Über die Einschätzung in der Technik von Risiken. *Technische Überwachung* 22, 9, S.353.
Kuhlmann, A. (1981). *Einführung in die Sicherheitswissenschaft,* Verlag TÜV Rheinland GmbH, Köln.
Putz, E. and Suttrop, F. (1981). Sicherheit elektrischer Anlagen. *Technische Überwachung,* 22, 3, S.124.
Rosin, O. (1983a). Allgemeine Sicherheitsbetrachtungen über Steurungen. *Technische Überwachung,* 24, 10, S.407.
Rosin, O. (1983b). Aufzugssteuerungen im Rahmen allgemeiner Sicherheitsbetrachtungen. *Technische Überwachung,* 24, 11, S.455.
Schaefer, E. (1979). *Zuverlässigkeit, Verfügbarkeit und Sicherheit in der*

Elektronic (Reliability, availability and safety in electrons). Vogel-Verlag, Würzburg.

Technische Regelm für Aufzüge. *TRA200 Lastenaufzüge Personenaufzüge; TRA101 Prüfung von Bauteilen.* Carl Heymanns Verlag KG, Köln.

Tobergte, E. (1980). Software-Zuverlässigkeit, ein integraler Bestandteil technischer Sicherheit. *Technische Überwachung,* 21, 7/8, S.335.

Part 8
Innovation

27

Latest developments in lift technology

Mr M. Kaakinen, KONE Lift Group, Helsinki, Finland

ABSTRACT

The microcomputer can be considered to be the most significant technological innovation into the lift industry. After a review of lift operations, this chapter examines traffic control systems, monitoring and command systems, drive systems and automatic doors. Each topic is examined with respect to the introduction of modern solid state technologies.

1. INTRODUCTION

Land is scarce and expensive in big cities, and hence buildings must be tall. From the building owner's point of view, the total purchasing cost of the lift installation is the sum of the cost of lift equipment plus the cost of space occupied by them. To survive in today's keen competition, the lift manufacturers have to supply lifts that provide the maximum performance inside the given building core space. In addition to lift performance expressed in terms of waiting time and handling capacity, the quality aspects (reliability, short down-time, ride comfort and low noise level) are playing a steadily increasing role as competition tools in the mid- and high-rise lift market segments. Also in the low-rise lift market similar requirements are emerging. In spite of the present falling trend in the cost of energy, the demand on lift systems designed to conserve energy will be growing.

Throughout the electronic era, which began with the invention of the transistor more than 30 years ago, the lift industry, with the exception of a few manufacturers, has been slow in adapting to the new technologies. The entire industrial world experienced a technological revolution when the microprocessor chip was invented in 1968 by Intel, the American electronics manufacturer. Today all the lift manufacturers seem to share the opinion that microcomputer technology is the technology that has to be adapted to all lift systems involving some sort of data processing or motion control. It is felt that when applied correctly, the new technology will not only meet future product challenges, but also equip the new products with advanced features and options that could not be thought of before the microcomputer became available.

What are the reasons behind the slow adaptation of the modern electronics in lift production? First, the general reluctance to increase the variety of spare parts and the training of installation and maintenance people, and second, the lack of engineers familiar with the new technology. The lift and its operation are so complex that the transfer to a new technology over the whole range requires vast hardware and software development investment. Also large errors have been made in the implementation of microcomputer technology: this concerns the software, the hardware and maintainability aspects. Lift makers with a long experience in the development, manufacturing and maintenance of a complete range of lifts, backed up by in-house computer design and manufacturing know-how, have been in a fortunate position in avoiding the pitfalls. Companies having to rely on outside sources have generally been less well placed.

The subject of this chapter is to outline the ways that development is going and to describe new features and benefits that the new generation lift systems offer to the building owner and the lift passengers as well as to the lift manufacturer and maintenance company.

2. THE LIFT ROUND TRIP TIME

2.1 The Elements of the lift round trip time

The lift round-trip time can be divided into three main elements, the flight time, the passenger transfer time and the door operation time. Figure 1 shows these elements as functions of increasing building height and lift speed.

According to the graphs, the flight time is about 50%, and the door operation time about 30% of the round-trip time. As they together account for 80% of the time, it is quite evident that it is worth studying how the modern microelectronics can improve performance.

What the graph does not disclose is the fact that the lift group control system is equally important in ensuring the maximum performance out of the installed lift capacity. In addition to improving handling capacity during intensive traffic peaks, a modern control system greatly improves the quality of passenger service by reducing the average waiting time as well as the probability and magnitude of long waiting times.

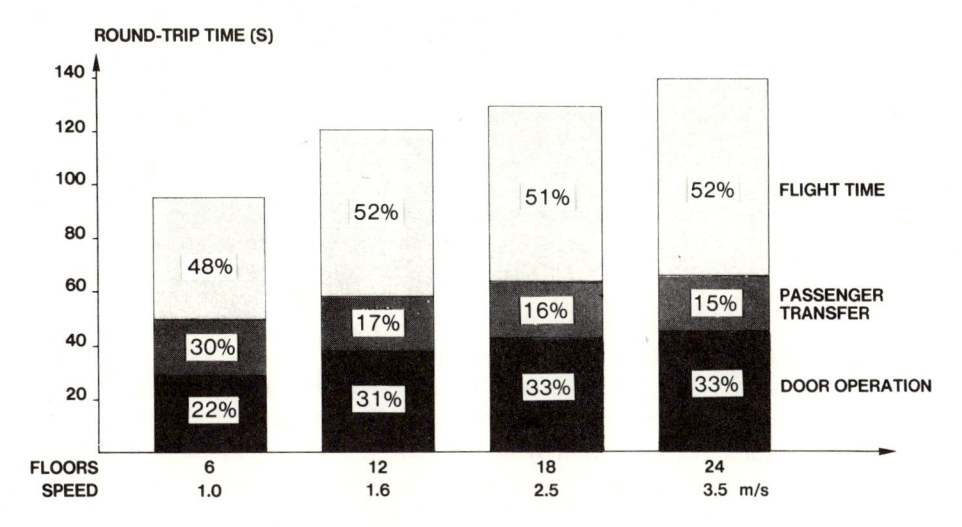

Ch. 27]   **Latest developments in lift technology**   267

Fig. 1 — Round trip time and its main elements.

2.2 Lift Cycle Time

The efficiency of the lift drive system and doors can be judged with the aid of the door to door cycle time, as defined in Fig. 2. The cycle time depends on:

(i) *the drive system*, and particularly the rates of acceleration, deceleration and jerk, start delay, levelling phase and use of intermediate speeds;

(ii) *the doors*, and in particular their opening and closing times, time losses in opening and closing the car door coupler, advance door opening and the inteligence of operation.

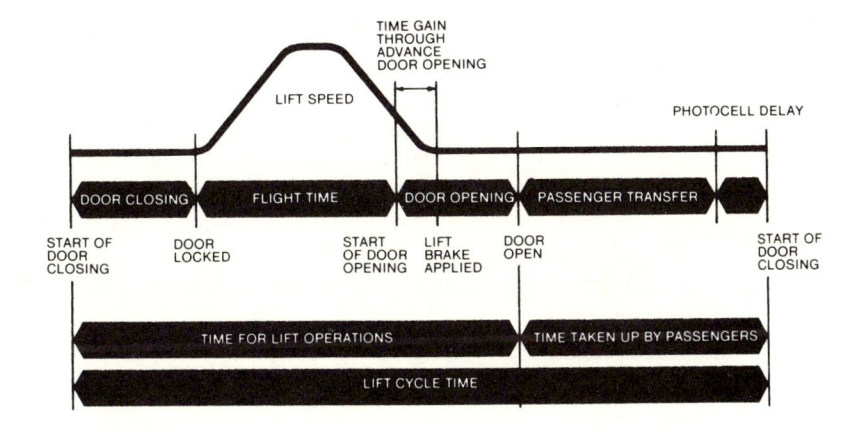

Fig. 2 — Lift cycle time.

The cycle time can be simply measured on site by a stopwatch. The result reveals mercilessly lift equipment efficiency or inefficiency. A cycle time less than 8 s proves that the installation is working efficiently, every second above 10 shows that capacity is being wasted. For instance a cycle time of 12 s (subject to lift speed being at least 1.5 m/s) implies that there is at least 4 s available for improvement. A good rule of thumb is that a second saved at each stop improves the capacity by approximately 5%.

3. CONTROL SYSTEMS

3.1 System Principles

In the past, the lift control systems were built of telephone relays and some companies even made their own relays. The relay technology was never completely superseded by the hard-wired electronic systems, which emerged in the early 1970s (e.g. KONE's Modulec ICS, Schindler's Aconic and Haughton's 1092 IC in the USA). A transition period is presently occurring from the earlier technological generations to microcomputer solutions and the status of implementation varies from company to company. The larger manufacturers have started from controllers for medium-sized or large lift banks, but the independent component manufacturers are concentrated on simpler systems.

There are two approaches to developing a microcomputer control system. One is to take an existing relay or hard-wired electronic logic and to translate its logic into computer language from switch to switch and circuit to circuit. The other way is to develop new operating principles and control algorithms that are especially suited to computers, and which exploit their computing power. This approach calls for a thorough knowledge of the lift traffic demand, deficiencies of old systems and how their efficiency and service quality can be improved.

There seems to be different approaches, with regard to the control system structure:

(i) The most popular choice for large lift groups has been to provide a separate microcomputer for the lift group control, and to have separate computers for the control of individual lifts. The reliability of group control operation can be improved by having a back-up computer for this purpose.
(ii) To use one single computer to control several lifts (in lower quality lift installations).
(iii) To decentralize the lift group control to the individual lift computers, thus omitting the group control computer (suitable for medium-sized lift groups).

The inter-computer communication has also been solved in different ways:

(i) A common data bus, to which all the computers are connected. For improved reliability, two parallel data buses can be used.

(ii) Separate data links from each computer to the others. The signal transmission can be secured by echoing the received signals to other computers through the other data links.

The system structure is vital from the reliability point of view. For instance, if one computer is controlling the registration of landing calls and their allocation, its breakdown will severely degrade lift group operation. The use of a back-up computer and decentralized control are superior in this respect.

Some customers have expressed concern about the guaranteed availability of software particularly in cases of local or international conflicts. Only international companies, can fulfil this requirement where software development and storage is taken care of by several factories around the world.

3.2 Operating Principles

The programmability of microcomputers has made it possible to build in control features which are the prerequisites for efficient control algorithm:

(i) Call time measuring.
(ii) Automatic detection of traffic pattern and modified operation to favour dominant traffic flow.
(iii) Capability to cut long waits by deviating lift routes from the collective principle.
(iv) Forecasting of service times to that by-passed calls will not become the worst cases.
(v) Considering the car load in order to balance number of stops between cars and to reduce passenger destination times.
(vi) Efficient dispatching of vacant lifts to determined floors to ensure short average waiting time.

There are other concepts that have been used, or proposed, to improve the quality of lift service:

(i) *Advance signalling*. The computer is used to determine which floors each car can serve before becoming full. Once the decision is made, the waiting passengers on the landings are told which car is coming. While this information will shorten the 'psychological' wating time, the decision may prove wrong, especially if unexpected delays occur or cars acquire full loads earlier than anticipated. The waiting passengers also tend to book passengers trying to exit from the car. During traffic peaks, the flexible, last minute decision principle gives better results.

(ii) *Lobby detectors*. These can be used to detect the number of passengers on the landing in order to enable the control system to act on more accurate information than just the call waiting time. They detect floors with heavy demand, so that sufficient capacity can be allocated to such floors. Available space in cars can be matched with the number of waiting passengers on various floors. However, the problem with these devices is their inability to determine how many persons want to go up and down.

(iii) *Advance registration of car calls*. On arriving in front of the lift group passengers register their destination floors with the aid of call buttons. The computer obtains more preicse information, and in principle it can make a better allocation decision. It can invite passengers with the same destination to board the same car, and it can consider coincidence calls in advance. This principle works well in computer simulations, but not so well in practice as people do not act systematically.

Compared with systems with less adequate operating principles, good control systems provide shorter waits at similar traffic flows or allow increases in the transportation capacity with similar waits.

4. EXAMPLES OF MICROMPUTER LIFT CONTROLLERS

To illustrate Section 3 a number of examples from the KONE 'Traffic Master' range is given below.

4.1 Low rise system

The TMS 200 system brings computer intelligence to single car to triplex installations in low and medium rise buildings with speeds up to 2.5 m/s. As the system is competing against low-cost relay controllers, cost-effective system solution was developed. Each lift controller houses a single PC-board computer with 8085 processor. Instead of the conventional serial data link, a new 'common memory area' principle was developed for high-speed intercomputer data transfer.

The PL/M programming language and a multi-tasking operating system are used. The engineers use the computerized Lift software production and Management System (LMS) with artificial intelligence to assemble the required software packages. The control algorithm includes features such as call time monitoring, a predicting procedure for call service optimization and many optional extras. The group and lift operation statistics and fault log modules are standard. One of the optional features is access to the collected statistics with a printer or via a telecommunication line.

4.2 Medium rise system

The TMS 516 controls lift groups with up to six cars and speeds up to 4 m/s and offers certain advantages not possible with the older techniques, such as lift operation advantages not possible with the older techniques, such as lift operation and waiting time statistics and maintenance diagnostics. Updated with the powerful 8088 processor, the system offers enhanced features and capabilities, including a Lift Monitoring and Command (LMC) system with colour CRT screen.

Each TMS 516 lift controller houses a microcomputer, which is connected to each other with serial data links (Fig. 3). The most important signals received by each computer are echoed to the other computers, thus one faulty communication line does not cause degradation of the group control operation. Each computer memory contains program modules written in the high-level PL/M language for lift control, group control, statistics and fault finding, running under the supervision of a multi-task operating system.

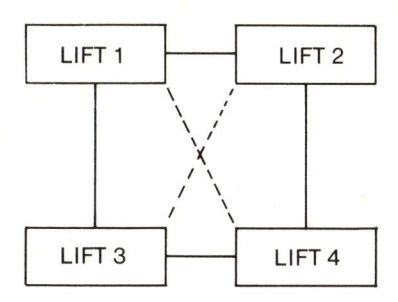

Fig. 3 — Controller interconnections.

The landing calls are wired to each controller. Therefore the complete loss of landing call information is not possible. The omission of the group controller also makes the group control function practically invulnerable — a faulty lift is simply disconnected from the bank, and any computer can act as the master.

The computer contains a unique external programming board, with the aid of which the lift installation parameters (main lift bank data, lift operation and option activation) can be tailored to suit the particular building without the need of program changes. Complete reprogramming is, of course, also possible. The parameter module simplifies spare part requirements and reduces the probability of engineering errors.

The group control software is designed to combine short average waiting times with a high handling capacity, and to eliminate long waits. To do this, it collects statistics of lift operation and landing calls for statistical evaluation of the traffic pattern, and uses computed call allocation criteria to operate the lifts accordingly. It monitors the call waiting times, and classifies them into ordinary, priority and timed-out calls with variable boards. The system allocates the necessary capacity for each call group, avoiding idle runs and maximizing the handling capacity. During traffic peaks only the timed-out calls with waiting times clearly in excess of the other calls in this group are served individually, to avoid waste of capacity.

The system performance can be verified with the waiting time statistics, which show the number of up and down calls on each floor, and their average and maximum values. To ease maintenance, the statistics module collects lift events, including starts and stops and a number of fault situations. It is also possible to write special fault finding programs into the computer memory using an easy-to-learn language.

4.3 High rise system

The TMS 900 system, intended for large high-rise lift installations (up to eight cars and speeds up to 8 m/s), also employs the powerful 8088 processor, but to cope with the increasing data processing demand, one computer is dedicated to group control. For the sake of reliability, an identical back-up group control computer is provided. The group control computers are located inside the lift controllers housing the individual lift computers. In high-rise installations, each lift car has a computer, communicating with the lift computer. The system includes a new concept, a separate Data Concentrator computer, which is linked to the group computers. The data buses between the computers are doubled for high reliability.

The Data Concentrator serves as a communications interface between the lift group and the building control centre and eases lift maintenance. It collects comprehensive group and lift operation statistics, which can be accessed and viewed by colour display terminals located in the Data Concentrator itself, at the building control centre and at some additional locations (Fig. 4).

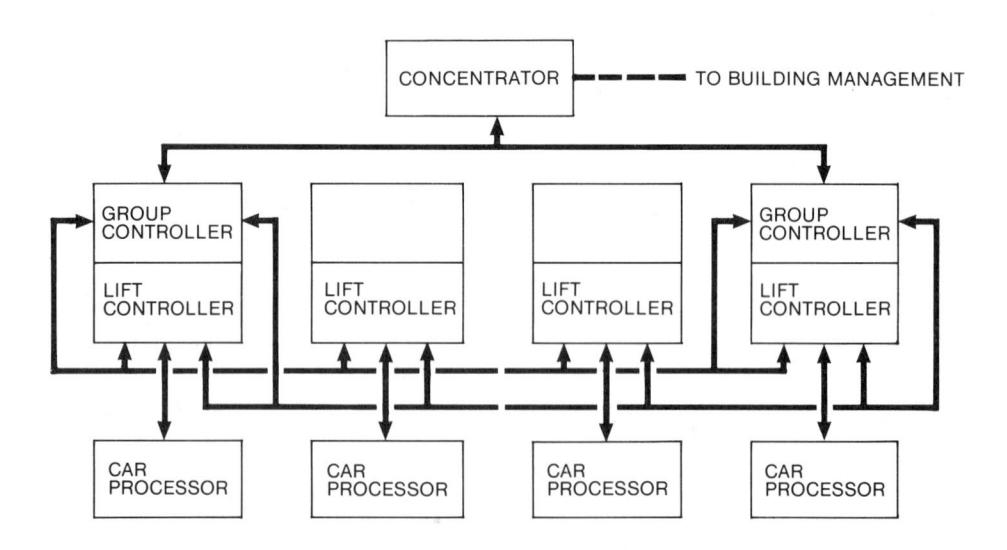

Fig. 4 — Controller communications.

The software includes all the features built into the TMS 516 and an external programming board is also provided. The group control software uses an optimizing algorithm, which selects the best lift routing from the feasible alternatives, and before deciding to by-pass any calls, calculates a prognosis for their probable service time.

A special version of the TMS 900 designated 'MO' has been developed to update existing relay controls with minimal disruption of building operation. The key to the solution are the adapter computers between the individual lift controllers and the group computer, translating relay signals to computer language and back. In essence the 'Modernization Overlay' means furnishing the old relay system with a computer 'brain', laid 'over' the relay equipment, and taking over the registration of landing calls and call allocation.

Incorporating all the TMS 900 control features, the MO system has proven to increase the handling capacity considerably. The average waiting times show a drop of 25 to 40%, and the longest waits are reduced usually by 50% or even more. All this can be achieved with an out of service a few hours or, at most, a few days per lift.

As the second phase, the lift components needing updating (controller, drive system, car, doors, signalling etc.) are replaced lift by lift. Owing to the overall performance improvement brought about by the group computer, the modernization can usually be carried out almost unnoticed.

4.4 Lift Monitoring and Command System

In microcomputer technology, the traditional lift group display panels with tell-tale lights are substituted by video displays. Equipped with a keyboard, the display can provide access to control and supervise the lift group operation. KONE have developed a special 'Lift Monitoring and CommandSystem, available in connection with all the TMS controls. This system forms an advanced interface between the lift system, building automation and lift maintenance. As the data collection, processing, video control and system communications are controlled by separate, dedicated computers, the system can manage up to 10 terminals.

The extensive menu structure covers a comprehensive traffic display and the group command, statistics, fault, test, parameter set-up and security modes, in colour. Options include remote transfer of traffic data and a tele-surveillance interface.

Recent experiences has shown that customers appreciate the usefulness of the enhanced Lift Monitoring and Command system, and specify it in their inquiries.

5. DRIVE SYSTEMS

Lifts have been equipped with one of the following drive systems, depending on the speed, desired quality of ride; and period of history

Before 1960
- hydraulic, up to 1 m/s
- single-speed a.c. motor, up to 0.63 m/s,
- two-speed a.c. motopr, up to 1.25 m/s,
- geared and gearless variable voltage drive with rotating motor-generator set above 2 m/s.

From 1960
- controlled a.c. drives up to m/s,
- improved variable voltage drives (e.g. electronically computed speed reference and thyristor control of generator excitation current).

Current
- electronically controlled hydraulic value,
- d.c. drives with solid-state thyristor converters, and
- variable voltage, variable frequency a.c. drives (inverters).

5.1 Controlled a.c. drives/ stator voltage control or eddy current braking
For controlled a.c. drives, two main solutions are used:

(i) two-speed motor with additional eddy-current brake. The disadvantages of this older solution are expensive construction and high inertia.
(ii) single or two-speed motor lift. The braking effect generated either by injecting d.c. to the motor windings, or by reversing the stator field. The disadvantages of the field reversing method are its high energy consumption and motor heating, which become problematic especially with higher lift speeds and at low maintenance speed.

The best controlled a.c. drives acelerate to the maximum achievable speed also within short distances, and feature fast, landing approach. There are, however, still a number of systems on the market which lack one or both of these quality features.

5.2 d.c. drive with Static Converter
The replacement of the motor generator used in the traditional variable voltage or 'Ward-Leonard' drive was a natural step after the implementation of thyristors in the control of the generator excitation current. This step has not been easy as evidenced by the fact that KONE is the only manufacturer who has been offering the static converter drive as the standard solution in connection with the microcomputer controls.

5.3 Controlled a.c. Drives with Voltage and Frequency Control
With frequency control, the rectified (or battery) d.c. voltage is inverted by pulsewidth modulation to variable a.c. voltage with variable frequency. Thyristors, GTO thyristors or power transistors are used as the power switch semiconductor. Because of their high efficiency, simple power circuitry, high switching speed and good power factor, the power transistor may well gain in popularity. Low starting current, high efficiency (although not more than 5 ever that of the static converter) good controlability and low-cost motors are the advantages of frequency control in lifts.

The first 'vvvf' drives being introduced on the market for speeds up to 4 m/s seem primarily to be provided with helical gears offered in place of variable voltage d.c. drive. In the long run, this gear solution, because of its inherent drawbacks as a lift machine, may well prove to be only an interim solution. Some companies are also using industrial inverters, which are not really suited for lift applications. Unable to reach a true zero-speed, their ride quality is rather closer to an improved two-speed drive. Also, as they use ballast resistors to absorb the regenerated energy, so considerable energy savings can be made.

Being a new drive solution, frequency control will require more time, before it reaches the same level of dependability as the best d.c. drives.

6. EXAMPLES OF DRIVES

The KONE ECD-drive system employs a single-speed a.c. motor for speeds up to 1.0 m/s. The system features, control of acceleration by thyristors, and eddy current braking with direct landing approach giving good ride comfort and excellent levelling accuracy. The brake problems of the single-speed lift and the time wasting levelling speed of the two-speed lift now belong to the history.

The KONE TAC drive system (Fig. 5) uses a two-speed a.c. motor for speeds up to 2 m/s. With its first generation introduced already 20 years ago, the TAC drive has been the subject of repeated technical improvements over the years. The last improvements include new generation electronics, packing the electronics from five PC-boards on to two boards, and the new computer calculated speed reference, which enables the system to perform very short floor runs with the basic construction. The TAC system uses optimal speed reference controlling the lift speed by simultaneous thyristor regulation of the a.c. torque and eddy current braking from start until stop. The optionally available relevelling is made by the hoisting motor without any inching gear.

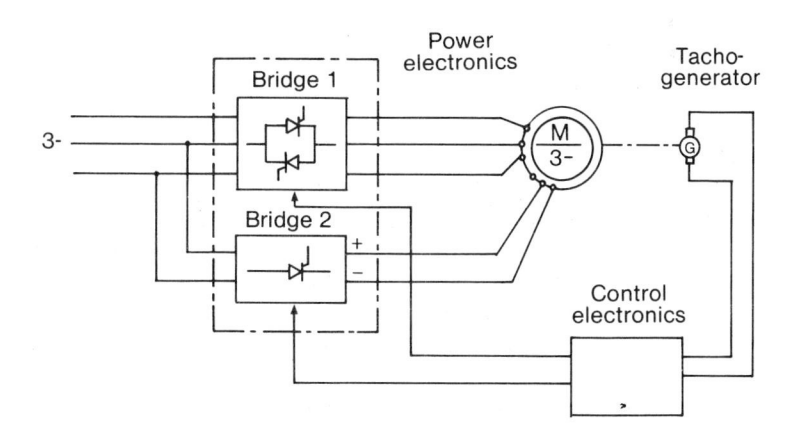

Fig. 5 — TAC drive system.

The KONE Dynaglide II drive solves the most distinctive problem of the solid state d.c. drives in lift applications; fuse and thyristor burn-out in power failure situations. A schematic diagram of Dynaglide II is shown in

Fig. 6. Among its advanced features are the special resonance and commutation filters, which allow high system reliability and elimination of the typical 300 Hz thyristor drive noise.

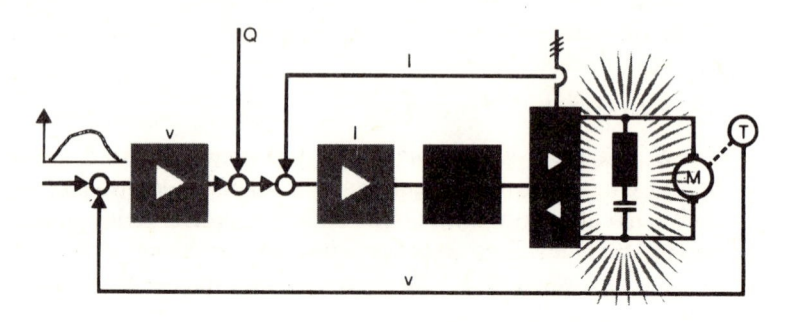

Fig. 6 — Schematic of Dynaglide II.

The Dynaglide II has proven its dependability in some two thousand installations all over the world. Furthermore, it offers excellent ride comfort, energy savings (up to 40% in comparison with the modern d.c. drives with mg-sets and up to 60% in modernizations) and optimum performance with selected acceleration and jerk rates. There is little doubt that the solid state d.c. drive will be the dominant high-speed lift drive during the next few years.

7. AUTOMATIC DOORS

7.1 Door Gear

Door operation and passenger transfer through the car entrance is becoming more and more important from the handling capacity point of view, because the performance of the new technology control and drive systems is now approaching the theoretical maximum limits dictated by the laws of probability and physical comfort.

The modern automatic door is expected to operate fast, without unnecessary delays, but at the same time smoothly and considerately. It is also expected to operate safely, which is in the interest of the passengers, but also of the building owner and manufacturers considering product liability legalities.

The traditional door mechanism comprises:

 (i) sinusoidal door movement by crank mechanism
 (ii) d.c. door motor with elementary speed pattern, because the door gear mechanism dictates the door speed pattern
 (iii) fixed speed patterns for opening, closing and nudging.

This specification was questioned by KONE engineers, who decided to develop a new generation automatic door, taking advantage of the latest achievements of the computer, controlled a.c. drive and surface mounted electronic technologies.

The first product decision concerned the type of door motor drive. Just as with lift drives, where the trend is from d.c. drives toward controlled a.c. drives such as inverters, an a.c. motor with two thyristor bridges controlling the motor speed in four quadrants was selected. The drive out-performs d.c. drives not using full speed and position feed-back, and is quieter and needs less maintenance.

The second choice was the type of speed reducer. Because the crank mechanism distorts the ratio between the motor rpm and door speed, a constant velocity drive with single step poly-V-belt transmission was selected.

To enable complete control of door speed at any time, the speed control is provided with two positively driven feed-back devices, one for door speed and the other for their position.

The controller which is mounted on the door gear includes everything necessary for the independent door operation, including control logic, speed program generation, speed control, thyristor firing and safety supervision. The controller communicates with the TMS computer, which can overrule the built-in logic instructions, to adapt the door operation to the prevailing traffic pattern. Locating the controller on the door gear, also ensures good safety and reliability.

The electronic controller incorporates the following features:

(i) four ready adjusted speed patterns for opening and closing;
(ii) computer-selection of speed programs as functions of traffic;
(iii) computer overruling of built-in door reopening, hold-open and nudging logics to match the detached traffic pattern;
(iv) use of maximum permitted or achievable speed within each door movement.

Fig. 7 illustrates the new door gear and Fig. 8 depicts door operating cycles in detail.

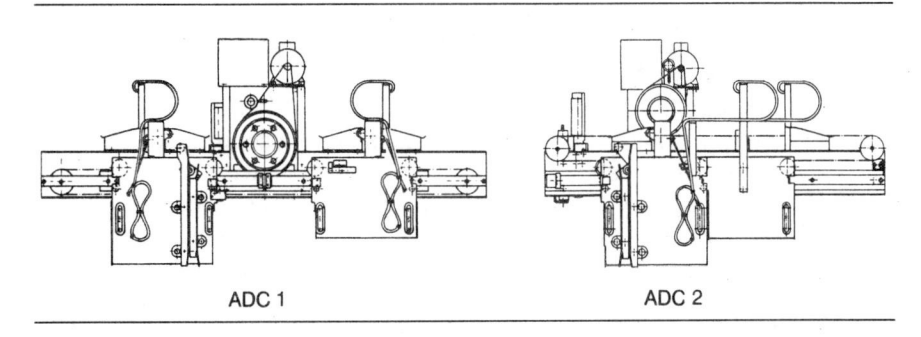

Fig. 7 — ADC automatic door gear.

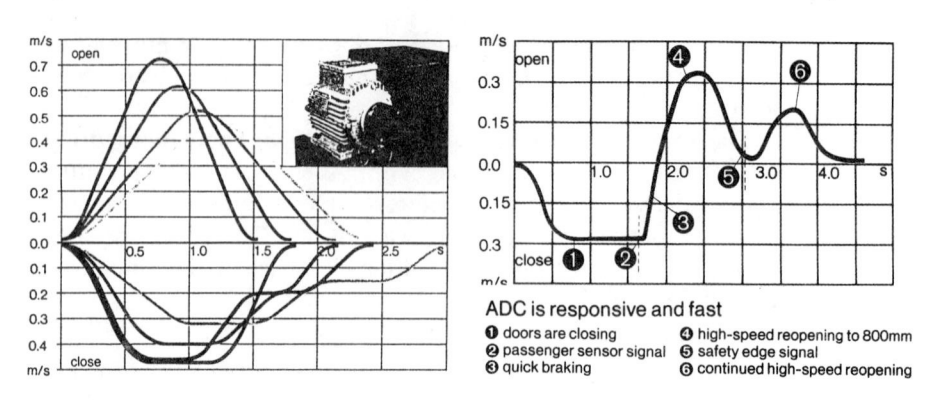

Fig. 8 — ADC doors operating cycles

ADC is responsive and fast
① doors are closing
② passenger sensor signal
③ quick braking
④ high-speed reopening to 800mm
⑤ safety edge signal
⑥ continued high-speed reopening

7.2 Passenger Detection

Over the years, many attempts have been made to design electronic door safety edges, and most of them have proven unreliable or insensitive in use. The ADC doors incorporate a new electronic safety edge, which detects obstructions with an image processing method, handling a picture of capacitance values created by multiple antenna sections. Mounted on a protruding safety edges on the car door panels, the sensing field extends more than 10 cm ahead of doors, and also towards the landing doors, greatly improving protection for passengers. As both the safety edges are in antiphase oscillation, high sensitivity can be maintained until doors are closed, unlike systems which reduce the sensitivity whilst the gap is closing. Thus the ADC door safety even protects fingers left between the door panels!

The same processing principle can be used for an optical passenger detector, which is mounted above the car lintel to detect approaching passengers. Operating on ambient illumination, the device does not emit any radiation. The device has proven its reliability in practice. It allows shorter door hold-open times without compromising passenger comfort, and reduces door damage during goods transit.

8. SUMMARY

Modern computer and microelectronics technologies, when applied to lift control, drive and door systems and their associated equipment, increase lift group performance, reliability and maintainability. They make it possible to introduce new features and capabilities, at the same or reduced cost in comparison with the earlier technologies. To do this calls for skilled and well trained personnel throughout the organization, and high levels of expertise in the research and product development laboratories. In the foreseeable future, the rapid trends will continue, imposing even higher demands on professional skills and research laboratories.

28

Skytrak: a new era of passenger transport?

Mr Michael Godwin, Lift Design Partnership, London, UK

ABSTRACT

In general people-moving systems operate in either the horizontal or vertical planes. Escalators used for mass transit cannot be compared with the personalized rapid transit system proposed here which exploits new degrees of freedom for architects and developers in building design.

One attempt has already been made to provide a system of transporting packets of people in cabins from ground level to an observation deck around a curved track. This was installed some thirty years ago on the Gateway Arch in St Louis, Missouri using a ropeway and multiple cabins.

The author proposes a cabin or cabins of perhaps six persons capacity which can travel in any of the three planes by means of a suitably curved track, and is ropeless! When travelling on an inclined plane or vertically it is intrinsically safe, although no conventional safety gear is provided. Car speeds can be varied to suit service schedules and 'Skytrack' could operate with efficiency in either outdoor or indoor environment.

1. INTRODUCTION

Combining horizontal and vertical transportation systems would seem more likely if the constraints and limitations of ropes were removed. Some ten years ago the author proposed a ropeless scheme for taking small packets of persons up a structure that was to be similar to the Gateway Arch at St Louis to an observation deck 230 m high.

At that time, however, the technology was hardly available to make such a proposition feasible, but today, the author believes that the technology is far enough advanced to make such a scheme viable. 'Skytrak' is a proposed transportation system consisting of a modular track that may be straight, curved or twisted on which small passenger carrying capsules can travel.

2. THE TECHNOLOGY

The past decade has seen one of the key elements of 'Skytrak', namely the material used for the magnetic field system undergo tremendous technological development. Present day rare earth magnets can achieve working flux densities of 0.5 Tesla because of their properties of high coercivity and high remanence, and thus provide the necessary energy storage element needed for the safe operation of 'Skytrak'. Given that the stability of the physical properties of rare earth magnets are guaranteed under normal conditions a motor can be designed such that the capsule is safe under free fall conditions without recourse to any external power source.

The basic costs involved in the processing of rare earth magnets have fallen consistently over the years whilst the energy product has increased. At present the world supplies of rare earth materials for processing into high energy permanent magnets are more than adequate to meet the existing demand so that the quantity required for this system presents no supply problem. Combining a moving magnet array with a wound fixed stator produces in effect a linear motor of the switched, synchronous type machine. Again present day state of the art transistor switching devices are now relatively cheap and this is important because at first glance there would seem to be many hundreds of these required, and cost per unit length of track could become a significant factor, especially as the current in the winding of each pole might be individually switched using transistors to connect them to the processor controlled inverters.

The third area where, over the past ten years, the development of the technology is now at an appropriate stage is the fibreoptic transmission system whereby the positional data of each passenger carrying capsule on the same track can be remotely monitored with high integrity. Clearly the bandwidth of fibreoptics is more than adequate for dealing with the data transmission which may require positional data every millimetre of track as well as the unique signatures of every capsule, including voice communication. Interference techniques using fibreoptics have already been applied to the machine-tool industry where it is achieving far higher resolution than is required in this application.

The timely evolution of powerful multi-tasking microprocessor computers means that the real time computing problems of control should not be difficult and therefore initial steps towards the development of 'Skytrak' seem appropriate.

3. SAFETY

Safety when applied to passenger carrying vehicles is concerned with the control of electro-mechanical forces which cause the vehicle mass to accelerate or decelerate. The ultimate test for 'Skytrak' must be concerned with limiting the capsule descent velocity under a free fall condition. This condition has always been traditionally dealt with on conventional lifts by the use of a safety gear first invented by Elisha Otis in 1857. In principle there is no reason why 'Skytrak' should not have something similar; however, the author argues, that such a device is unnecessary and might cause more problems than it solves. The essential point to grasp is that whereas in rope driven systems there is nothing that prevents free fall of a lift car should the ropes break, in the case of 'Skytrak' it contains its own energy package on board each vehicle in the form of permanent magnets which energy cannot be degraded or converted under normal conditions. Provided the capsule remains mechanically connected to the track and there are the huge attractive forces of the magnet array assisting braking force an electro-magnetic force will oppose gravitational acceleration to limit the descent velocity of the capsule assembly to approximately 1 m/s.

As with safety gears on conventional lifts the mechanism for triggering it in anger must be reliable and as far as possible very simple. As springs in compression are simple and have become accepted as reliable the author has in mind that should an over-speed condition be detected by some mechanical pick-off, springs are relaxed to force a metal plate to lift clear of the array of magnetic switches positioned along the track. It is the operation of these normally open magnetic switches which lift the fail-safe contactors and thus steer power into the stator windings from an array of inverters. Preventing the magnetic switches from operating by lifting, the metal plate ensures that the fail-safe contactors cannot lift and by remaining in their normally closed position each of them provides a path for the current in the individual windings of the stator caused by the emf generated by the descending magnet/capsule assembly. As there are many of these fail-safe contactors per metre of stator track all of which are electrically interlocked, this represents a huge redundancy factor so that under no cirmcumstantes is the safety of 'Skytrak' dependent upon the correct operation of a single element such as one fail-safe contactors or magnetic switch. It becomes a matter of routine for the computer not only to check the electrical interlocking of such contactors but also the safety of a complete track by dispatching a capsule with a dummy load under power to the highest point of travel, removing the power and litting it free fall whilst checking its speed of descent and hence the uniformity of the braking force. With such ease of checking combined with the assured safety of 'Skytrak' the fitting of conventional safety gear would seem to be superfluous, after all what could be more unnerving to passengers than to be stuck in a capsule perhaps hundreds of feet in the air waiting for power to be restored, or for some alternative mechanical switch to lift or lower the capsule, the whole operation seemingly taking an age! Clearly the dynamic safety system as described will always operate first and hence the redundant safety gear never being required to operate in anger represents only additional dead weight for each passenger-carrying capsule which 'Skytrak' could usefully do without. Needless to say each capsule carries some form of energy absorbing buffer at its front or back to absorb the impact of another capsule coming into contact. The important inherent design features that ensure the safety of passengers on 'Skytrak' can be summarized thus:

(i) The energy stored in the permanent magnet field system attached to the moving capsule when combined with the wound stator circuit limits its free-fall velocity to less than 1.0 m/s.
(ii) The fail-safe contactor that is normally de-energized and thus short circuits its small section of stator winding to which it is connected.
(iii) The huge redundancy factor inherent in the array of normally closed de-energized fail-safe contactors and magnetic switches, situated along the track.
(iv) The enormous attractive force of the permanent magnet to the stator iron keeps the vehicle glued as it were to the track.

Figure 1 shows the braking curve for a 1000 kg capsule and load. Above the

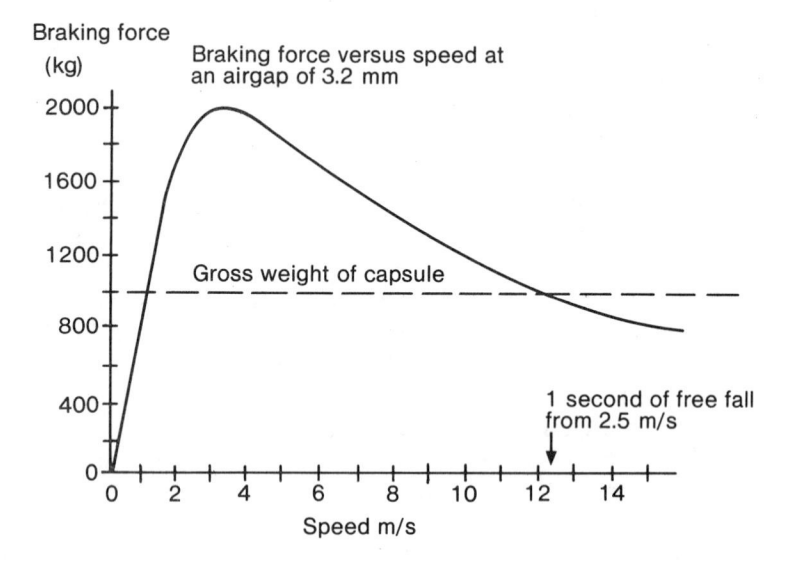

Fig. 1 — Braking curve.

line drawn at 1000 kg deceleration takes place. By design the maximum braking can be increased by reducing the air gap and shape of the curve changed by altering the stator winding resistance. A design can be chosen that will ensure that braking force is operational at the highest velocity predicted under the free fall condition.

4. POWER SYSTEM

No optimum configuration of the magnet/stator arrangement has yet been finalized. To some extent it is a matter of careful mechanical design in packaging the articulated magnet frame so that it can follow the curves of the track with minimum change in the air gap. As all the energy of the system is in the air gap, the size of this is crucial. Obviously the smaller the air gap, the greater the thrust obtained for a given size stator and magnet assembly. An air gap of approximately 5 mm is considered to be both mechanically feasible and not unreasonably large from the viewpoint of motor design. Table 1 is an example of a motor design based on an all-up weight of passenger capsule assembly including load of 1000 kg.

Table 1 — Single-sided motor design.

Characteristic	Value
Gross lift weight	1000 kg
Peak braking	2000 kg
Speed	2.5 m/s
Pole pitch	45 mm
Airgap (generating)	3.2 mm
Length of magnet member	3.4 m
Length of each stator section	0.85 m
Air gap (motoring)	5 mm
Width of stator section	0.184 m (0.139 m active)
Depth of stator section	0.09 m
Weight of each stator section	67 kg
Width of magnet member	0.139 m
Depth of magnet and backing iron	0.03 m
Weight of magnets	40 kg
Total weight of magnet member (magnets + backing iron)	114 kg
Efficiency	90.5% P.F. 0.89
i/p power	27 kW
Dissipated watts per m^2 of stator when motoring	1920
Battery voltage	275 V
Current drawn from battery when motoring	80 amps
Inverter current/phase	50 amps
Total force of attraction on magnets when motoring	2289 kg
Total force of attraction on magnets when braking	8117 kg

From Table 1 it would seem that the size of motor is determined by ensuring sufficient braking force to limit the descent velocity under free fall to approximately 1.0 m/s and this may always be the determining factor. To increase the braking efficiency of the field magnet array it would be feasible to allow it to mechanically collapse directly on to the face of the stator iron which could be suitably covered by a thin sheet of appropriate material so that the effective air gap is only perhaps 2 mm. Braking force would thus be vastly increased for a given size of motor. The mechanism to cause the collapse might be a fail-safe vacuum cylinder linked to the tracked wheels such that when vented by the operation of the over speed governor the magnet array being no longer restrained at 5 mm air gap moves into contact with the stator assembly to give the smaller air gap of 2 mm. Quite clearly from the point of view of powering the capsule, the machine could be very intermittently rated owing to the distributed nature of the stator windings. In so far as powering the motor is concerned a heavy current bus could form an integral part of the modular track fed at a suitable d.c. potential from a low impedance supply such as a battery of nickel cadmium cells. Small inverters, fed by the common bus using power transistors, are stacked along each section of modular track to provide appropriate current and voltage into the windings. As already mentioned, the control data pick-off is via a fibre-optic transducer to the main computer which determines from the speed and position of each capsule the appropriate power and frequency that should be fed into the stator. .The static inverters are of course four quadrant devices capable of recovering a high proportion of the energy used in providing the vertical thrust.

5. CONCLUSIONS

Whilst there have been various ideas put forward in patents, etc., applying so-called linear motors to lifts, these have as far as the author knows, used ropes for the suspension of the passenger carrying vehicle. 'Skytrak' represents a new departure and requires some radical re-thinking of the traditional concepts of safety and traffic circulation. At first glance it may seem that 'Skytrak' belongs more to Disneyland providing fun rides than to serious commercial applications such as airports and high-tech buildings of the twenty-first century, and it is difficult to foresee where the principal application of such a system will lie, mainly because it is so novel. The first application for it does however exist, and a serious development programme is about to be undertaken.

It can be further imagined a track switching network in the horizontal plane where passenger capsules can change tracks by a simple points system similar to railways. Such a versatile system will take a long time to commercially exploit, but will doubtless find its niche in the market-place some time in the future. Above all it is a transport system that can give an extremely high quality of ride equal to the best presently attainable in lifts and escalators and should not therefore be classed as a form of mechanical hoist or similar where the quality of ride is not considered important. The nearest analogy is to consider it as an unwrapped s.c. gearless elevator without the discontinuities and noise produced by the friction drive and hoisting ropes. Static variable frequency four quadrant drives for lifts are already well established using conventional rotating machines, so that a linear version using this form of drive follows naturally. Of course, much development work is required to achieve the lightweight cabins and support assemblies, no doubt employing some of the new materials used in the aircraft industry. Most probably therefore its development will spring naturally as an offshoot of the air-frame industry rather than the general heavy industry associated with conventional lifts.

To this end the author invites constructive criticism and comment concerning this novel concept and reasonably expects that the persons rightly concerned with the safety aspects of transportation systems will consider it with an open mind and without prejudice, not forgetting the quotation: 'For the idea which does not at first glance seem crazy, there is no hope'.

Part 9
Escalators

29

The kinematics and dynamics of escalator steps and safety

Eng. Dr Lubomir Janovsky, Technical University, Prague, Czechoslovakia

ABSTRACT

Analysis of the motion of escalator steps in the event of a failure of the mechanical linkage between the drive unit and the top shaft has been carried out with considerations of the operation of the escalator emergency brake.

Two methods of calculation of the dynamics and kinematics of the steps were adopted. All static, kinematic and dynamic quantities are tabulated. Speed/distance and speed/time diagrams for different initial conditions reveal a rapid increase of the escalator speed in the event of emergency brake failure.

In conclusion the fundamental requirements for an escalator emergency brake are specified.

1. INTRODUCTION

Escalators must be equipped with a safety device for bringing the steps to a standstill in case of the kinematic linkage between the drive motor and the top drive shaft being interrupted. This device is represented by an emergency brake, which is mechanically operated and located on the top shaft.

An escalator accident that took place in a Prague underground station in Sepember 1980 revealed the importance of a well-functioning emergency brake. A duplex drive chain transmitting the power from the drive machine to the top shaft broke and the step running upwards transporting a number of persons (afternoon peak time) stopped quickly. An undesired change of the running direction took place allowing just a few people to leave the steps. The emergency brake failed and the speed of the steps increased rapidly so that only few passengers could alight safely at the bottom landing. Many people fell down and two of them suffered severe injuries resulting in disability.

The emergency brake must be so designed as to stop the fully loaded escator irrespective of its running direction. The braking torque must be induced mechanically. Before the brake is actuated the main electric circuit must be disconnected.

When a drive chain forms the last element of the mechanical linkage between the motor and the top drive shaft the impulse for emergency brake operation is usually drived from the motion of a chain tensioning-device or a spring-loaded arm that takes place in the event of chain breakage. If the drive mechanism is of different design a safety device including a speed governor is provided. It cuts out the motor circuit in case of the speed of the steps being in excess of its nominal rated value by 15–20% and actuates the emergency brake when the nominal rated speed is exceeded by more than 30–40% or when an undesired change of running direction occurs.

2. KINEMATICS AND DYNAMICS OF THE STEPS

2.1 Methods of calculation

Graduated calculation by resolving the motion of the escalator into a number of phases. Each phase responding to a one step shift (400 mm). The load is assumed to be constant during each phase and consequently the motion of the steps is uniformly accelerated or retarded respectively.

Solution for the continuously changing (decreasing) load on escalator steps.

Assumptions for calculation:

(i) At time $t = 0$ (the time of drive chain breakage) the load is equally distributed along the inclined section of th escalator; horizontal sections on both ends of the escalator are empty.
(ii) The escalator is descending.
(iii) No passengers enter the escaltor after the chain breakage.
(iv) Escalator emergency brake is actuated readily after the breakage without any time delay.

The driving force (F_{ho}) at the beginning ($t = 0$):

$$F_{ho} = m_1 \cdot g \cdot L_i \cdot \sin \alpha, \tag{1}$$

where m_1 is the mass of persons on the length of 1 metre of the inclined section (kg/m), g is the acceleration of gravity (mm/s^2), L_i is the total length of the inclined section of the escalator (m) and α is the angle of inclination(°).

Frictional resistance against the motion is proportional to the overall mass of all movable parts of the escalator, including passengers on the steps:

$$F_{\text{st o}} = \text{const} \cdot (m_1 \cdot L_i + m_s), \qquad (2)$$

where m_s is the mass of all movable parts of the escalator (kg).
At time $t > 0$ (see Fig. 1) the driving force is:

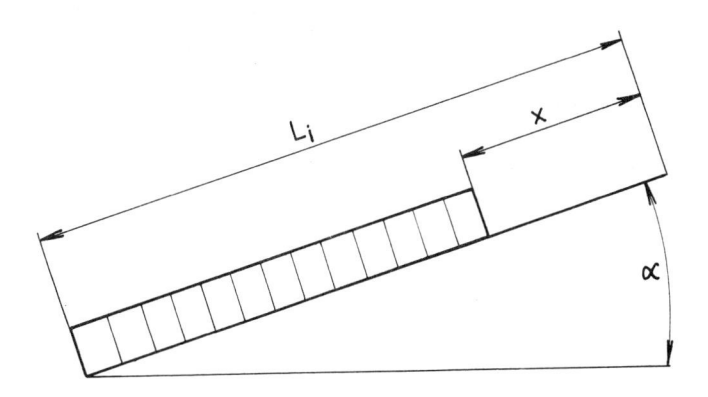

Fig. 1 — Representation of forces.

$$F_h = F_{ho} \cdot \frac{L_i - x}{L_i}, \qquad (3)$$

where x is the distance (the travel of the steps) at time t (m). The mass of passengers on the steps:

$$m = m_1 \cdot (L_i - x) \qquad (4)$$

Frictional resistance:

$$F_{st} = \text{const} \cdot [m_1 \cdot (L_i - x) + m_s] \qquad (5)$$

At time t the loaded steps are in linear motion with the acceleration $\ddot{x}$ and the mass of the movable system m is decreasing. The fundamental equation of motion is of the following form:

$$F_h - F_{st} - m \cdot \ddot{x} = 0 \qquad (6)$$

and after substitution:

$$F_{ho} - \frac{F_{ho}}{L_i} \cdot x - \text{const.} \, (m_1 \cdot L_i + m_s) + \\ \text{const.} \, m_1 \cdot x - m_1 \cdot L_i \cdot \ddot{x} + m_1 , x \cdot \ddot{x} = 0 \qquad (7)$$

which is a non-homogenous second-order second-degree differential equation and may be written in a simplified form:

$$C_1 + C_2 \cdot x + C_3 \cdot \ddot{x} + C_4 \cdot x \cdot \ddot{x} = 0 \tag{8}$$

where C_1, C_2, C_3 and C_4 are constants.

To solve this equation a convenient numerical method must be used.

Comparison of the results achieved by both methods was carried out showing only a slight difference and therefore the first method was given preference as it proved to be simpler and its assumptions corresponded better to the actual operation conditions.

2.2 Technical data and fundamentals of calculation

A detailed calculation is not presented here, but an analysis has been included as well as the survey of the results obtained by the calculation of a heavy-duty escalator, with the following technical data:

Length of step	1000 mm
Width of step	400 mm
Rise	6.8 m
Angle of inclination	30°
Nominal speed	0.7 m/s

The theoretical length of the inclined section of the escalator is 13.6 m and the number of steps in this section 34. Lengths of horizontal sections are 1255 mm at the upper end and 1320 mm at the bottom (in accordance with the appropriate drawings).

The fundamental equation of motion

$$F_h - F_{st} - m \cdot a = 0, \tag{9}$$

where the symbols are of the same significance as in Section 2, only the acceleration is designated a instead of $\ddot{x}$.

The driving force F_h is represented by the sine component of the load on escalator steps in the inclined section. The mass of each person was supposed to be 80 kg.

The frictional resistance F_{st} comprises resistances of the steps, handrails, drive sprockets and tensioning device, including frictional resistance of all bearings. Its value was estimated for the purpose of this calculation in relation with the results of practical tests which were made in September 1980. When the drive chains of a heavy-duty escalators with rises of 4.5 m and 4.2 m, respectively were disconnected and the emergency brake released the steps were set in motion by 8 persons stepping on the inclined section. It is possible to calculate the frictional rsistance as it is known to be equal to the driving force that sets the steps in motion. Hence

$$F_{st} = 10 \cdot 80 \cdot g \cdot \sin \alpha = 3924 \text{ N} \tag{10}$$

With the load increasing the resistance will also increase, but with adequate accuracy it may be assumed to remain proportional to the total mass of the movable system including persons on the steps. For each load resistance F_{st} was calculated as a proportional number.

3. CASE STUDIES

3.1 Case 1: Emergency Brake Failure

The breakage of drive chain was supposed to take place when the escalator was descending. As the load is variable during the motion, the operation was divided into individual phases, each of them responding to the shift of 400 mm (one step). The load was constant during each phase and consequently the steps were either uniformly accelerated or uniformly decelerated. For each phase the calculation of kinematic values was carried out using the following formulae:

$$a = \frac{F_h - F_{st}}{m} \text{ (m/s}^2\text{)} \tag{11}$$

$$v = \sqrt{v_0^2 + 2s \cdot a} \text{ (m/s}^2\text{)}, \tag{12}$$

where v is instant velocity of the steps (m/s), v_0 is the initial velocity of the steps (m/s) and s is the travel of the steps (m).

$$t = \frac{v - v_0}{a} \text{ (s)} \tag{13}$$

where t is time of each individual phase (s).

In the case of a fully loaded escalator there might be 68 persons on the inclined section. This number is theoretical only as the capacity depends considerably on the speed of steps. Observations revealed that with fast-moving escalators the capacity was not proportional to speed and quite a number of steps were not fully loaded on arrival at the top (or bottom) landing. A graph showing variation of capacity with speed is displayed in Fig. 2. Curve 1 shows the results of a survey caried out by London Transport, curve 2 was given by Bovin and curve 3 by Hitachi, Japan. The hesitancy of passengers to board an escalator running at high speed results in the decrease of capacity in comparison with its theoretical value. A straight line showing the increase of the theoretical capacity has been also incorporated in the graph.

According to the rated speed of 0.7 m/s 45 persons were supposed to be standing in the inclined section of the escalator at the instant of failure.

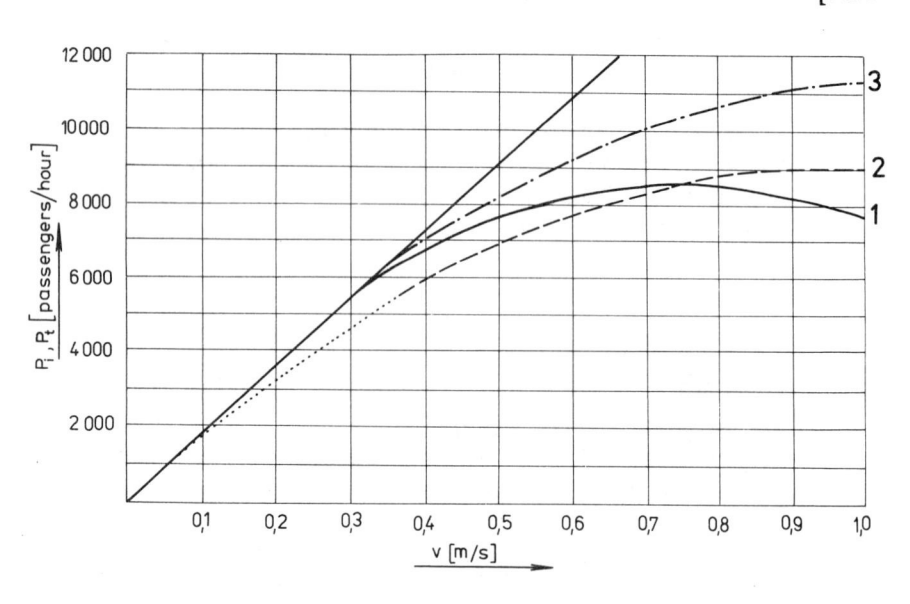

Fig. 2 — Variation of capacity with speed.

The following distribution of passengers was assumed:

Lower horizontal section	Inclined section	Upper horizontal section
3 steps	34 steps	3 steps
1 2 1	45	2 1 2

Calculations proved that after the travel of 400 mm (1. phase of escalator motion) the speed increased to 1.26 m/s. Another two passengers were supposed to step on at the upper end and the second phase followed with a slightly changed distribution of passengers on the step band. At the end of the second phase the speed reached the value of 1.64 m/s; it was assumed that even at this rather high speed another two persons stepped on. Since the speed increased to 1.93 m/s during the third phase of motion no other person could board the escalator. During the two following phases the upper horizontal section gets empty and starting with the sixth phase the distribution of passengers in the inclined section will be represented by the model tabulated in Table 1.

An unfavourable distribution was assumed for the purpose of this calculation, in fact it is likely to be more favourable. Each step was occupied by one person at least, two persons on one step occurred mostly in the upper part of the inclined section. The load was supposed to act upon the step until the step disappeared under the comb-plate at the bottom landing. This small inaccuracy did not influence the results.

The dependence of the step velocity upon the distance and time respectively are featured in Fig. 3 and Fig. 4 respectively.

Maximum speeds of steps, namely 5.18 m/s, is achieved in less than 4 s after the travel of 13.2 m. Along the distance of another 400 mm the speed

Ch. 29] Kinematics and dynamics of escalator safety

Table 1 — Distribution of passengers

No. of step	1	2	3	4	5	6	7	8	9	10	11	12	13	14	15	16	17
No. of persons	1	1	1	2	1	1	1	2	1	1	1	2	1	1	1	2	1
No. of steps	18	19	20	21	22	23	24	25	26	27	28	29	30	31	32	33	34
No. of persons	1	1	2	1	1	1	2	1	1	1	1	2	2	1	2	2	2

Table 3 — Distribution of passengers

No. of step	1	2	3	4	5	6	7	8	9	10	11	12	13	14	15	16	17
No. of persons	1	1	1	2	1	1	1	2	1	1	2	1	1	2	1	1	2
No. of steps	18	19	20	21	22	23	24	25	26	27	28	29	30	31	32	33	34
No. of persons	1	1	2	1	1	2	2	1	2	1	1	2	1	1	2	2	1

Table 4 — Energy & inertial, static and braking torque

Distance s (mm)	Velocity v (m/s)	Mass of movable system (kg)	Kinetic energy (J)	Time (s)	Inertial component	Static component	Total
0	0.7	7440	1822.8	0	5946.05	4139.856	10085.9
400	1.2664	7440	5966.02	0.4068	5946.05	4139.856	10085.9
800	1.6424	7520	10142.5	0.6818	6010.0	4108.287	10118.3
1 200	1.9316	7520	14028.85	0.8933	6010.0	4108.287	10118.3
1 600	2.2011	7440	18022.8	1.0868	5946.05	4139.856	10085.9
2 000	2.454	7360	22161.38	1.2586	5882.1	4327.668	10209.8
2 400	2.6912	7200	26073.2	1.4141	5754.24	4390.405	10144.6
4 000	3.4503	7800	40475.53	1.9339	5434.56	3763.03	9197.6
6 000	4.127	7320	53821.52	2.4604	5050.94	3010.586	8061.5
8 000	4.6145	7840	62177.34	2.9167	4667.3	2258.14	6925.4
10 000	4.9547	7280	64809.49	3.3338	4219.77	1536.86	5756.6
12 000	5.1485	7800	63616.92	3.7288	3836.16	626.97	4463.1
14 000	5.1722	7240	56713.5	4.1164	3388.6		

Table 2 — Kinematic values

Phase	Distance s (mm)	Number of persons — Inclined section	Number of persons — Total	Acceleration a (m/s^2)	Velocity v (m/s)	Partial time t (s)	Total time t (s)	Driving force F_h (N)	Mass of movable system m (kg)	Frictional resistance F_{st} (N)
1	400	45	53	1.3924	1.2664	0.4068	0.4068	17 658	7440	7298
2	800	45	54	1.3671	1.6424	0.275	0.6818	17 658	7520	7377
3	1 200	45	54	1.3671	1.936	0.2115	0.8933	17 658	7520	7377
4	1 600	45	53	1.3924	2.2011	0.1935	1.0868	17 658	7440	7298
5	2 000	46	52	1.4714	2.454	0.1718	1.2568	18 050	7360	7220
6	2 400	46	50	1.5254	2.6912	0.1555	1.4141	18 050	7200	7063
7	2 800	45	49	1.4991	2.9055	0.1429	1.557	17 658	7120	6984
8	3 200	44	48	1.4716	3.1015	0.133	1.69	17 266	7040	6906
9	3 600	43	46	1.4715	3.2858	0.1252	0.8152	16 873	6880	6749
10	4 000	41	45	1.3848	3.4503	0.1187	1.9339	16 088	6800	6671
11	4 400	40	44	1.3547	3.6039	0.1133	2.0472	15 696	6720	6592
12	4 800	39	43	1.321	3.7472	0.1088	2.156	15 304	6640	6533
13	5 200	38	41	1.32	3.8856	0.1048	2.2608	14 911	6480	6357
14	5 600	36	40	1.2262	4.0098	0.1013	2.3621	14 126	6400	6278
15	6 000	35	39	1.192	4.127	0.0983	2.4604	13 734	6320	6200
16	6 400	34	38	1.1572	4.2377	0.0956	2.556	13 342	6240	6121
17	6 800	33	36	1.1488	4.3447	0.0931	2.6491	12 949	6080	5964
18	7 200	31	35	1.0463	4.44	0.091	2.7401	12 164	6000	5886
19	7 600	30	34	1.0076	4.5299	0.0092	2.8293	11 772	5920	5807
20	8 000	29	33	0.9676	4.6145	0.0974	2.9167	11 380	5840	5729
21	8 400	28	31	0.9533	4.6964	0.0859	3.0026	10 987	5680	5572
22	8 800	26	30	0.8414	4.7675	0.0845	3.0871	10 202	5600	5493
23	9 200	25	29	0.7962	4.834	0.0835	3.1706	9 810	5520	5415
24	9 600	24	28	0.75	4.8955	0.082	325268	9 418	5440	5336
25	10 000	23	26	0.7284	4.9547	0.0812	3.3338	9 025	5280	5179
26	10 400	21	25	0.6036	5.0032	0.0803	3.4141	8 240	5200	5101
27	10 800	20	24	0.552	5.0471	0.0795	3.4936	7 848	5120	5022
28	11 200	19	23	0.4984	5.0865	0.079	3.5726	7 456	5040	4944
29	11 600	18	21	0.4664	5.123	0.0782	3.6508	7 063	4880	4787
30	12 000	16	20	0.3268	5.1485	0.078	3.7288	6 278	4800	4709
31	12 400	15	19	0.2661	5.1691	0.0774	3.8062	5 886	4720	4630
32	12 800	13	18	0.1183	5.1782	0.0769	3.8831	5 101	4640	4552
33	13 200	12	16	0.07	5.1837	0.0785	3.9616	4 709	4480	4395
34	13 600	11	15	0	5.1837	0.077	4.0386	4 316	4400	4316
35	14 000	9	13	−0.1478	5.1722	0.0778	4.1164	3 532	4240	4159
36	14 400	7	12	−0.3206	5.1474	0.0773	4.1937	726747	4160	4081
37	14 800	6	11	−0.4039	5.1159	0.0779	4.2716	2 354	4080	4002
38	15 200	4	9	−0.5703	5.0703	0.0785	4.3501	1 570	3920	3845
39	15 600	2	7	−0.772	5.009	0.0794	4.4298	785	3760	3688
40	16 000	0	6	−0.981	4.93	0.0805	5.51	0	3680	3510
41	16 400	0	6	−0.981	4.8497	0.0818	4.5918	0	3520	3453
42	16 800	0	4	−0.981	4.7681	0.0831	4.6749	0	3360	3296
43	17 200	0	2	−0.981	4.685	0.0847	4.7596	0	3200	3139
	18 000		0		4.5143	0.174	4.9336			
	20 000				4.0564	0.4667	5.4003			
	22 000				3.5399	0.5265	5.9268			
	24 000				2.9337	0.618	6.5448			
	26 000				2.1644	0.7846	7.3294			
	28 000				0.8711	1.3179	8.6473			

Fig. 3 — Step velocity and distance.

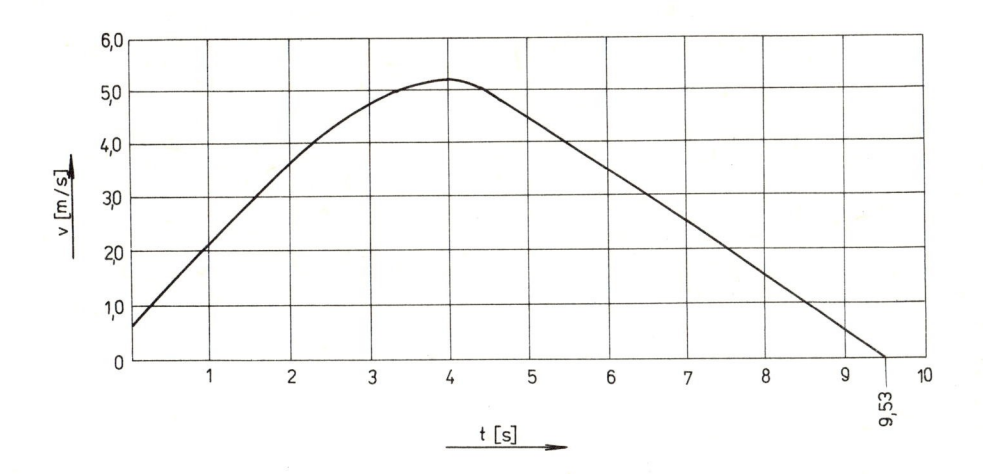

Fig. 4 — Step velocity and time.

will be constant as a result of the equality of the driving force and frictional resistance. A period of retardation will follow afterwards, as the braking force represented by frictional resistance will exceed the decreasing driving force. As the braking force was small then the rate of retardation will also be small. At the total distance of 15.6 m the speed will still be as high as 5 m/s. At this point the driving force decreases to zero and the motion of steps will be uniformly retarded. The total escalator travel will be 23.4 m and the total time 9.53 s.

Frictional resistance does not effect the speed of steps to a large extent during the period of acceleration as it is small in comparison with the driving force, but its influence is substantial after the driving force has decreased to zero. Even a small increase of friction may shorten the running distance considerably.

Further analysis and calculations would confirm that there is only a small difference between the downward and upward motion of escalator steps from the kinematic viewpoint. Let us assume that the breakage of drive chain took place when the escalator was ascending. If the distribution of passengers remains the same the escalator will stop after a very short distance, owing to both the sin component of the load in the inclined section of the escalator and frictional resistance. The rate of retardation will be relatively high (3.35 m/s^2) and the stopping distance short (73 mm) as well as the braking time (0.208 s).

Having stopped the steps are set in reversed motion. It may be assumed that passengers travelling on the two former steps at the top landing left the escalator. Consequently only one passenger remains standing on horizontal steps at the upper end of the escalator. After the travel of 400 mm this passenger appears on the inclined section. A model of load distribution tabulated in Table 3. was accepted for calculations of kinematic values.

In Fig. 5(a) velocity–distance diagram is displayed. The curve is very

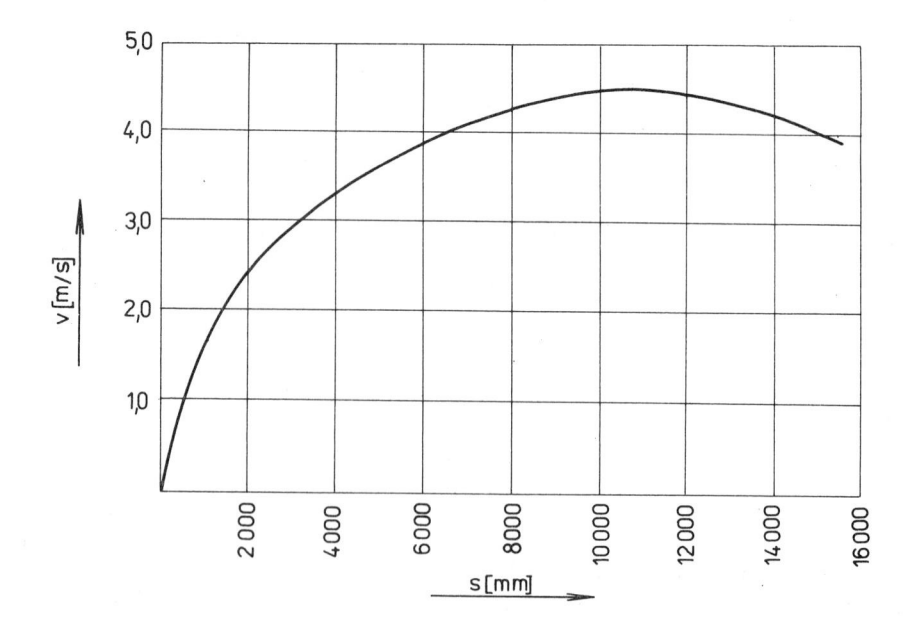

Fig. 5 — Velocity–distance diagram.

similar to that shown in Fig. 3. The maximum speed of escalator steps will be rather lower (4.25 m/s) and achieved at a shorter distance (10.8 m).

The initial velocity is not of primary significance, but it is very important that the time delay between the drive chain breakage and the operation of escalator emergency brake be as short as possible. During this time the steps are accelerated in downward direction and their speed increases rapidly. If the time delay is long the emergency brake starts to operate when the kinetic energy of the movable system is considerable. The increase of energy is shown in Table 4 together with the inertial, static and braking torques in particular phases of the step motion.

3.2 Case 2: Emergency brake operation
3.2.1 Calculation of the braking torque
The braking torque is composed of two parts: the static component for keeping the movable system at standstill and the dynamic component necessary for stopping the system in question. The value of the dynamic (inertial) component is restricted by the maximum permissable rate of retardation. Both componens are quoted in Table 4 as well as the total value of the braking torque. Calculations reveal that the total value should be about 10 000 Nm. If it is smaller the braking time will be very long and quite a number of passengers will have to step off at high speed.

3.2.2 Calculation of kinematic quantities
Calculations were carried out for the escalator descending with the same model of load distribution as that tabulated in Table 1. Since the exact value of the time interval between the drive chain breakage and the beginning of emergency brake operation was unknown three different values were determined for the purpose of this calculation:

(i) *A long time delay*: 1.26 s. During this interval the steps ran a distance of 2.0 m and their speed increased to 2.45 m/s (see Table 2). The braking torque M_b was supposed to be 10 000 Nm and hence the braking force

$$F_b = \frac{2M_b}{D} = 23\ 524\ \text{N}, \tag{14}$$

where D is the pitch diameter of the step drive sprockets; $D = 850.2$ mm.

The rate of retardation of the steps was calculated from the equation of dynamic equilibrium of forces acting upon the movable system and the remaining kinematic quantities were given by the equations of the uniformly retarded motion, namely

$$a = \frac{F_b + F_{st} - F_h}{m}\ (\text{m/s}^2) \tag{15}$$

$$v = \sqrt{v_0^2 - 2a \cdot s} \text{ (m/s)} \tag{16}$$

$$t = \frac{v_0 - v}{a} \text{ (s)} \tag{17}$$

The meaning of particular symbols is identical to those used in Section 3.1. All calculated values are tabulated in Table 5.

Table 5 — Kinematic quantities: long time delay. Braking Force F_b = 23/524

Distance s (mm)	Velocity v (m/s)	Retardation a (m/s²)	Partial t (s) time	Number of persons
2000	2.454	—	—	—
2400	2.151	1.74 125	0.174	2
2800	1.784	1.80 477	0.203	1
3200	1.2987	1.86 988	0.26	1
3600	0.3587	1.94 767	0.482	2
3631	0	2.07 45	0.173	—
		Sum =	1.292	

The braking distance was 1.6 m. The retardation reached the highest value at the end of braking. Six passengers left the steps during the braking operation, but only three of them at a dangerous speed.

In order to see what might happen in the event of the braking torque being smaller than 10 000 Nm the braking force of 15 00 N was taken for further calculation. The results are shown in Table 6 and are favourable. The braking distance was 3.7 m and 12 persons had to step off the escalator during the braking period, eight of whom at a dangerous speed.

The inertial force F_i acting upon a passenger of 80 kg in weight achieved the following maximum values:

With braking force of 15 000 N F_i = 94.5 N
With braking force of 23 524 N F_i = 166 N

It is most desirable that this force is reasonably small as it might cause a 'cascade' accident through passengers being thrown off blance.

(ii) *A short time delay* — 0.41 s. With the braking force F_b = 23 524 N the braking distance is 455.24 mm and only two persons must leave the steps at a low speed. With the braking force of 15 000 N the braking

Table 6 — Kinematic quantities: long time delay. Braking Force F_b = 15 000

Distance s (mm)	Velocity v (m/s)	Retardation a (m/s²) time	Partial t (s) time	Number of persons
2000	2.454	—	—	—
2400	2.361	0.557	0.1669	2
2800	2.256	0.607	0.1729	1
3200	2.136	0.659	0.1821	1
3600	1.998	0.708	0.1949	2
4000	1.826	0.821	0.2095	1
4400	1.622	0.877	0.2326	1
4800	1.372	0.935	0.2673	1
5200	1.043	0.994	0.3309	2
5600	0.44	1.117	0.5398	1
5682	0	1.181	0.3725	—
		Sum =	2.6698	

distance is much longer — 1.29 m four passengers must leave the steps at the bottom landing, but only two of them at a speed in excess of 1 m/s.

(iii) *No time delay — instantaneous emergency brake operation.* under these theoretical conditions the stopping distance would be very short and the rate of retardation quite acceptable as given below:

Braking force (N)	15 000	23 524
Retardation (m/s²)	0.623	1.79
Braking distance (mm)	393	138

The brake would stop the escalator before any passengers have to step off at the bottom landing.

The comparison of kinematic quantities calculated in this Section emphasize the necessity of a rapid action — the time delay should be as short as possible.

4. CONCLUSIONS

(i) If the breakage of an escalator drive chain occurs and the kinematic linkage between the drive machine and the top shaft is interrupted, the speed of the steps loaded with a number of passengers will increase considerably in a very short time.

(ii) The time delay between the drive chain breakage and the commencement of emergency brake operation should be minimized. This requirement is of prime importance. The control mechanism should be simple and the mass of its movable parts small.

(iii) Bearing in mind all operational conditions the braking torque should be proportional to the load. Unfortunately this requirement seems to be theoretical only. If the braking torque is of a constant value and calculated for stopping the loaded escalator when descending, the inertial force effecting the balance of passengers on the steps will be very high. With only few persons on the steps the danger of a 'cascade' accident will occur. If the braking torque is small the stopping distance will be long and quite a number of passengers will have to step off the escalator at high speed. With some escalators emergency brakes with progressive braking torque are employed. This design does not solve the problem satisfactorily as the rate of retardation as well as the inertial force increase during the braking. The danger of falls and consequent injuries of passengers is only delayed to the end of the braking period.

(iv) An escalator emergency brake must be so designed as to facilitate the control, maintenance and replacement of all components.

30

Heavy duty escalators for public service

Dipl.-Ing. Emil Braun, Orenstein and Koppel AG, Dortmund, West Germany

ABSTRACT

Nowadays escalators can be used without great problems in commercial buildings and department stores. There are external influences on escalators for public use and on Metro excalator installations besides the transport of passengers e.g. all influences aggrevate the task of the escalator and reduce its availability. Appropriate counter-measures are special protection against corrosion, selection of special materials, rich lubricants, strong gear components, automatic control systems and above all preventive maintenance.

The development aims at an improvement of these measures, an improvement of the service life of parts subject to wear, energy-saving drive systems and the application of computers and processor technology.

1. CAPITAL INVESTMENT AND CORROSION PROTECTION

Public service escalators are exposed to completely different influences than encountered in stores of offices. These influences arise, from the very nature of the installation and include:

(i) the unsupervised mode of operation;
(ii) rough treatment, even vandalism;
(iii) high incidence of dust and dirt, both during installation and in operation;
(iv) environmental influences (dampness, snow, salt spread to thaw ice, heat, cold) aggressive cleaning agents;
(v) operating cycles of 20 hours a day.

The influences of uncontrolled passenger traffic flows also contributes to the step chain running conponents being exposed to excessive wear, the steps and handrails becoming dirty quickly, panelling parts being dented, damaged or even destroyed etc. On the whole, escalators are subject to abnormal stops and an increase in the repair requirements. Such experiences, led to the basic considerations in the design of public service escalator of pricing as well as ease of servicing. Thus, a modular assembly programme was developed. The principal features are:

(1) A uniform lattice truss construction for all escalator models. The all-welded structure, made of MSH rectangular hollow profiles with a good moment of inertia to weight ratio can be competely hot-dip galvanized in one piece as shown in Fig. 1.

Fig. 1 — Hot-dip galvanizing an escalator truss.

(2) A water and dust-protected compact positive drive unit with a form-locked shaft for step band and handrail drives. All major drive components, such as sprockets, worm wheels and roller bearings, are run maintenance free in an oil bath. The common chain drives, with their wear susceptibility and high maintenance requirements, have been relinquished (Fig. 2).
(3) A step guide system is fitted with plastic rails in which the step chain (the actual step moving mechanism) is returned (from inclined to horizontal) so that the chain rollers are prtoected agianst overstressing. The step

Fig. 2 — Compact positive drive unit.

chains are supplied with oil from an automatic lubricating system which ensures a constant film of oil between individual chain links, even when they are exposed to water and high humidity. Stainless steel panelling covers prevent water, dirt and other aggressive media from falling directly into/onto the step guide system.

(4) Compact steps made of corrosion-resistant, die-cast aluminium (GD Al Si 12), with cleated risers, so that the resulting gap between consective steps which could prove a danger by trapping objects (Fig. 3). The closed ribbing of the steps means that, although of extremely light weight (approximately 16 kg), each step has a maximum distortion resistance and is extremely foot-sure. The low weight of the steps contributes to the circulating deadweight of the step chain being kept to a minimum. The steps have been type-tested statically and dynamically (Fig. 4.). The step rollers are lifetime lubricated and have watertight bearings (two ball-bearings with RS seal and a grease chamber in between, and, a second grease chamber between the seal and a nilos ring).

Fig. 3 — Cleated meshed steps.

(5) Handrails with a steel-cord axial strengthening (to prevent stretching during operation) and Nylon radial braiding to prevent shrinkage due to the influence of water. The handrail guideways are of stainless steel to prevent any rust formation during escalator stand-still that could lead to handrail damage upon the escalator being restarted.
(6) Heating installations for step chain and return stations. The robust gilled-tube radiators keep the accessible areas of the escalator free of snow and ice during winter operation. This system also makes it easier and cheaper (by saving power) to start up the escalator after it has stood still over a cold night.
(7) Electric motors with IP 65 protection, according to DIN 40 050 (protection against water jets and dust).

2. OPERATIONAL READINESS AND SAFETY

The measures described in section 1 cannot completely prevent breakdowns, irrespective of servicing. The service departments of various municipal underground transport corporations have, for example, noticed that 90% of breakdowns are due to delinquency/vandalism, 5% to objects falling into the systems, and the remaining 5% to failures/defects in the actual escalator running components. Whatever the cause of an escalator or

Fig. 4 — Dynamic test.

control system failure, a return to operational status is required as quickly as possible. To ensure this, the following possibilities are available:

(1) The connection of each escalator within each underground station, along a whole underground route, to a central control room. Upon a failure/defect occurring, the relevant monitor lamp on a control panel goes out (the signal transmitted is for example, a collective fault indication). The service engineer can then be sent to the respective escalator. A disadvantage of this type of system is that the fault and time consuming fault location then has to be carried out.
(2) The fault indication is undertaken in the same manner as (1) above, but with an extra facility in the form of a fault indicator panel on the escalator itself. This panel has an indicator connected to each safety

contact so that it is possible for the service person to localize the source of 'trouble' without any delay (Fig. 5). A further possibility in this respect is the installation of indicator lamps in two colours to enable a member of the station personnel (not the service persons) to see whether the escalator can be put back into operation or whether the presence of a service mechanic is required.

(3) Individual fault indications from each monitored escalator can also be passed directly to the central control room. The mechanic then has an indication of the symptoms concerned before leaving.

(4) Closed-circuit TV monitoring of the escalator, relayed to the control room. Such a system would then be combined with a collective fault indication in order to recognize intentional stoppages (usually due to emergency stop buttons being pressed) and to start up the escalator again via remote control. The opinions expressed by the operators of high traffic flow escalators differ widely with regards to remote control. On one hand there is the question of expense and the risk of damage by vandals; on the other hand the uncertainty as the cameras cannot cover each individual point of the escalator, or in case of human error in the control room. One fundamental safety aspect is that the escalator must be completely empty when being restarted.

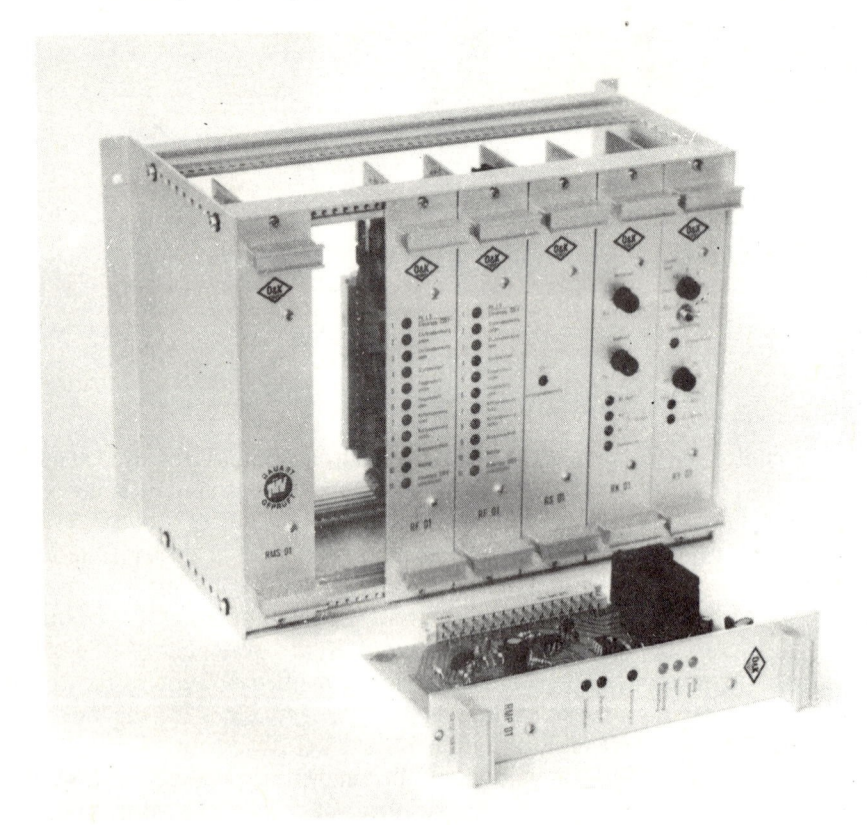

Fig. 5 — A fault indicator panel.

For the future, several noteworthy solutions have been drawn up and already partially put into operation in pilot installations where they have achieved a considerable increase in escalator availability (i.e. overall running time). They include:

(1) Optical monitors grouped as a display in the control room. Preprogrammed computers then assess all possible faults and defects and give a print-out of solutions and suggestions for repair. The optical monitors indicate the defective component on the escalator.
(2) An escalator self-monitoring system that checks an emergency STOP has tripped. Checking is carried out by a system of infra-red photocells positioned in the area above the step band (Fig. 6). Whilst the escalator is checking itself all starting switches are locked (i.e. rendered inoperative). Then following an interval of 10–30 seconds after the system has registered no objects on the stepband, the escalator switches back to its standby status of 'ready for operation'.

Fig. 6 — Photocells behind the skirtpanel.

3. SAFETY DEVICES

Experience has shown that the risk of accidents on unsupervised escalators increases primarily on account of careless and incorrect use, as well as through stupidity. Some 70% of accidents are due to falls on the steps and 30% for objects becoming jammed in the unavoidable gaps moving and fixed components. In order to reduce the risk factor as far as possible, suitable measures have been introduced:

(1) The meshing of tread and riser surfaces of consecutive steps in order to eliminate the continuous gap in which the toes and tips of shoes could become wedged.
(2) Specially rigid walls in the area of the moving steps (balustrade skirting) with an anti-friction coating to reduce the danger of objects being caught in the gap (maximum 4 mm) between step and skirting panel.
(3) Brush obstruction, with safety contacts, to protect the gaps at the points where the handrails run into the balustrade newels.
(4) Safety contacts, to safeguard the points at which the steps enter the combs.
(5) Installation of a braking system that ensures a smooth stop of the step band with a uniform retardation under all weather conditions! In this respect, load-dependent brakes (working on the 'flywheel effect' have proven themselves particularly suitable on long escalators (vertical rises of 10 m).

4. STEP BAND SPEED AND DRIVING PROPERTIES

The actual speed of an escalator step band has, surprisingly, to be different for users in different parts of the world. Some user groups require slower transport whilst others cannot be moved fast enough. Russian high-rise escalators for instance in the underground stations of Moscow, Leningrad, Prague, and Budapest have a speed of 0.9 m/s. The underground system in Stockholm, uses escalators with step band speeds of 0.75 m/s. In the Frankfurt underground, the downward speed was reduced from 0.65 to 0.5 m/s. An initial trial with a speed of 0.65 m/s is currently being undertaken in the Stuttgart underground tram system. Currently it seems as though a gradual increase from 0.5 to 0.65 m/s speed is becoming acceptable in West Germany. Whatever the speed, it is clear that all interconnecting escalators installed in one system should run at the same step band speed.

Another important factor towards user comfort is an absolutely smooth run with the maximum possible horizontal run (at least two step widths) at the landing points and with large-radius transitions from the inclined section into the horizontal, especially in the downward direction.

And last (but not least) the angle of inclination should not be greater than 30°; first to ensure optimum travel lines; secondly to avoid any 'pit' impression and feelings of vertigo when looking into a deep, downward travelling step band. O & K have installed new escalators with 23° inclination, which guarantee comfortable walking on the running escalator.

31

Escalator safety in the UK

Mr B. G. James, Health and Safety Executive, Bootle, UK

ABSTRACT

The use of modern escalators incorporating rigid, low-friction skirtings and maintaining the clearance between step-treads and skirtings to 4 mm maximum (3 mm at opposite side) has not eradicated the trapping risk.

In 1983, the Health and Safety Executive decided that additional safeguards should be provided at UK escalators, particularly those travelling downwards. In particular, a deflector device was recommended which would prevent trapping, rather than a device designed to stop an escalator automatically very soon after trapping has occurred.

London Regional Transport (formerly London Transport Executive) have many years experience of escalator usage and had long ago incorporated deflector devices in the balustrades of their escalators. They had recently adopted a bristle-type of deflector device which could be fitted retrospectively to the metal balustrades of their modern escalators.

Consequently, the Health and Safety Executive issued guidance for safety at escalators which included the recommendations that suitable deflector devices be fitted to all new escalators and to existing down-travelling escalators. In practice, many owners of large numbers of escalators have been conditionally permitted to complete the retro-fitting exercise over a period of four years.

1. INTRODUCTION

In the early years of universal use of lifts and escalators, lift accidents were often very serious, sometimes causing loss of life. Escalator accidents rarely resulted in serious injury of death, and it is not surprising, therefore, that lift safety attracted the greater interest of legislators and standards authorities. In the UK this was particularly so for lifts used in factories, offices, shops, where persons worked, and, such lifts form a large proportion of the lift 'population', which itself exceeds the escalator 'population'. Until the advent of the Health and Safety at Work, ect. Act, 1974, health and safety inspectors had rarely been required to consider the subject of escalator safety. Escalators, apart from those installed at public transport undertakings, were usually to be found in department stores and multiple retailers' stores, and were intended almost wholly for use by members of the general public rather than by the workforce. Since 1974 escalator safety has received a greater degree of attention from safety authorities, escalator owners and Members of Parliament.

The largest single owner and user of escalators within the UK has long been the London Underground and their experience of escalator safety over the past 70 years or so is naturally relevant. However, it is the safety of escalators used in public areas other than public transport undertakings which has been the main cause of concern within the UK in recent years.

2. CURRENT POSITION

An escalator forms an alternative in the two-part system of mechanized transportation most frequently provided within our society for transferring members of the general public from one floor level to another. The other part of that system is of course the lift (elevator). Often, both these parts of the transport system are available for use within the same premises, and it is then that the distinction is most clearly defined.

When using a lift, passengers enter a 'box' which is then sealed before it ascends or descends to a different floor level. During such time the passengers are not exposed to trapping or falling risks etc., other than those which might be caused by failure or deterioration of lift components. There is little that passengers can normally do, either wilfully or inadvertently, to interfere with the safe operation of the lift.

However, passengers using escalators are exposed immediately to risks, which are present as long as the escalators are in the use. Such risks exist even when all components are functioning as intended. These risks are created by the very nature of escalators (i.e. moving stairs) and in some cases because it has long been accepted that exposed running clearances are an inevitable and tolerable feature of escalator design.

Escalators have now been in use for more than 80 years and whilst advancing technology has produced certain improvements in design etc., it seems to have been accepted that particular types of accidents will continue to occur at tolerable frequency rates. This clearly seems to be the situation in the case of falling and trapping accidents. Whilst reference is made in the ensuing sections to both falling and trapping of persons, attention is focused upon particular trapping hazards and recommended protection against such hazards.

3. FALLING OF PERSONS

Movement of persons between different levels is a potential source of falling accidents, even when inclines (gradients) are provided for persons to negotiate under their own power and at their own speed, e.g. fixed stairs or ramps. Predictably, the use of moving stairs (escalators) increases the likelihood of falling accidents, particularly when escalators are being used more and more by young children (Fig. 1). However the increase in actual accidents may not be as great as anticipated.

Rated speeds and angles of incline of escalator treadways have generally become standardized, and are unlikely to be changed drastically in the foreseeable future. It may be anticipated that members of the public including adults with accompanying children will continue to act as they do now when using escalators, so that the frequency rate for accidents caused by falling is unlikely to be greatly reduced.

Such factors as better illumination, increased slip-resistance of step treads, clearer demarcation between stationary landings and moving treadways, etc., will help to reduce accidents. Also, experience has indicated to some authorities that greater radii of curvature in the transitions of the treadway from the horizontal to the inclined are needed, as well as an extended length of horizontal treadway travel at the exit and entry ends of escalators adjacent to landings.

4. TRAPPING OF PERSONS

Trapping risks exist because of the existence of running clearances between adjacent parts of an escalator, although advances in technology have reduced such clearances (Table 1) to a seemingly ultimate practicable level, as indicated by national and international standards (codes).

Whilst these escalator clearances are exposed to every passenger, it has been known for many years that children and young persons are at greater risk from such clearances, because of the greater likelihood that they will be wearing particular types of 'soft' footwear (e.g. training shoes, 'sneakers', wellington boots), or because they are unable to read or assimilate written or pictorial warnings.

The UK experience of foot and leg trapping accidents indicates that the type and/or condition of footwear being worn are contributory factors

Fig. 1 — (Reproduced by permission of London Underground Ltd.)

towards such accidents. However, accident experience within USA suggests that footwear is not so often a contributory factor, even when trapping of legs occurs between steps and balustrade skirtings.

5. PASSENGER SAFETY

When passenger safety is considered, it is fair and logical, to make a comparison between safety at lifts and safety at escalators. Statisticians may always argue over such factors as accidents per passenger-mile, accidents per passenger-journey, overall numbers of lifts/escalators, type/severity of accidents, etc., but, it cannot be denied that today, members of the general

Table 1 — Summary of standards

Standards	Step/skirt clearance (any side — maximum)		Step/skirt clearance (total — both sides maximum)	
	in.	mm	in.	mm
BS 2655:4:1969	0.19	5.0	0.25($\frac{1}{4}$)	6.5
ANSI.17.1–1982	0.187($\frac{3}{16}$)	4.8	0.375($\frac{3}{8}$)	9.6
EN 115:1983	0.16	4.0	0.28	7.0
ANSI.17.–1925	None Quoted		None Quoted	
ANSI.17.1–1960	0.187($\frac{3}{16}$)	4.8	0.25($\frac{1}{4}$)	6.5
ANSI.17.1–1965	0.187($\frac{3}{16}$)	5.8	0.25($\frac{1}{4}$)	6.5
ANSI.17.1–1971	0.375($\frac{3}{8}$)	9.6	0.75($\frac{3}{4}$)	19.2

	Skirting panels, rigidity etc.	Skirt obstruction device
BS 2655:4:1969	'Of rigid construction'	None
ANSI.17.1.–1982	Shall not deflect more than $\frac{1}{16}$ in. (1.6 mm) (Force: 150 lbf (667 N)	Yes At upper/lower comb plates
EN 115:1983	4 mm—Force: 150 N (337 lbf) over 25 cm²	None

public are more at risk when travelling on escalators than they are when travelling on lifts.

Since the 1920s the London Underground have considered it necessary to fit deflector devices to the balustrade skirtings of their escalators. The emergence of the modern designs of escalators has not induced London Underground to modify their policy regarding deflector devices. Over the years, various types of deflector devices were incorporated at the manufacturing stage into the balustrades of escalators (Fig. 2). More recently, some London Underground escalators have been retro-fitted with a new type of deflector device comprising a continuous strip of bristles fitted along each balustrade skirt. This bristle-type deflector device is now being incorporated into the balustrades of new London Underground escalators at the manufacturing stage. (A typical example can be seen at the Marble Arch London Underground station.) A detail is shown in Fig. 3.

Deflector devices have been fitted in the past to public transport escalators in at least one European capital city other than London, although the devices have been different to those mentioned above.

6. ACTION TAKEN WITHIN THE UK

Safety standards — particularly those intended to prevent trapping — have been reviewed, and the effectiveness of the Euronorm (EN 115) in this respect has been carefully assessed.

Fig. 2 — (Reproduced by permission of London Underground Ltd.)

The Health and Safety Executive consequently published guidance in 1983 which 'summarizes risks associated with the use of escalators and recommends standards of safety to be met by those in charge of premises in which they are installed'.

The experience gained by London Underground in the use of deflector devices on escalators is reflected in this guidance, which should now ensure that from 1 January 1984:

(i) All *new* escalators will be provided with suitable deflector devices before being taken into use.
(ii) Over a period of four years, all existing down-running escalators will be retro-fitted with suitable deflector devices.

7. CONCLUSION

It is not claimed that deflector devices provide the ultimate protection against certain trapping hazards, and there is certainly no reason for complacency, now or in the foreseeable future. It was recognized that something positive had to be done quickly, but the opportunity exists for perhaps better solutons, which may eventually emerge.

Naturally, it is preferable for hazards to be removed by design rather than hazards being left in and in need of protection when an escalator is installed and taken into use.

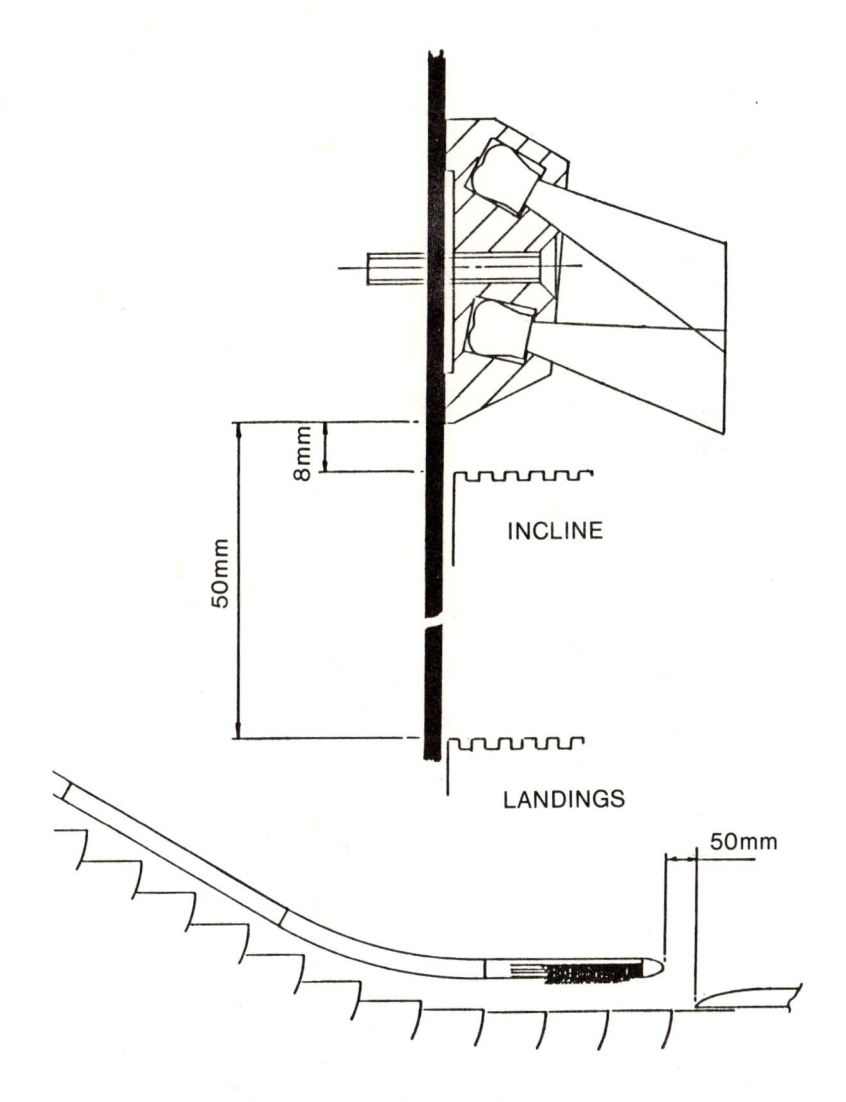

Fig. 3 — Detail of bristle-type deflector — (Reproduced by kind permission of Kleeneze Industrial Ltd.)

32

Wheelchair escalator

Mr E. Watanabe, Mr S. Yokota, Mr. M. Yonemoto and Mr M. Asano, Mitsubishi Electric Corporation, Inazawa, Japan

ABSTRACT

In recent years, the outdoor activity of persons confined to wheelchairs has been increasing. Mitsubishi Electrip Corporation has developed the wheelchair escalator by adding a special function to an ordinary escalator following a long study of escalator technology.

This escalator is equipped with special steps to make it available for a person on a wheelchair. On this escalator, a wheelchair will ascend or descend with the particular step, safely keeping the wheelchair in a comfortable position. The escalator is also available as an ordinary escalator whilst it is not used by a person on a wheelchair. Mitsubishi Electric Corporation has completed eighteen units in Japan.

This chapter describes the structure, principles of operation and basic specification of Mitsubishi wheelchair escalator.

1. INTRODUCTION

Outdoor activity of persons confined to wheelchairs has been increasing in recent years. It was resolved, by the conference in 1981 International Handicapped Person's Year, that various kinds of facilities and establishments for handicapped persons were to be improved continuously within ten years and the necessity of these facilites has been recognized by persons in every field across the world.

Under these circumstances, Mitsubishi Electric Corporation has developed a wheelchair escalator available for the person in a wheelchair by adding the special function to an ordinary escalator as a result of a long study of escalator technology.

2. OUTLINE OF WHEELCHAIR ESCALATOR

The Mitsubishi wheelchair escalator will be an escalator equipped with special steps to make it available for a person in a wheelchair to get onto the steps. On this escalator, a series of steps will ascend or descend with a particular step formation which keeps the wheelchair safely in position. The escalator is also available as an ordinary escalator whilst it is not used by a person in a wheelchair.

The characteristics of the wheelchair escalator are as follows:

(i) It does not require any particular equipment other than the escalator for wheelchair operation.
(ii) Planning of the building construction is easier, because the escalator enables both ordinary passengers and persons on wheelchairs to use the same route.
(iii) The administration of the system is simplified.
(iv) The mechanism for the use of the wheelchair is contained within the escalator, and it can be operated automatically by the starting switch.

3. BASIC STRUCTURE

The escalator is equipped with three special steps in series for carrying the wheelchair. These three steps have the same function as that of other steps when the escalator is under ordinary operation. Only when the escalator is operating for a person in a wheelchair, will the special steps perform the special function. The function and the structure of the special steps for carrying a person in a wheelchair are as follows (refer to Figs. 1 and 2).

As shown in the figures, from the lower side of the escalator, the special steps are named as 'wheel stopper step', 'fork step', and 'inclination step'.

When the escalator is under wheelchair operation, the tread and the riser of the 'wheel stopper step' will be supported by the adjacent 'fork step'. The 'wheel stopper step' and the 'fork step' will make two steps of the same level and these steps will travel from the lower landing to the upper landing or from the upper landing to the lower landing.

Only during wheelchair operation, will the wheel stoppers automatically emerge from the tread of the 'wheel stopper step'. The 'inclination step' at the highest position of the special steps will have special guide rollers, and these rollers will engage with the special guide tracks arranged in the truss.

The centre part of the 'inclination step', whose width is 750 mm, will incline with the movement of the step to secure the space for the wheelchair.

The basic specification for the wheelchair escalator is given in Table 1.

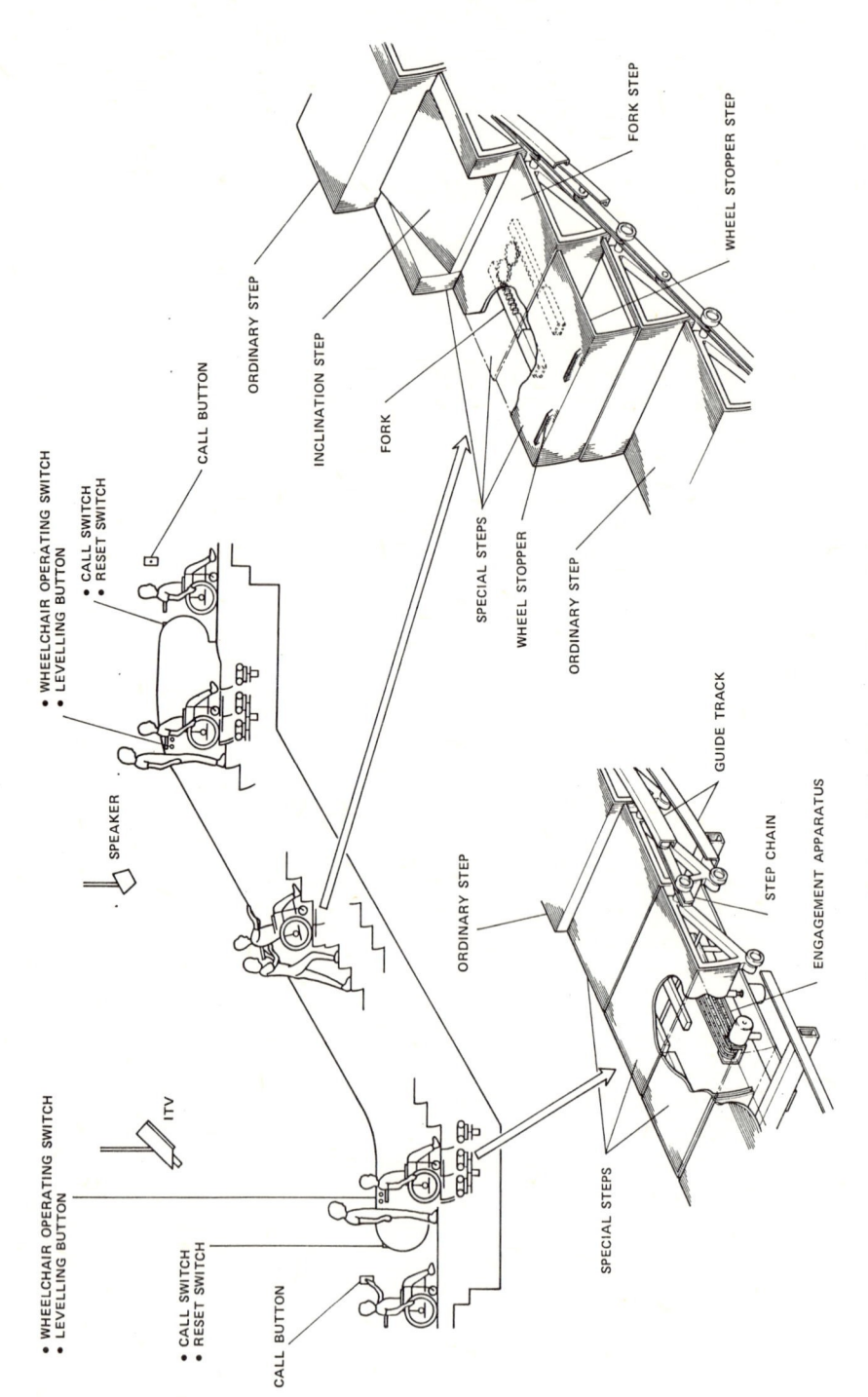

Fig. 1 — Outline of Mitsubishi wheelchair escalator.

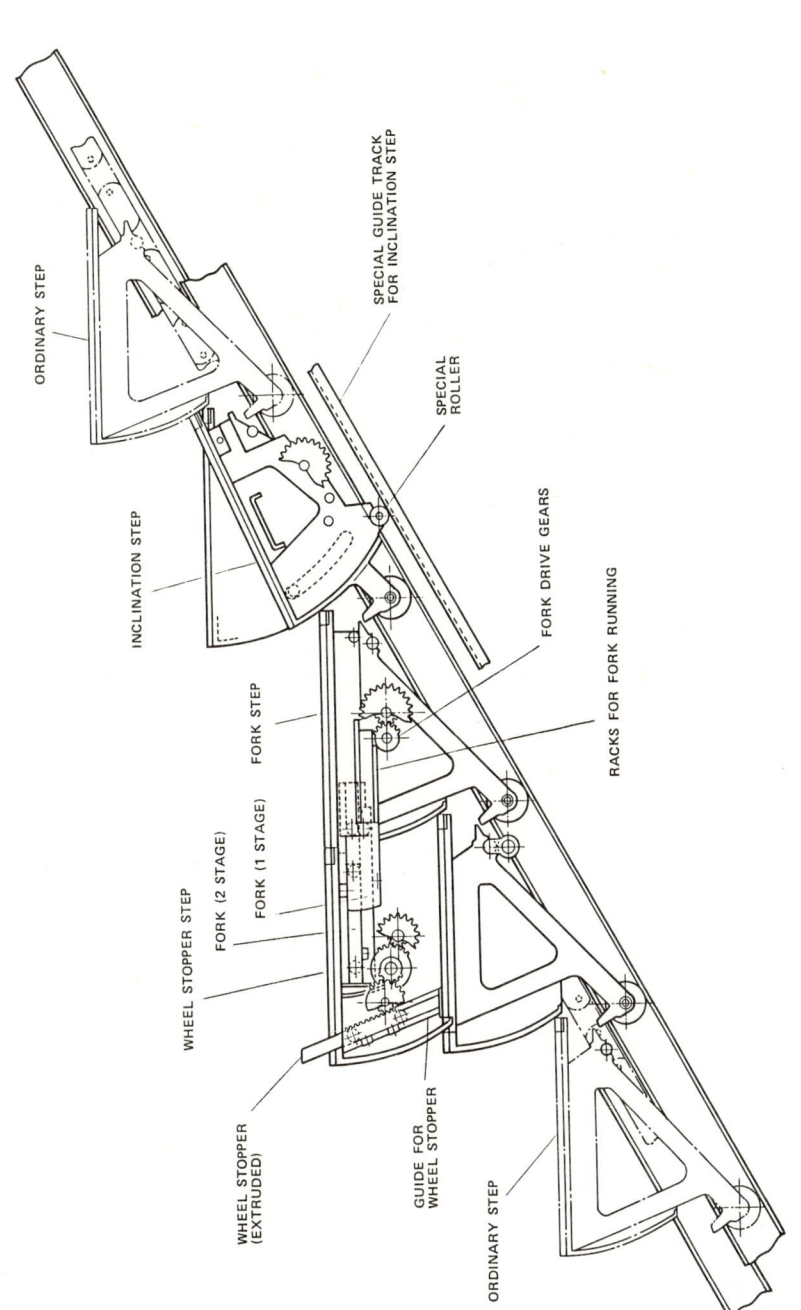

Ch. 32] **Wheelchair escalator** 321

Fig. 2 — Special steps at 30° inclined portion (side section).

Table 1 — Basic specification for wheelchair escalator.

Nominal width	1200 mm
Nominal step width	1000 mm
Speed	30 m/min at ordinary operation
	15 m/min at wheelchair operation
Bulustrade	Panel type
Call button	To be pushed by the person on the wheelchair to call the operation personnel. (Installed near the upper and the lower landing.)
Operating switch	Starting switch
	Signal switch
	Emergency stop switch
	(above three switches are the same as those of an emergency escalator.)
	Call switch
	Wheelchair operating switch
	Levelling button
	Reset switch

Operational peripheral apparatus	
CCTV monitor	Consisting of a camera and monitoring TV. (To monitor escalators and the surrounding area of the entrance landings.)
Announcement	An announcement can be made through the speaker installed close to each landing using the auto announcement apparatus.
Indicator panel	The indicator panel can be installed in the operations room and indicates the operating condition of the escalator.

4. OPERATING PROCEDURE

This is best explained as a sequence of steps and reference to Fig. 3.

(1) The call button close to the entrance landing of the escalator will be pushed by the person in the wheelchair to call an attendant.
(2) The attendant will come to the calling escalator and inform the passengers on the escalator by natural or synthesized voices: 'Operation is going to be changed for wheelchair operation'. After the attendant has confirmed that the ordinary passengers are off the escalator, then the switch to call the special steps for the wheelchair will be set.
(3) The special steps will stop at the entrance landing. The person in the wheelchair will get on the special steps and engage the brake of the wheelchair.
(4) The attendant will push the operating switch for the wheelchair after confirming wheelchair safety. The escalator will start in the upward operation after the wheel stoppers come out.

In the downward operation, when the special steps stop at the entrance landing, the wheel stoppers come out. In the downward operation, when the special steps stop at the entrance-landing, the wheel stoppers come out.

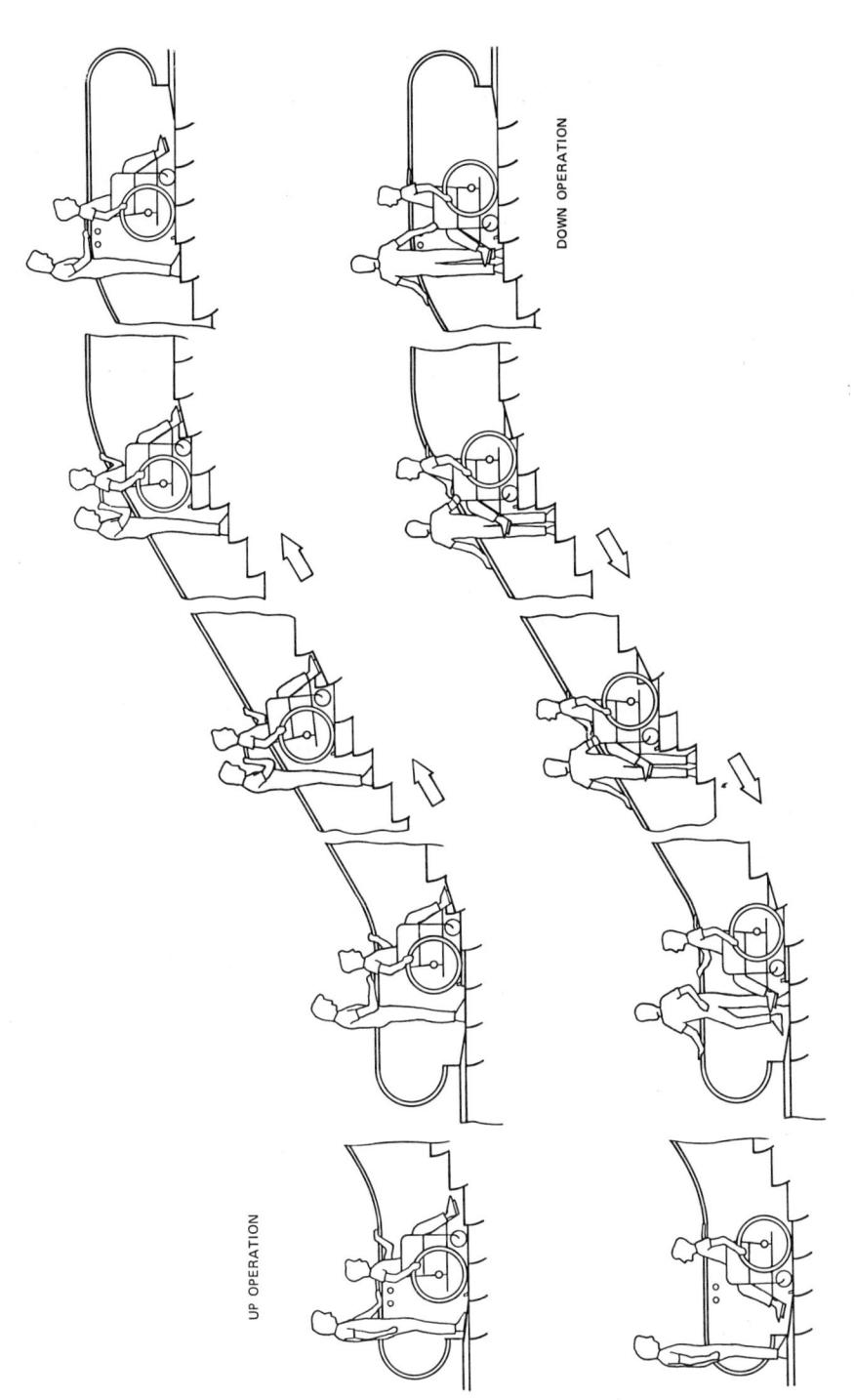

Ch. 32] **Wheelchair escalator** 323

Fig. 3 — Movement of wheelchair.

(5) As shown in Fig. 3, the person in the wheelchair will ascend or descend accompanied by the attendant.
(6) When the special steps arrive at the other landing, the escalator will stop automatically. The person in the wheelchair and the attendant will get off the escalator.
(7) After the person on the wheelchair leaves the escalator, the wheel stoppers will be put back. Then the attendant will operate the 'starting switch', and the escalator will be returned to normal operation.

5. CONCLUSION

The wheelchair escalator is already in service at Yokohama, Tokyo and Nagoya, Sapporo, Japan, and is acquiring a good reputation.

33

Spiral escalator

Mr S. Goto, Mr H. Nakatani, Mr T. Kaida, Mr M. Tomidokoro and Mr R. Saito, Mitsubishi Electric Corporation, Inazawa, Japan

ABSTRACT

The idea of a spiral escalator has a long history. However, since it has a complex of three-dimensional curves, the idea has not been able to turn into realization due to a number of technological difficulties.

But now, Mitsubishi Electric has succeeded in developing the world's first spiral escalator that moves up and down along spiral lines, through the full application of the computer analysis technology and NC machine tools.

The most distinctive feature for realizing this escalator is the application of the innovative mechanism that varies the centre and radius of curvature in accordance with the postions of steps.

This chapter introduces the drive theory, the drive mechanism, structural features and safety of the Mitsubishi spiral escalator.

1. INTRODUCTION

Elevators and escalators have been playing an important role as an effective element in the interior space of the building as well as a means of transportation. Modern architectural designs have been more and more popularly employed in open interior spaces for amenity, exhibition or show, and this recent trend requires elevators and escalators to have something more distinctive as an interior element. Mitsubishi Electric has successfully developed the world's first spiral escalator, which goes up and down along a spiral curve, to meet the demand of modern architecture.

Escalator engineers have desired the realization of a spiral escalator ever since the appearance of the world's first escalator at the International Exposition in Paris in 1900. A few years after the Exposition, an idea of a spiral escalator was presented in an application for a patent in the USA, and some attempts were made. However, the complexity of spiral structure comprising various three-dimensional curved surfaces has been making it very difficult to develop into practicality.

The technological problems have been solved by maximally employing computers for analysis and advanced NC machine tools for precision manufacturing operation, and by the development of innovative mechanisms incorporating a number of original ideas.

It is expected, that this new escalator will have a significant effect upon the world of architecture and the escalator industry as an epoch-making product, which makes it possible to create a new architectural space. Since spring 1985, four Mitsubishi spiral escalators have been operating at Tsukuba Shopping Centre (Fig. 1) and Oska International Exhibition Hall and have achieved a good reputation.

Fig. 1 — Appearance of spiral escalator.

2. STRUCTURE AND DRIVE PRINCIPLE

2.1 Overall structure

The spiral escalator is obtained by bending a conventional linear escalator into an arc shape. It consists of upper and lower horizontal zones, an intermediate inclined zone where the steps keep a fixed rise and the transit zones where the steps gradually change in height and which smoothly connect the upper and lower zones. The steps move up and down across these zones along a three-dimensional curved line.

Ch. 33] **Spiral escalator** 327

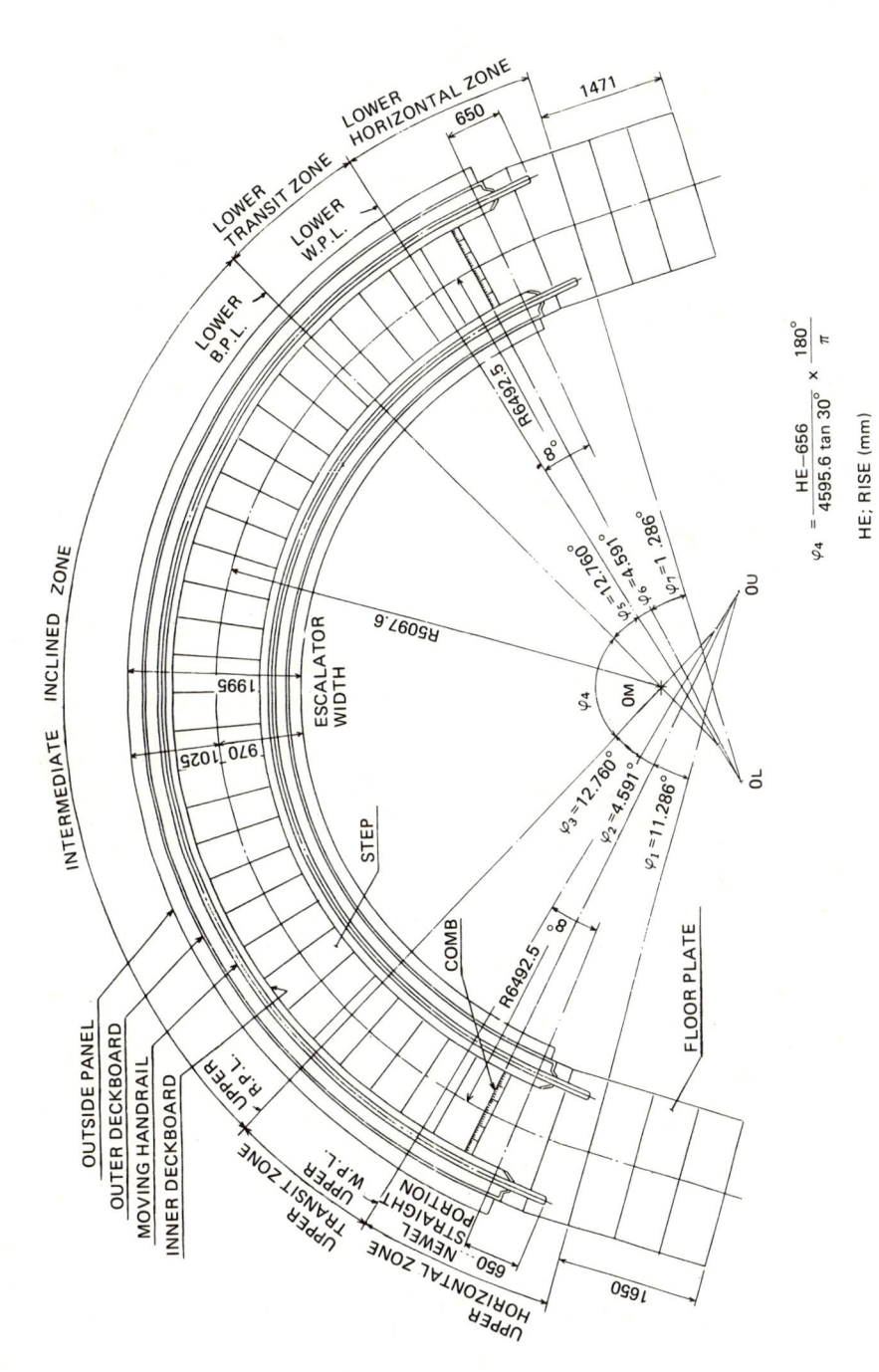

$$\varphi_4 = \frac{HE - 656}{4595.6 \tan 30°} \times \frac{180°}{\pi}$$

HE; RISE (mm)

Fig. 2 — Plan view.

Fundamentally, the escalator has to operate so that the tread surface of all the steps is kept horizontally, the clearance between adjacent steps is kept constant and the steps move synchronously with moving handrails over the whole range of the operating zones. That is, in an arc-shaped spiral esculator, the steps and the handrails must move at equal angular velocity with respect to the centre of curvature. However, it has been very difficult to turn this principle into practicality since the escalator has a complicated combination of the horizontal, inclined and transit zones described above and since the inclination angle varies across the inside and outside peripheries. Computers were employed to help develop a configuration incorporating several new concepts.

Figure 2 shows the plan view of the spiral escalator. Each zone of step movement has its centre of curvature as 0U and 0L for the upper and lower horizontal zones and 0M for the intermediate inclined zone. The radius of curvature varies between the inclined zone and horizontal zones as shown. The centre of curvature of the transit zone shifts from 0L to 0M in the lower section and from 0M to 0U in the upper section. The radius of curvature of this zone accordingly changes gradually. The locus of centre in this zone is determined, based on a numerical analysis which corresponds to the rise between steps.

This concept of a centre movement system has made it possible to keep the angular velocity of the steps and handrails constant with respect to the centre of curvature over the whole range of escalator movement. It is possible to introduce a variable velocity mechanism as the driving system for the steps and handrails, in order to make both the centre and radius of curvature fixed for the whole range of operation, and to allow the exterior configuration to be simple. However, a variable velocity mechanism requires a very complicated drive system, which results in inferior reliability.

2.2 Drive mechanism

Figure 3 shows a general structure of the drive mechanism. The drive unit (driving machine and motor) located inside the upper portion of the truss drives the endless step-chains through the drive-chains. At the top and bottom ends, the step-chains engage with the inner and outer sprockets which are different in diameter. The step guide tracks and sprocket teeth near the turning zone are designed to have a special configuration so that the steps turn smoothly and make a conical curved surface. The moving handrails are driven by the upper sprockets synchronized with the steps. In this way, a complicated three-dimensional drive movement is controlled by a mechanism which is basically similar to that of the linear escalator, and which has been realized by a simple and realiable drive system. All of this has been accomplished by adopting the centre movement system.

2.3 Steps

As shown in Fig. 4, a step consists of a tread, a riser and other parts. The tread has a sector shape opening from the inside to the outside periphery and has concentric cleats on it. The riser plate is made from a part of a conical

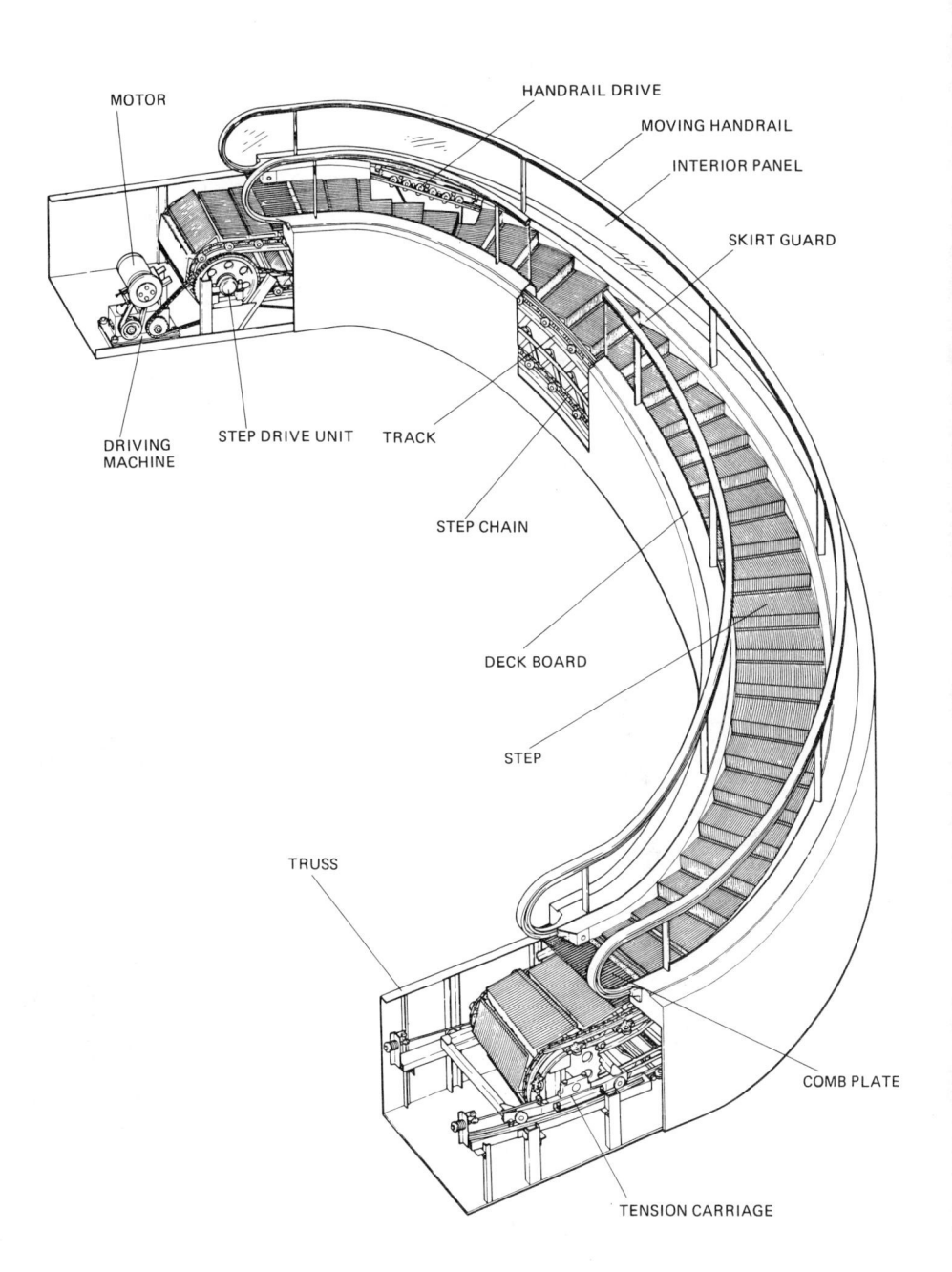

Fig. 3 — Central structure.

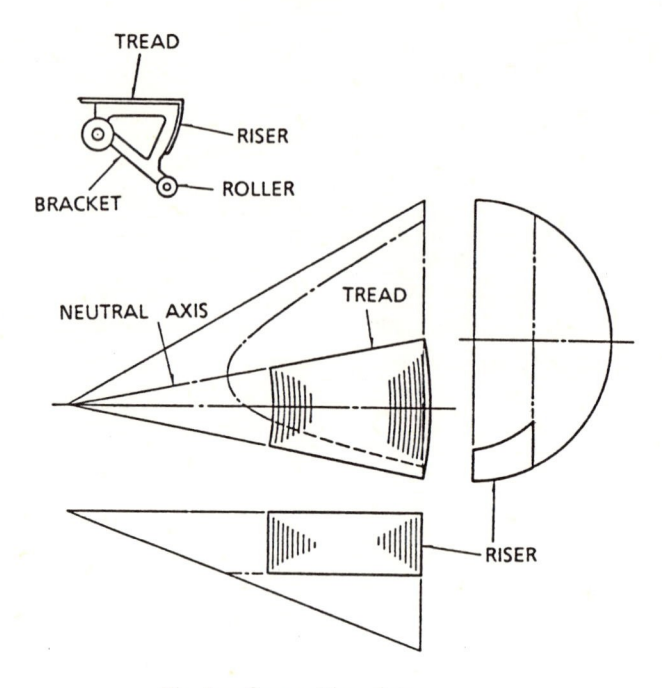

Fig. 4 — Composition of step.

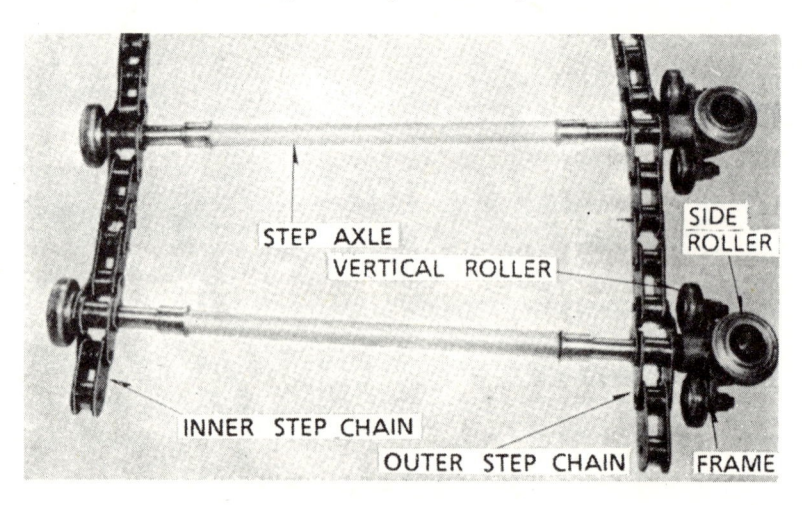

Fig. 5 — Step-chains and axles.

curved surface and has cleats of which the curvature increases in an arithmetic progression from the inside to the outside periphery. Therefore, the engagement of the tread and riser of adjacent steps, which move relatively to each other in the upper and lower transit zones, and that of the tread and comb-plate at the top and bottom landing point, occur in the same manner as that of linear escalator. Also, the combining effect of the cleats provides safety at the same level.

2.4 Step-chains

The step-chain and axle, as shown in Fig. 5, comprise inside and outside roller chains, which are different in pitch, and step-axles and rollers which are mounted on to the chains at a fixed interval. The step-chains incorporate spherical bearings to make the steps travel smoothly in a three-dimensional direction along a spiral track. Another structural feature, which ensures smooth three-dimensional movement is given by the axle end on the outside periphery. This end of axle is provided with a set of two vertical rollers and a side-roller. The side-roller is to support the force, which acts in the direction of arc centre, caused by the tension of the step-chains and also functions as a guide roller of step in the horizontal direction. A pair of vertical rollers run on the guide track, while restricting the inclination of the side roller for a smooth guide movement with minimum slippage.

2.5 Truss

The truss supports the total weight of whole equipment and passengers. The whole structure consists of five or six blocks which are joined with high tension bolts. each block is of welded construction mainly using shape steel and has an approximate arc shape of which nodal points are connected with straight lines.

Since the truss is subjected to large torsional moment caused by the eccentric load, it must be supported at an intermediate point as well as at the top and bottom ends. Computers were used for the three-dimensional structural analysis and strength check in order to analyse the complicated configuration.

3. CONCLUSION

A number of features and accomplishments can be earmarked.

(i) The spiral escalator is the world's first configuration which features three-dimensional curved surfaces and can help to create innovative and unique architectural space.
(ii) The panoramic effect which changes every moment while riding gives passengers a fresh feeling and riding fun.
(iii) The constraint conditions required in respect of the building are almost the same as those for a linear escalator. This means a wide applicability to a variety of buildings.
(iv) Complicated three-dimensional movement has been realized by a highly reliable simple drive mechanism. The advanced manufacturing technology has made it possible to achieve precision for three-dimensional bending and forming.
(v) The spiral escalator provides the ultimate in safety between adjacent steps, step and comb-plate and step and skirt quard equally with linear escalator.

Part 10
Epilogue

34

Practising engineering, we each make a difference

Mr V. Quentin Bates, Lerch, Bates and Associates, Littleton, USA

ABSTRACT

This chapter is an edited version of a keynote addresses. It deals with ingenuity and the sharing of knowledge and ideas. The author suggests that an organization such as the genesis International Association of Elevator Engineers is a step in the right direction.

1. INGENUITY

We had some unexpected snow last fall in Colorado; snow in quantity in October before everyone had snow tyres mounted and shovels ready. One evening as I approached the house, under such conditions, I noticed my neighbour trying to get up the driveway in his compact pickup. He was about halfway up with the wheels spinning. A I got out of my car I thought, 'I should go help him', followed almost instantly with the second thought, 'It would take more than just one person to get him up the drive'. I went over anyway, my neighbour rolled down the truck window as I approached. His response to my offer was similar to the thought I had had, 'I don't think it would do any good, I'll just leave it here'. I suggested we try, so he backed down to the road and with me pushing from behind, we started up the drive again. I would like to report that we made it — we did not. The truck got to about the same spot as before and began to spin out. My neighbour stuck his head out of the window, 'Let's give up — thanks for the try'. But now, I was not ready for defeat. 'Let's try once more, but first I'll clear some tracks up the drive to the garage'.

Having tramped down the snow in front of the tyres, we again returned to the street and started up the drive. We got to the spot where we had spun out before and the tyres started to spin, but ... we kept inching forward. I continued to push, the car kept creeping forward and finally, the truck was in the garage.

As I walked home several thoughts passed through my mind — 'If I had not gone over, nothing would have happened'; Success does not necessarily come easily on the first try, sometimes you have to consider alternate ways to solve a problem; and, 'one person can make a difference'.

Those thoughts have application to us. One person can make a difference; each of us can make a significant contribution by sharing an idea, by communicating understanding, by being a friend, by expressing a view, by taking some action. Nothing happens until someone does something. Success may not be immediate.

A line in Proverbs says, 'There is nothing new under the sun'. As we contemplate the revolution in equipment and application which we have seen and are discussing at this conference, we might well keep that in mind. Most progress comes by better application of the known. Most inspiration is perspiration; and only occasionally does any individual or event burst like a skyrocket giving absolutely independent illumination of a problem or process. The real genius of the average man, available to each of us, is the desire to improve upon what has been done before, no matter how significant the task.

Each of us bear or aspire to the title, 'engineer'. Why else should there be an inaugural gathering of the 'International Association of Elevator Engineers'? We might well ask, What is an engineer? In simplistic form, it is a noun. An engineer is a person trained in, skilled at or professionally engaged in a branch of engineering. The Latin derivative, ingenium, means 'talent'. As engineers, we engage in engineering. I believe this is important. If we are engineers, we engage in *engineering*.

> Application of scientific principles to practical ends as the design, construction and operation of efficient and economical structures, equipment and systems.

There are several key words in that definition, but let me focus on three: *practical, efficient, economical*. As we seek, as engineers, to improve what has been done before, we should emphasize the practical, the efficient and the economical. Not at the expense of safety or quality or aesthetics, but concurrently with them.

Several years ago, *Elevator World* featured the World of 'Elevator Consultants' in an annual issue. I was asked what quality I felt was essential for a successful consultant. I immediately thought, 'technical competence', and answered, 'ingenuity'. Ingenuity is a quality available to each of us. I think of this quality as something which should be regularly exercised. Using imagination (rather than dogmatically repeating what has been done before) allows a person to seek that heart of a proposition and evaluate alternative solutions.

2. TWO HEADS ARE BETTER THAN ONE

I would like to talk now about a situation that developed in our office in respect of a small hotel. We take pride in large jobs which receive major attention and are highly visible, but there is a satisfaction in small details, well done, which every craftsman feels and understands. Every job cannot be a World Trade Center or an Eiffel Tower; in fact, the small job with limited budget, may call for the greatest amount of imaginative effort.

The hotel is relatively small, 220 rooms and six floors. The architect's plan (Fig. 1) indicated space for two elevators grouped together. One had a

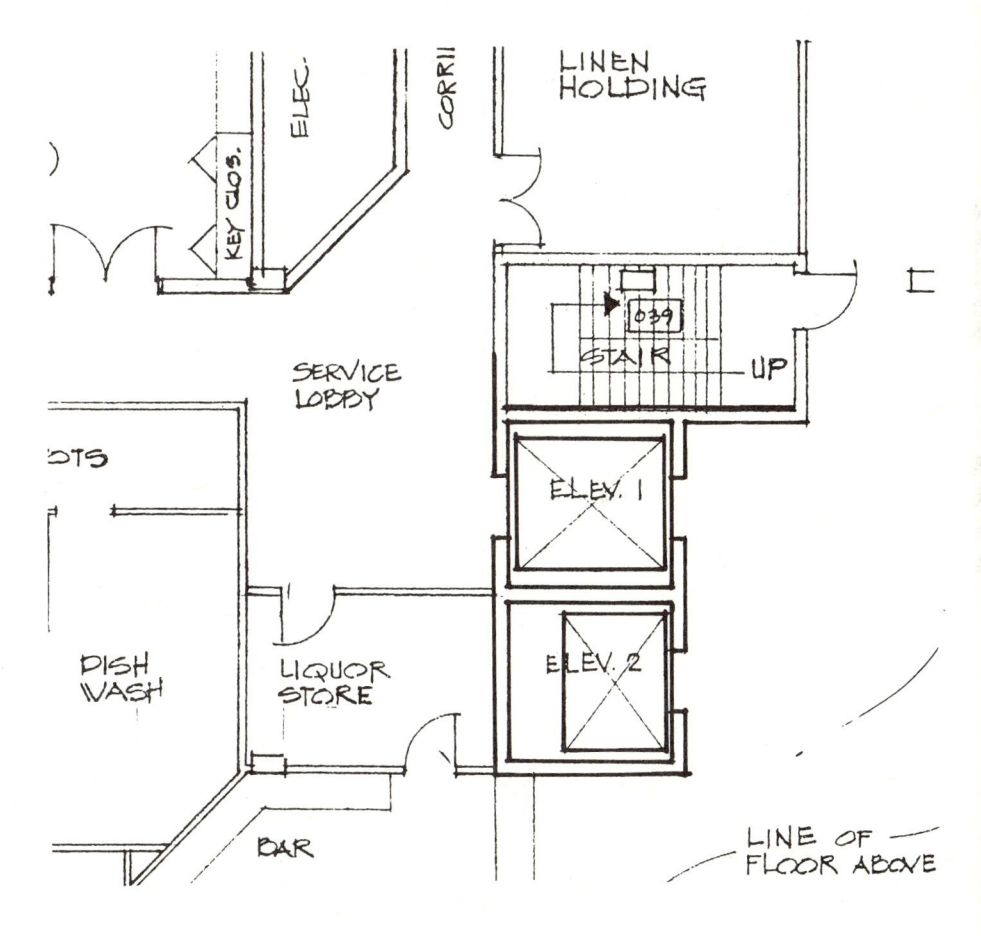

Fig. 1 — The architect's plan.

rear door accessing the service corridor leading to the kitchen. The drawing was $\frac{1}{8}$ in scale so there was little detail but dimensionally the hoistway box approximated 18 ft wide by about 10 ft deep. The owner's project manager asked us to provide elevator car and hoistway dimensions for the structural engineer's guidance. He told me, 'I need the information this morning'. So as soon as I hung up the telephone, I began to doodle on a scratch-pad.

One of the other consultants passing by asked me what I was working on; I briefly explained. He said, 'I've got a layout of just that arrangement, I'll get you a copy' (Fig. 2).

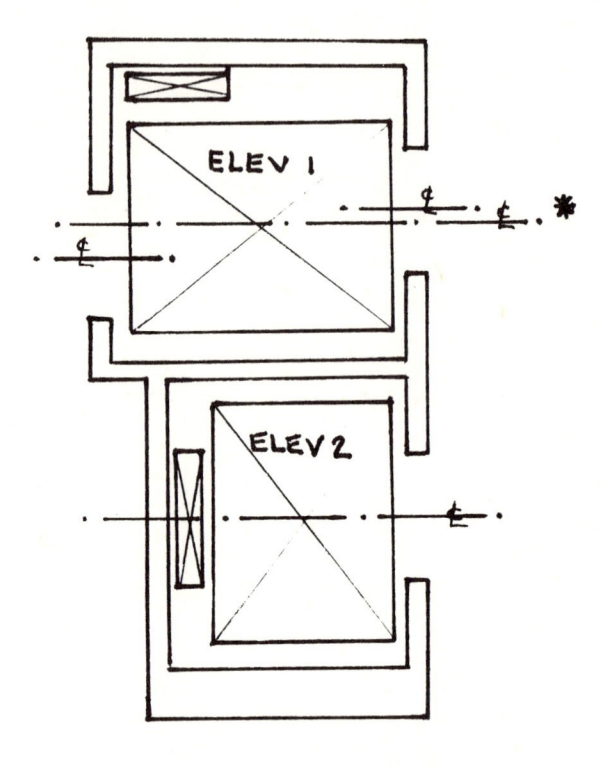

Fig. 2 — The previous plan: door and car not centralized *.

The passenger car was *practical, efficient and economical* — perfect for my engineering sense. The car arranged with rear doors was similarly straightforward. Sized somewhat deeper to facilitate loading carts and handling freight, it had the counterweight on the side and entrances arranged as shown. 'Boy', I thought, 'that's perfect; just what I need'. Then the imagination began ... looking at the drawing I realized that someone had put thought into the layout. Dimensions were somewhat standard and the entrances were almost symmetrical on the hoistway. The plan shown would be different on my job though, because the walls were 8 in thick and touched the hoistway like a stair riser. The front wall on one side would be 8 in wider than the other because of this sidewall thickness. As I studied the drawing, I decided that balancing the front wall would be important to the architect. The elevators were a prominent feature in the lobby; balance would be harmonious with other design elements. The first revision with the front entrances balanced is given in Fig. 3.

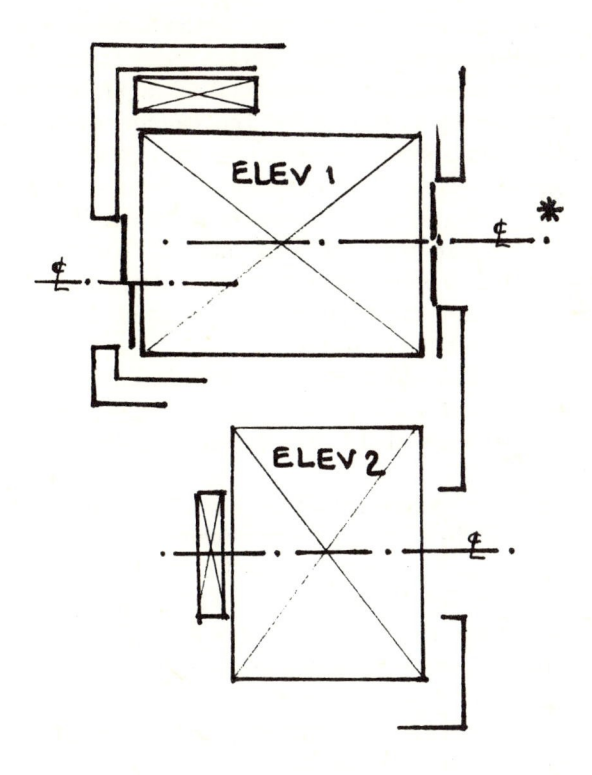

Fig. 3 — First revision: front entrances balanced *.

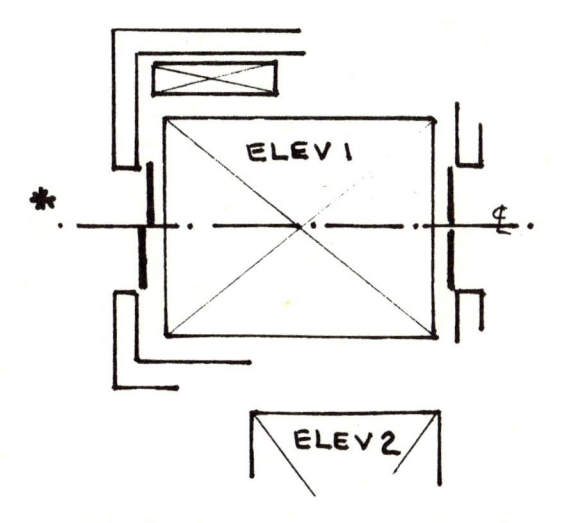

Fig. 4 — Second revision: the ultimate — front and rear entrances aligned *.

I looked some more. If balancing the hoistway entrances was desirable, so would equalizing the front returns inside the service car. It did not take long to find a position for the car that accomplished both objectives. Having gone that far, I got a little cocky. The rear door was two-speed, side opening, in order to maximise the opening width for service use. But why not try to completely balance the car by centring the rear door. I was having trouble achieving this goal until the thought occurred, 'Why do we need a 4 ft wide door on the rear entrance?' Everything that comes in has to exit by the front doors which are only 3 ft 6 in wide. There it is, Fig. 4, the second revision, the perfect elevator: hoistway entrances symmetrical, car entrances, front and rear centred on the platform.

I called the architect and gave him the dimensions. I do not know if he recognized that a lot of thought and I believe, ingenuity, went into that layout. I do know that I felt the craftsman's pleasure at achieving a good detail.

That could be the end of the story, but it is not. I gave my sketch to a draftsman to do a drawing for the architect. He came back the next day with the drawing and a question, 'Why did you put a two-speed door on the back of the car?' My response was that it was the only way to avoid the counterweight. He laid Fig. 5 on my desk.

There is the car, same arrangement, same balance, but now the entrances are not only dimensionally the same, they are both centre opening. All that was required was shifting the counterweight behind the car guide-rail instead of adjacent.

Here is another thought. No matter how good you are (or think you are), two heads are sometimes better than one. Never hesitate to seek advice, to brainstorm, to let that imagination roam with others.

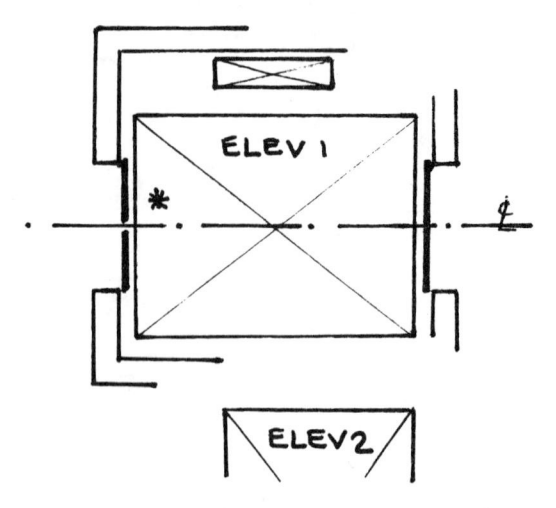

Fig. 5 — Final arrangement: centre opening doors *.

Ch. 34]   **Practising engineering, we each make a difference**   341

But, back to that hotel. There are a few more subleties the two cars would be balanced with two conventional care like the passenger by adding 1 inch to the hoistway between the elevator entrances. Relocating the rear entrance allowed service loads to be more easily manoeuvred from the corridor, and rearranging walls to conform to the required hoistway, saved space for a larger liquor storage room.

3. ONE MORE FLOOR

The Lincoln Center Building is an attractive 25-storey office building in downtown Denver. It is noteworthy because the first seven levels above the lobby are parking floors with 17 office floors above. The exterior finish is broken concrete (concrete ridge broken with air hammers to expose rough texture and aggregate).

Several years ago, a new owner found that almost 8000 sq. ft. of leasable space could be developed on the 26th floor. The only problem was that the six passenger elevators stopped at the floor below. This problem had been discussed with the elevator manufacturer and a structural engineer. Planning had proceeded on the basis that two of the six cars would be extended to serve the 26th floor. Necessary overhead would be gained by eliminating the secondary deck, extending the hoistway of the two cars about two feet above the existing machine room floor level and using reduced speed approaching terminal floors which allowed reduced stroke buffers. Planning reached the point of obtaining firm prices and drawing up contracts for the work.

We were asked to make a review. Our investigation revealed:

(1) Existing elevator service was marginal and needed to be improved.
(2) The 25th floor was occupied by a single tenant. That tenant was not agreeable to allowing people from the floor above to use his space for their access or elevator transfer.
(3) The potential tenant for the 26th floor space was not agreeable to service from only two out of six cars; the lease, already presented, called for service comparable to the rest of the building.

Well, all cars would have to serve the 26th floor; that is simple. Simple! The cost for the elevator work tripled; cost for building work more than doubled; elevator shutdown time would be measured in months rather than weeks; machine room code clearances could not be maintained and, worst of all, the building could not afford to stretch the marginal elevator service to cover another floor.

It appeared the space could not be utilized. Cost and problems were too great. In a meeting reviewing this final conclusion, we pointed out the main problem was the need to extend the hoistways and raise the machine room level. Maybe, rather than moving the machine room up, it might be easier to move the 26 floor elevator lobby down and use a couple of steps at each end of the new lobby to get to the regular floor level.

original bid. We got a plus out of the final plan. Though parking levels are served by separate elevators, the building elevators stopped at the highest parking floor. By eliminating this stop, overall service was improved and building security tightened. Elimination of this stop also allowed substitution of material and functions. Entrances and hall signals were removed and reinstalled at the upper floors. All car signal and operating fixtures were reusable with adjustment of floor designations, and controller and selector components were shifted from the parking level to the 26th floor without major additions and modifications.

How about dropping the elevation of the 26th floor lobby? That proved to be the easiest part of all. It was accomplished over a single weekend by installing angle supports below the floor, sawing out the portion to be depressed and lowering it gently into place. The only change on the 25th floor was in the design of the lighting fixtures in the elevator lobby. The tenant on this floor thought the owner was improving the appearance by redecorating.

4. IN CONCLUSION

In telling these stories, I hope similar experiences have been recalled to your mind and that you are encouraged to share your experiences with each other. That is the genius of any gathering — sharing ideas and experiences, exercising that imagination. There is nothing to delight an elevator engineer's heart like a 'war' story. That sharing may help to solve a problem or improve an idea. It will certainly establish or improve a friendship.

On person can make a difference. But only if we make an effort, only if we communicate our ideas, only if we did our imagination to the dialogue. Leonardo De Vinci may be the greatest innovative genius who ever lived. In a book I have that sets out one philosophers' listing of the 100 most-influential individuals that have graced the earth, Leonardo is mentioned only in a footnote. His greatness was never realized because his ideas were not shared.

Perhaps none of us has the wisdom of a Solomon, the imagination and talent of a Leonardo or the intuitive brilliance of an Einstein. But each one of us can resolve to contribute to progress by improving on the known, each one of us can add lustre to our title of 'engineer' by practising 'engineering' — practical, efficient and economical. each one of us can refuse to accept repetition of the past without review. We certainly do not want to reinvent the wheel with each assignment, but consideration of alternatives, flexing of imagination and ingenuity, may lead to great personal satisfaction in knowing we have made a special contribution. And finally, each one of us can make a difference to this genesis organization — the International Association of Elevator Engineers. We have to support freely, add our ideas and share our experiences.

Index of contributors

N. A. Alexandris, BSc, MSc, PhD
Piraeus Graduate School of Industrial Studies, 40 Karaoli and Dimitriou Street, Piraeus, Greece

Dr N. A. Alexandris was born in Greece and studied for his BSc in Mathematics in the Department of Mathematics and Physics of University of Athens. He took his MSc in Computation at UMIST. He received his PhD degree from the Control System Centre at UMIST. His PhD project was 'Statistical Models in Lift Systems'. He has published articles on lift systems and in other areas. He has also contributed to books in mathematics in his country.
   He is now working as an Associate Professor in the Department of Statistics and Informatics in Piraeus Graduate School of Industrial Studies.

Ewald H. Allaart, ME
Tebodin, PO Box 1029 – 2500 BA, The Hague, Netherlands

A mechanical engineer and graduate of Technische Hogesschool Delft, Mr Allaart has worked at Tebodin for many years. He is mainly concerned with all types of mechanical handling systems for material flows and passenger transport. Married with two sons one of whom is studying at Technische Hogeschool Delft.

G. C. Barney, BSc, MSc, PhD, C.Eng, FIEE
University of Manchester, Oxford Road, Manchester M13 9PL, UK

G. C. Barney after some years in industry graduated as an electrical engineer from Durham University in 1959. Subsequently he obtained an MSc in control theory (again from Durham University) and a PhD in control applications at Birmingham University. Dr Barney joined the Control Systems Centre in 1967 as a lecturer specializing in continuous system simulation and digital computer systems. He subsequently was promoted to Senior Lecturer in 1971. Dr Barney has led a research team since 1968 investigating the design and control of elevator systems, graduating some 25 students in this field. Currently he is the Director of the Network Unit at the University of Manchester Regional Computer Centre. Dr Barney is a Fellow of the Institution of Electrical Engineers.

V. Quentin Bates
Lerch Bates & Associates Inc., 8089 South Lincoln Street, Suite 201, Littleton, Co. 80122, USA

Mr Bates is a graduate of Brigham Young University and of West Point. After a period of naval service Mr Bates entered industry and joined Charles Lerch is consulting firm some 25 years ago. He is currently President of Lerch Bates and Associates.

Paul H. Beard, BSc(Eng.), C.Eng, MIEE
Wessex Regional Health Authority, Highcroft, Romsey Road, Winchester SO22 5DH, UK

Paul H. Beard is a chartered engineer. After an apprenticeship in the electrical supply industry, he graduated in 1954 with a BSc(Eng.) External degree of London University after study at Northampton Polytechnic (now City University). Following naval service, he returned to the power industry. He spent eight years with Esso Petroleum Company at Fawley Refinery as a project engineer in petrochemicals. For the past 17 years, he has been with the Wessex Regional Health Authority designing building services for hospitals and health premises, and is responsible for the design and specification of lift systems in the region.

Jonathan R. Beebe, BSc, MSc, PhD, C.Eng, MIEE, MBCS
Lift Innovations Ltd, Unit 11, Hartford House, Weston Street, Bolton BL3 2AW, UK

In 1976 Dr Beebe graduated as an electronic engineer from the University of Manchester Institute of Science and Technology and subsequently gained an MSc in digital electronics whilst developing automatic data entry devices for a central records computer monitoring experiments in a scientific laboratory. Whilst still at UMIST, he studied the application of computers to lift management, for which he was awarded the degree of PhD. In 1980 Dr Beebe joined Lift Design Partnership and was responsible for the development of a high performance, computerized lift control and monitoring system. He is now Managing Director of Lift Innovations Ltd, a company specializing in the design of lift management systems.

Index of Contributors

Dipl.-Ing. Emil Braun
O & K Orenstein and Koppel, Postfach 1702 18, D-4600 Dortmund 1, West Germany

Dipl.-Ing. Emil Braun, a graduate of the Technical University at Stuttgart, is now Executive Vice-President Engineering in the Escalator Division of O & K Orenstein and Koppel, Dortmund, Germany and is responsible for the design and construction of the firm's escalators and autowalks.

Wojciech Cholewa, PhD
University of Mining and Metallurgy, Wire Rope testing Laboratory, Al. Mickiewicza 30, 30-059 Kraków, Poland

Dr Cholewa is a lecturer at the Interbranch Laboratory for Wire Rope Testing and Rope Transport Equipment at the University of Mining and Metallurgy in Cracow, Poland. He was awarded a Doctorate in 1981 in the field of the reliability diagnosis of steel wire ropes during operation. He has interests in the following problems: the reliability diagnosis of rope transport equipment, discarding criteria of steel wire ropes, analysis of the frictional contact between rope and traction pully.

Eng. Félix de Crouy-Chanel
Ascinter Otis, 141, rue de Saussure, boite postale 727, 75822 Paris Cedex 17, France

Félix de Crouy-Chanel has been with Ascinter Otis (France) since 1959, where he has held various responsibilities as sales engineer, quality insurance manager, marketing manager, product development and engineering manager. Since 1984, he has been Director of Codes for Otis European and Transcontinental Operations. He is the Chairman of the ISO/TC178 and CEN/TC10 Technical Committees, dealing with lifts, escalators and passenger conveyors.

Dr-Ing. Carlo Distaso
Volpe Edtore, 20090 Segrate (Milano), Via Paeinotti 4, Italy

Carlo Distaso graduated at Rome University. He began his career as a teacher, covering the roles of vice-headmaster, technical office director and mechanical-working laboratory director. At ENPI (The State Institute of the Security-test of elevators), Milan, he has been organizing and supporting meetings for elevator technical staff. After starting as columnist for *Elevatori Moderni* he became chief editor and a member of the UNI commission, deputy for the UNI to the ISO/TC178. he is a consultant for the elevator security-test and carries on an intense activity of study and research for the IGV S.r.l. of Segrate Milan.

Index of Contributors

James W. Fortune
Lerch bates & Associates Inc., 8089 South Lincoln Street, Suite 201, Littleton, Co. 80122, USA

James Fortune holds an AA degree from Pasadena City College and a BS degree from California State Polytechnic University. He has extensive experience in the elevator industry having worked at Westinghouse as a sales engineer. For the last 15 years he has been with LBA and is currently a Vice-President. Fortune has an extensive speaking and publishing record in thefield of elevators.

Alfred Garshick P.E.
BIW Cable Systems Inc., 65 Bay Street, Boston MA 62125, USA

Alfred Garshick is Vice-president of Research and Developments of BIW Cable Systems Inc. and Registered Professional Engineer, State of Massachusetts. A member of the Institute of Electrical and Electronic Engineers' Power Engineering Society and Chairman of the Committee on Tests and Measurements of the Insulated Conductors. Has been active for 30 years in the design and performance of cables in adverse environments for the television, industrial, utility, petroleum and elevator industries.

A. M. Godwin, BSc
Lift Design Partnership, Bush House, Aldwych, London WC2B 4PY, UK

Adrian Godwin graduated from Salford University with a BSc in Electrical Engineering. After some experience running a small company making electronic equipment he joined LDP. Currently Managing Director he is interested in applying modern technology to lift systems.

Michael Godwin
Lift Design Partnership, Bush House, Aldwych, London WC2B 4PY, UK

Michael Godwin has been involved with the lift industry for most of his working life. Leaving William Wadsworth of Bolton in 1974 he founded Lift Design Partnership a consultancy firm specializing in the design of innovative lift systems. Currently Chairman of LDP he has a wide range of interests in applying high technology to lift systems.

Shigeru Goto
Mitsubishi Electric Corporation, Inazawa Works, 1-Hishi-Machi, Inazawa, Aichi 492, Japan

Shigeru Goto was born in Japan and obtained his mechanical engineering degree from Nagoya University in 1965. He joined Mitsubishi Electric Corporation (Japan), and is engaged in various phases of escalator engineering. He is currently manager of the mechanical engineering section in the Inazawa Works, Mitsubishi Electric Corporation, Japan.

Eduard Hadorn
Betriebsokonom Dipl. OEK, Beringer Hydraulik, Neuheim, Switzerland

Eduard Hadorn studied Betriebsokonom, Dipl. OEK, at Zurich with a subsequent two years practical study in the Republic of South Africa. From 1980–1982 Mr Hadorn represented the sales activities of a Swiss Company which produces power distribution components and substations for the Near and far East. In 1984 he joined Beringer Hydraulik GmbH, which manufactures hydraulic components and system applicaitons for the elevator and industrial market. At the end of 1985 Eduard Hadorn took over the responsibility of the marketing department of this company.

Jozef Hansel, DSc
University of Mining and Metallurgy, Wire Rope Testing Laboratory, Al, Mickiewicza 30, 30-059, Kraków, Poland

Dr Hansel is the Director of the Interbranch Laboratory for Wire Rope Testing and Rope Transport Equipment at the University of Mining and Metallurgy in Cracow, Poland. A Doctoral dissertation in 1972 in the field of influence of pulley lining on fatigue life of steel wire ropes, was followed by a postdoctoral dissertation in 1977 in the field of reliability and prediction of lifetime of steel ropes. Currently he coordinates scientific works carried out in Polish scientific and industrial centres in the field of rope transport equipment, rope construction, analysis of rope wear during its operation.

Bernard G. James, C.Eng., MIMechE
Health and Safety Executive, Magdalen House, Stanley Precinct, Bootle L20 3QZ, UK

Mr James is employed as one of HM Senior Engineering Inspectors by the Health and Safety Executive (HSE) UK. He is a Chartered Mechanical Engineer and has specialized for 16 years in safety matters relating to lifts (elevators), escalators, and other kindred equipment. During that time, he has represented the HSE at several relevant British Standards' and International Standards' drafting committees.

Dr-Ing. Lubomir Janovsky
Technical University of Prague, Suchbatarova 4, 166 07 Praha 6, Czechoslovakia

Dr-Ing. Lubomir Janovsky is a graduate of the Technical University of Prague (1957). Currently employed as a Senior Lecturer at the Technical University of Prague, mainly concerning vertical transportation, but also hoisting equipment, conveyors and mechanical handling machines. In addition to his pedagogical activities he has been involved in research, standardization and consulting activities. He published eight university textbooks, numerous papers and a 700-page book on the design and operation of elevators and escalators, and their application to buildings. He has been a Foreign Correspondent of *Elevator World*.

Matti Kaakinen
KONE Lift Group, Box 6 05851 Hyvinkaa, Finland

Matti Kaakinen, MSc(Eng.), was born in Helsinki, Finland, and graduated from the Technical University of Helsinki in 1960. He served the Finnish Waertsilae Corporation until 1963, when he joined the KONE Corporation. Since then, he has held various positions within the KONE organization in Finland and abroad, involving managerial, manufacturing, product planning, sales and marketing tasks. Currently he is the manager of market engineering of the KONE lift group, being responsible for international sales support, such as lift traffic analysis, surveys and simulation services. he is the co-author of the article 'New Formulae for Elevator Round-trip Time Calculations', published by *Elevator World*, which solved the probability problems of up-peak traffic situation.

Richard Laney
Siecor Corporation, PO Box 1237, Highway 301N, Rocky Mount, NC 27801, USA

Richard H. Laney is Engineering Manager with Republic Wire and Cable; a division of Siecor Corporation, Hickory, NC. He has held various engineering positions with Siecor during the past 15 years and has been actively involved with the manufacture of elevator cable products for the last eight years. Mr Laney graduated from The Citadel in 1968 with a BS degree in Electrical Engineering.

Eng. Ami Lustig
Eng.S. Lustig Ltd, 5 Beer Tuvya Street, PO Box 3298, Tel Aviv, Israel

Ami Lustig, Consulting Engineer, was born in Israel. Graduate of Tel Aviv University, Israel. In 1972, he joined the elevator and electrical consulting firm of S. Lustig, Consulting Engineers. In addition to his work as a consulting engineer, he is responsible for the development of various computer programs, among them the LSP. He was also group leader for the team that designed the LDL 2000 data logger built round a microcomputer.

E. M. McKay, MSc, PhD, C.Eng, C.Phys, MIEE, MIERE, MInstP
Rubicon Technical Services, 11 Rambling Way, Potten End, Berkhampsted HP4 2SF, UK

From 1944 to 1958 Dr McKay worked mainly in telecommunication engineering with the British Post Office, including two years research into measurements and life testing methods. He then become a lecturer, initially in technical colleges and later at Bristol University. He taught mathematics, applied physics and electronic and radio engineering. His research interests during this time were in high vacuum measurements (subject of MSc) and pulse code modulation methods of data transmission.

In 1966 he joined the Building Research Station where his research topics included acoustics and vibration control, economic design of water services (subject of PhD), lift system traffic engineering and environmental control systems for buildings.

Since 1985 he has been self-employed as a consulting engineer concerned primarily with energy controls and instrumentation.

Jallal Salihi
US Elevator, 10726 US Elevator Road, Spring Valley, CA 92078-2097, USA

Dr Jalal T. Salihi received the degree of PhD. From the University of California, Berkeley, and the degree of BSc from the University of Leeds, England

He is presently the Vice-President of Research and Development at US Elevator. Previously, he was Director of Product Research at Otis Elevator, where he worked for eleven years. His experience includes seven years with the General Motors research laboratories and two years with the division of advanced automotive power systems of the US Environmental Protection Agency.

Dr Salihi has 22 United States patents in the field of power conditioning and electric drives and he is the author of 12 papers.

William McCallum
Siecor Corporation, PO Box 1237, Highway 301N, Rocky Mountain, NC 27801, USA

William J. McCallum is a Product Design Engineer with Republic Wire and Cable; a division of Siecor Corporation, Hickory, NC. He has been involved in the development of elevator compensating cables and is currently responsible for the design and engineering of elevator travelling cable and related products. Mr McCallum graduated from the University of North Carolina at Charlotte in 1982 with a BS degree in Mechanical Engineering.

Dipl.-Ing. (FH) Paul Schick
R. Stahl Switchgear, D-7118 Kvenzelsan, West Germany

Dipl.-Ing. (FH) Paul Schick is the manager of the sales department of R. Stahl Schaltgeräte GmbH and concerned with the explosion protection of electrical equipment. After graduating from the Technical University of Konstanz as an electrical engineer he gained experience in engineering sophisticated electrical installations for industrial plants at AEG and Metzenauer & Jung. Since his joining R. Stahl in 1961 he has taken an active part in the application of explosion protected electrical equipment and installations and their penetration in the domestic and international markets.

Dr.-Ing Joris Schroeder
Schindler Management AG, CH-6030 Ebikon/Lucerne, Switzerland

Dr Schroeder studies mechanical enginnering in Stuttgart and Berlin and obtained a PhD from Berlin Technical University in 1954. He has worked in the elevator industry since 1952 in the following positions: chief engineer, Flohr-Otis Berlin; engineering manager, Otis Europe; vice-president engineering, Otis New York; vice-president product development, Otis New York. He returned to Europe in 1977 and joined Schindler Management AG as a director, Dr Schroeder's main activities are in R & D, with several inventions in the elevator and escalator field. The author of several articles in English and German language publications.

Dipl.-Ing. H. Streng
7000 Stuttgart 75, Hocholzweg 18, West Germany

Harro Streng studied Electrical Enginnering at the University of Stuttgart, 1952–1957 and following some specialized studies in feedback control of electrical drives he started in the elevator business in 1957 with R. Stahl, Stuttgart. From 1962–1965 he was with Brown, Boveri & Cie in Switzerland and returned to R. Stahl (later Thyssen Aufzüge) in 1965 as export manager for elevators. He became managing director in 1974 and one year later became a member of the board of 'Stahlbau und Förder-technik'. Since 1983 he has been an independent consultant and chartered engineer for elevators and electrical equipment. A member of the German Elevator Code Committee and German delegate in TC10/WG1 of the European code-making committee CEN since 1978.

D. Allan Swerrie
244 San Felipe Way, Novato CA 94947, USA

Mr Swerrie graduated from the University of California (Berkeley) in 1950 and is a registered professional safety engineer. After some 15 years with Otis Elevator in San Francisco he joined the State of California in 1965 as a safety engineer. Dee Swerrie is a member of a number of regulatory authorities.

Malcolm Wareing, BSc, MSc, PhD (elect)
University of Manchester Institute of Science and Technology, PO Box 88, Manchester M60 1QD, UK

After studying at Salford University (UK) Malcolm Wareing gained an honours degree in Physics and was awarded the final year course prize (1981).

Moving to UMIST in 1981, he collaborated with a lift consultancy practice LDP and became involved in practical lift status monitoring. In 1983 a prototype network for the remote status monitoring was completed and he was awarded an MSc for this work. From 1983 Wareing investigated possible indices of lift traffic performance and was awarded a PhD for this work in 1985. Currently a programmer in the Bursar's department at UMIST Dr Wareing continues to take an active interest in lift technology and has been involved in establishing a data-base of lift publications.

Eiki Watanabe
Mitsubishi Electric Corporation, Inazawa Works, 1 Hishi-Machi, Inazawa, Aichi 492, Japan

He was born in Osaka, Japan and received his Electrical Engineering degree from Osaka University in 1963. After graduation, he joined Mitsubishi Electric Corporation, Japan and has been mainly engaged in the development of elevator control systems. He is presently Deputy Manager of the Engineering Department of Inazawa Works, Mitsubishi Electric Corporation, Japan.

Hans Westling, MSc, C.Eng, BA(Econ.)
Promandat AB, Oxenstiernsgaten 31, 11527 Stockholm, Sweden

Hans Westling, MSc (Civ. Eng.), BA(Econ.) is chairman of the 'lift group' of the Swedish Council for building Research and partner of Promandat AB. He has 20 years' experience in the building industry as contractor, developer, system building and project manager of various building projects — housing, administrative, industrial buildings, power plants. Mr Westling is now running his own consulting firm in project management and procurement and is especially involved in installation-intensive projects.

Index

A
accidents   238
advance signalling   269
automatic doors   276
average highest reversal floor   8, 22, 28, 63
average number of stops   6, 7, 22, 31, 63
average passenger waiting time   26
average waiting time   12

B
basement service   78
breakdown maintenance   197
busy lifts   27
busy period   36
buffer, hydraulic   232

C
cable twist   90
cable
   compensating   83
   travelling   90
CAD (*see* computer-aided design)
call backs   202
car dimensions   222
   load   10
CEN standards   225
compensating cable   83
   dense aggregate   85
computer-aided design   12
   (*see* simulation)
consumption of energy   120, 190
control system policies   13
control systems   268
control
   drive   139
   group   140, 203, 207
   hydraulic   111
   microprocessor   136, 157, 270
   positioning   138
   speed   132
   torque   132
   valve   111
corrosion protection   303
cycle time   171, 267

D
data logger   56, 63, 65
data logging   62, 143
dense aggregate compensating cable   85
diagnostic techniques   141
doors, automatic   160, 276
down peak traffic   34
   CAD analysis   14
drive controls   139
   system   139, 162, 180, 273
dynamics, escalator   289

E
emergency brake   289, 293
energy consumption   189
energy equation   193
engineering practice   335
escalator
   dynamics   289
   emergency brake   289
   heavy duty   303
   kinematics   289
   safety   289
   spiral   325
   wheelchair   318
expected number of stops   7, 28
explosion protection   244

Index

F
falling of persons   313
floor populations   9
frequency controls   116, 126, 274
friction drives   100
   factor   103

G
generalised formulae   22, 32
group control   140, 203, 207

H
H (*see* average highest reversal floor)
handling capacity   4, 63, 166
heavy duty escalators   303
highest reversal floor   6, 7
hospital lift traffic   47
hydraulic buffer   232
   lift   183, 240
   proportional valves   111, 274

I
index of lift traffic performance   36
inspection   241
interfloor   29
   CAD analysis   15
interval   10, 52, 55, 165
inverter   116, 127, 274
ISO standards   221

K
kinematics, escalator   289

L
lift control systems   51
lifts, hydraulic   183
lift-in-service indicator   207
lift management   196
lift reliability   161
lift system demand   41
lift traffic in hospitals   47
lift traffic performance   54
linear motor   280
lining of ropes   100
link chains   85

M
maintenance   196, 241
   breakdown   197
   performance guaranteed   201
   planned   197
   preventive   197

   replacement   198
management   196
management, lift-in-service   207
mathematical methods   1
microcomputer controllers   270
microprocessor control   136
modelling, traffic   48
modernisation   156, 174
monitoring, remote   207
MTBF   205

P
passenger arrival process   23
percentage population   1, 7, 169
performance guaranteed maintenance   201
performance index   39
planned maintenance   197
plastic lining   100, 102
population   9
positioning controls   138
power consumption   120
power system   282
preventive maintenance   197
probability of failure   253
progressive safety gear   229
proportional valves, hydraulics   101
protection
   corrosion   303
   explosion   224

Q
quality of service   10, 24, 52
queue length   26, 78

R
refurbishment   156
reliability   161, 200, 252
remote monitoring   207
remote signalling   210
replacement maintenance   198
residential buildings   174
response time   60
rope life   101
rope lining   101
round trip time   1, 22, 52, 165, 266
RTT (*see* round trip time)
running time   2

S
S (*see* average number of stops)
safety   179, 228, 236, 252, 280, 306, 311, 314
   codes   257
   devices   258, 308
   gear   228

Index

progressive   229
shuttle service   151
signalling codes   212
   remote   210
simulation   12, 37, 62, 67, 72, 73, 78
   analysis   12, 42
   techniques   1
sky lobby design   149
   criteria   150
speed control   132
spiral escalator   325
standards   219, 228, 244
   CEN   225
   ISO   221
standing time   2
statistical evaluation   22
system response time   55
system waiting time   70

T

testing cables   90
testing ropes   103
theoretical performances   165
time
   average waiting   12
   cycle   171, 267
   response   60
   round trip   1
   running   2
   standing   2
   system response   55

system waiting   70
torque control   132
traffic design   1, 58
   formulae   22
   modelling   48
   performance   54
   sizing   1
trapping of persons   313
travelling cable   90
twisting effect   99

U

up-peak CAD analysis   12
   handling capacity   1, 7
   interval   1
   traffic   3, 33
UPPHC (*see* up-peak handling capacity)

V

variable voltage and frequency (VVF)   116, 126, 274
variable voltage variable frequency (VVVF)   116, 126, 274

W

waiting time   204
wear resistance   106
wheelchair escalator   318